Metal Nitrosyls

Metal Nitrosyls

GEORGE B. RICHTER-ADDO

PETER LEGZDINS

Department of Chemistry
The University of British Columbia
Vancouver, Canada

New York Oxford
OXFORD UNIVERSITY PRESS
1992

Oxford University Press

Oxford New York Toronto
Delhi Bombay Calcutta Madras Karachi
Kuala Lumpur Singapore Hong Kong Tokyo
Nairobi Dar es Salaam Cape Town
Melbourne Auckland

and associated companies in
Berlin Ibadan

Published by Oxford University Press, Inc.,
200 Madison Avenue, New York, New York 10016

Library of Congress Cataloging-in-Publication Data
Richter-Addo, George B.
Metal nitrosyls / George B. Richter-Addo and Peter Legzdins.
p. cm. Includes bibliographical references and index.
ISBN 0-19-506793-2
1. Nitroso compounds.
2. Organometallic compounds.
I. Legzdins, Peter. II. Title.
QD305.N8R53 1992 547′.041—dc20
91-42087

1 3 5 7 9 8 6 4 2

Printed in the United States of America
on acid-free paper

Preface

This book originated from a 1988 *Chemical Reviews* article by the authors which deals with the then current aspects of organometallic nitrosyl chemistry. In preparing this earlier article, we soon realized that while a number of reviews over the years have been devoted to different aspects of nitric oxide and its complexes, there was no single source which provided an in-depth overview of the field. Hence, when Oxford University Press approached us in the fall of 1989 to prepare a volume to remedy this deficiency, their request fell on receptive ears.

The coverage in this book is comprehensive, the formal cut-off date for pertinent literature citations being mid-1991. Some particularly relevant work reported later in 1991 has been included where appropriate during the final stages of polishing the text. The level of presentation is intended to be comprehensible to a first-year graduate student or a non-specialist practising chemist. We also hope that the established researchers working in this or related fields will find this volume helpful both as a source of valuable data and as a fount for new research ideas. Each of the chapters in the book is intended to stand alone since cross-referencing to relevant material elsewhere in the text is ubiquitous. Hence, selections of various chapters may be used as the basis for a timely graduate course in inorganic chemistry, as we are in fact doing at U.B.C. at the present time.

This project would not have reached fruition had it not been for the dedicated efforts of a number of talented individuals. First, our heartfelt thanks go to V. Yakoleff for her insights and editorial comments concerning the biological aspects of nitric oxide chemistry and to G. Howard and the public relations staff of S.C.A.Q.M. in El Monte, CA for providing us with the photographs shown on page 340 and for helpful discussions concerning air quality in southern California. We are also grateful to our colleagues at U.B.C. who read and constructively criticized various portions of the manuscript. The efforts of J.E. Veltheer, M.J. Shaw, P.J. Lundmark, J.D. Debad, K.J. Ross, E.B. Brouwer, M.A. Young, and W.S. McNeil in this regard were above and beyond the call of duty. Finally, we thank E.A. Varty and S. Rollinson at U.B.C. for their professionalism while preparing the illustrations which complement the text. Nevertheless, it goes without saying that it is the authors alone who are responsible for any errors or omissions remaining in the final draft.

In closing, we acknowledge with deep gratitude our wives and families whose unwavering support and constant encouragement sustained us during the many hours devoted to this project.

Vancouver, B. C., Canada George B. Richter-Addo

October, 1991 Peter Legzdins

Contents

4 Organometallic Compounds 149

1

Introduction

The binding of nitric oxide to a metal center imparts unique chemistry both to the metal center and the nitrosyl ligand itself. In order to obtain a full appreciation of the uniqueness of metal-nitrosyl chemistry, however, an understanding of the bonding within the isolated nitric oxide molecule and its nonmetal derivatives is necessary. This is especially so, since nitric oxide and many of its nonmetal derivatives have well-established chemistry of their own, and are used as nitrosylating agents for the syntheses of metal nitrosyls. It is for this reason that this chapter contains an overview of (i) the bonding in free nitric oxide, (ii) the properties and reactions of the nitrosonium cation, (iii) the formation and properties of the XNO compounds (X = H, halide, pseudohalide), and (iv) the principles of metal-nitrosyl bonding. This background knowledge is essential, and allows for a more complete understanding of the physical and chemical properties of metal-nitrosyl complexes.

1.1: Nitric Oxide

1.1.1: Electronic Configuration of Nitric Oxide

Nitric oxide (NO) is the simplest, thermally stable paramagnetic molecule known. It is a colorless, monomeric gas (b.p. -151.8 °C) that is thermodynamically unstable (ΔH_f^o = 90.2 kJ mol^{-1}) with respect to N_2 and O_2. The molecular orbital diagram of NO is shown in Figure 1.1, and according to this rationale of the bonding, the odd electron is based in a π^* antibonding orbital. In valence bond terms, the bonding in NO may be represented by the resonance structures **A** and **B**. The resonance form **A** is

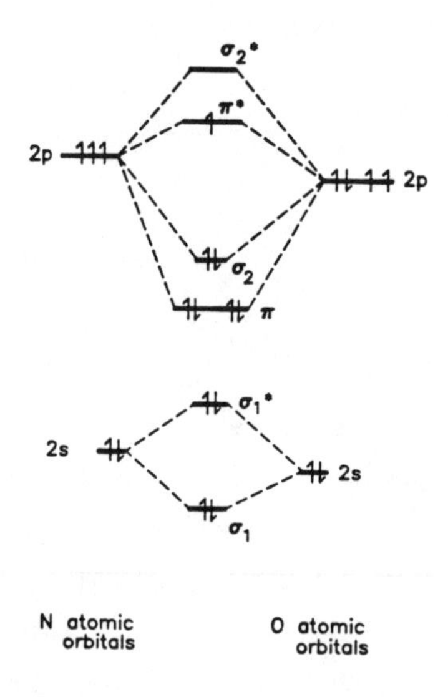

Figure 1.1 The molecular orbital diagram of nitric oxide.

slightly polarized due to the higher nuclear charge on oxygen, but the negative charge on the oxygen is cancelled by the positive charge in resonance form **B**.

$:\dot{N}=\ddot{O}$ ⟷ $\bar{\ddot{N}}=\dot{O}^{+}$ $\qquad :\dot{N}\doteq\dot{O}$

A **B** **C**

This accounts for the small dipole moment observed for NO (0.158 D), which is calculated to be of N^{-}-O^{+} polarity (*1,2*). Nevertheless, the NO molecule is also adequately described by structure **C**, which is consistent with the N-O bond length (1.15 Å) being between that of a double (1.18 Å) and a triple (1.06 Å) bond. The NO bond order is usually described as being 2 1/2 (*3,4*). It has been determined by ESR, however, that about 60% of the spin density is concentrated on the N atom of nitric oxide (*3*).

1.1.2: Preparation of NO

The preparation of NO on the industrial scale is achieved by the catalytic oxidation of ammonia over heated Pt (800 °C). On the laboratory scale, however, NO can be prepared by a variety of routes that involve the mild reduction of acidified nitrites or the disproportionation of nitrous acid *(3)*.

1.1.3: Oxidation and Reduction of NO

The simplest reactions of NO are those that involve the removal of the electron in the π^* antibonding orbital (oxidation) or the addition of an electron to this orbital (reduction). The ionization potential of NO is 9.26 eV (*5*), and oxidation of NO to generate NO^+ (the *nitrosonium* cation) is readily accomplished by a variety of methods. These methods include photoionization (*6,7*), electrochemical oxidation (*8*) or reaction with electron acceptors during the formation of charge-transfer complexes (*9*). As expected, the removal of the odd electron in the π^* antibonding orbital results in the strengthening of the N-O bond (Table 1.1).

Table 1.1 Comparison of the N-O Bond Lengths and IR Stretching Frequencies for NO^+, NO and NO^-.

	NO^+	NO	NO^-
N-O distance (Å)	1.06	1.15	1.26
ν_{NO} (cm^{-1})	2377	1875	1470

NO has a low electron affinity of 0.024 eV (*10*). The acceptance of an electron generates the NO^- anion (also known as the *nitroside* anion), and the addition of this electron into the π^* antibonding orbital weakens the N-O bond (*10,11*). Nevertheless, the nitroside anion may form weak Van der Waals complexes with rare gas atoms such as He and Ar (*12*) or may combine with molecules such as N_2O via Lewis base-Lewis acid interactions (*13*) to generate the $NO^-.He(Ar)$ and $NO^-.(N_2O)_{1,2}$ complexes, respectively.

1.2: The Nitrosonium Cation

As mentioned in the previous section, the odd electron of the neutral NO molecule is easily removed

$$NO \rightleftharpoons NO^+ + e^- \quad (1.1)$$

and the reversible redox potential (E^o) of this process in a variety of solvents has been determined. Interestingly, there is a marked solvent dependence on the value of E^o, a dependence which correlates with the previously established donor abilities of the solvents used (*8*). Thus, whereas the E^o of reaction 1.1 is measured to be 1.48 V (vs SCE) in dichloromethane, it is 1.34 V in nitromethane and 1.28 V in acetonitrile.

1.2.1: Methods of Preparation

A large number of synthetic routes to NO^+-containing salts are currently available, and these include the action of NO gas (*14,15*), N_2O_4 (*16-18*), nitrosyl halides (*19,20*) and organic nitrites (*21*) on various metal-containing complexes, e.g.

$$NO + OsF_6 \longrightarrow [NO][OsF_6] \quad (1.2)$$

$$6\ ClNO + Pt \longrightarrow [NO]_2[PtCl_6] + 4\ NO \quad (1.3)$$

$$FeCl_3 \xrightarrow{Cl_3CNO_2} [NO][FeCl_4] \quad (1.4)$$

$$Ln \xrightarrow{HF/NO_2} [NO]_x[LnF_{x+3}] \quad (1.5)$$

(Ln = rare earth element; x = 1, 3/2 etc)

Many nitrosonium salts containing MX_n-type anions are now known, where M is a main group element (*22-25*), a transition metal (*14,16,17,19,21-23,26-32*), a lanthanide or actinide element (*15,17,18*) or even a non-metallic element (*22,33-38*). However, the common nitrosonium salts employed for synthetic purposes are $[NO]BF_4$ and $[NO]PF_6$ or their ^{15}N-labelled analogues (*34,35*). The majority of these nitrosonium salts are ionic and crystalline. However, complexes such as $[NO]OTeF_5$ are covalent in the gas phase (*24*). Due to the reactive nature of the nitrosonium cation in these species, it is always advisable to sublime them freshly before use (*22*).

1.2.2: Reactions of the Nitrosonium Cation

The gas phase reactions of NO^+ include photofragmentation (*39*) (eq 1.6) and dissociative recombination (*40*) (eq 1.7), and such reactions are important in the chemistry of the upper atmosphere (see Chapter 8).

$$NO^+ \longrightarrow N + O^+ \quad (1.6)$$

$$NO^+ + e^- \longrightarrow N + O \quad (1.7)$$

Of more interest to the synthetic chemist is the fact that the NO^+ cation is electrophilic, and hence reacts with a very large number of basic molecules both in the gas phase and in solution by simple adduct formation (eq 1.8) or by charge transfer reactions (eq 1.9).

$$NO^+ + L \longrightarrow NO^+(L) \quad (1.8)$$

$$NO^+ + L \longrightarrow NO + L^+ \quad (1.9)$$

However, the distinction between these two reactions is not always unambiguous, since the electron affinity of NO (9.26 eV) is of the same order of magnitude as those of typical bases (for example, the electron affinity of benzene is 9.24 eV). Thus, whether the charge is localized on NO or on L depends largely on the ionization potential of the base (*41*). In any event, NO^+ reacts with a large number of molecules such as N_2, NO_2, CO, MeI and MeH in the gas phase (*42-62*) and the rare gases Kr, Ar and He (*46,63-66*) via reactions 1.8 and/or 1.9. The NO^+ cation also reacts with a variety of arenes to form intensely colored charge-transfer complexes of the form shown in eq 1.10 below (*8,67,68*):

$$NO^+ + \text{arene} \rightleftharpoons [NO^+,\text{arene}] \quad (1.10)$$

The NO stretching frequencies of these $[NO^+,\text{arene}]$ charge-transfer complexes is in the 1800 - 1880 cm^{-1} range, which is substantially lower than that for the NO^+ cation (2377 cm^{-1}). Also, the N-O bond distances in these charge-transfer complexes (as judged from X-ray crystallographic studies) range from 1.09 - 1.43 Å (*8,68-70*), reflecting considerable electron delocalization and weakening of the N-O bond in the η^6-centrosymmetric structures of these complexes. The ability of arenes to form charge transfer complexes with the NO^+ cation correlates with the strong contribution of the

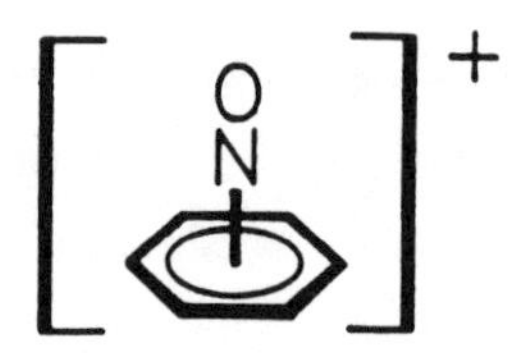

arenes (via electron donation) to the stabilization of the nitrosonium moiety. Indeed, such stabilizations are similar to those provided by various solvents, as evidenced by the data presented in Table 1.2. The data in Table 1.2 illustrate the powerful stabilizing ability of various solvents and arenes on the NO^+ group. Thus, the E° of NO^+/NO decreases with increasing solvent donicity (DN), and this point is illustrated graphically in Figure 1.2A. Interestingly, the oxidation potential of the free uncoordinated arene (E°_{Ar}) is directly related to the magnitude of the donor effect on the reduction of NO^+, as shown in Figure 1.2B.

Table 1.2 The Nitric Oxide/Nitrosonium Redox Potential Dependence on Various Solvents and Arenes.

cation	solvent	E^{o} or E_{c}^{o} [a]
NO^{+}	CH_2Cl_2	1.48
	$MeNO_2$	1.33
	MeCN	1.28
$[NO^{+},C_6Me_4H_2]$	MeCN	1.13
$[NO^{+},C_6Me_5H]$	MeCN	1.10
$[NO^{+},C_6Me_6]$	MeCN	1.02

[a] E_{c}^{o} is the redox potential of the charge transfer complex. [Data taken from reference 8a.]

The nitrosonium cation also forms adducts with crown ethers (*8,71,72*) and even β- and β''-aluminas (*73,74*). Nevertheless, it is a powerful

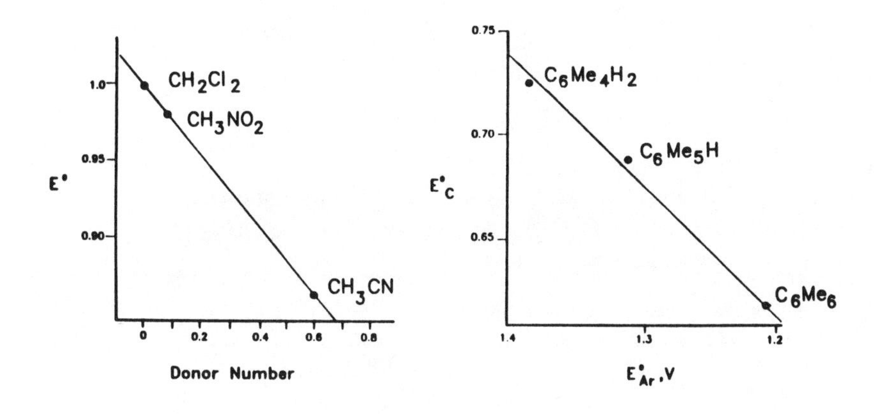

Figure 1.2 Free energy correlations of the reduction potentials (A) E^{o} of NO^{+} with the Gutmann solvent donor number DN and (B) E_{c}^{o} of the NO^{+}/arene complex with the oxidation potential (E_{Ar}^{o}) of the uncoordinated arene. [Redrawn from reference 8a.]

one-electron oxidant (*8*) and is used as an oxidative dopant for polymers (*75-83*). It is also used as a dopant for steel (*84*) in order to improve its corrosion resistance.

1.3: The Interaction of NO with Nonmetals

1.3.1: General Considerations

The ground state equilibrium geometries for the nitrosyl hydrides, halides and related compounds correspond to the X-NO forms (X = H, halide, pseudohalide) as opposed to the X-ON forms. This observation may be contrasted to that observed for the thiazyl halides which possess ground state equilibrium X-SN geometries.

The X-NO and X-ON dichotomy has been the subject of much theoretical work (*85-91*), and the X-NO geometries have been found to be more stable than their X-ON counterparts (which are referred to as *nitrosyl hypohalides* for X = halide). For example, the ground state HNO isomer is found to be about 24 kcal mol^{-1} more favorable than the NOH isomer (*88*), although both forms are stable compared to their individual NO and H fragments (*91*). Such theoretical work has been inspired by the consideration that the chemically unstable HNO compound is probably one of the most important intermediates in various ammonia oxidation processes.

Table 1.3 Calculated rate constants (s^{-1}) for the FNO -> FON and FON -> FNO isomerization reactions as a function of temperature.

T (K)	FNO -> FON	FON -> FNO
100	0	2×10^{-3}
500	0	2×10^{7}
1000	2×10^{-2}	2×10^{8}

Theoretical work on nitrosyl fluoride (*85,86,88*) predicts the F-NO isomeric form to be more stable than the F-ON form, with a barrier of 8.3 kcal mol^{-1}

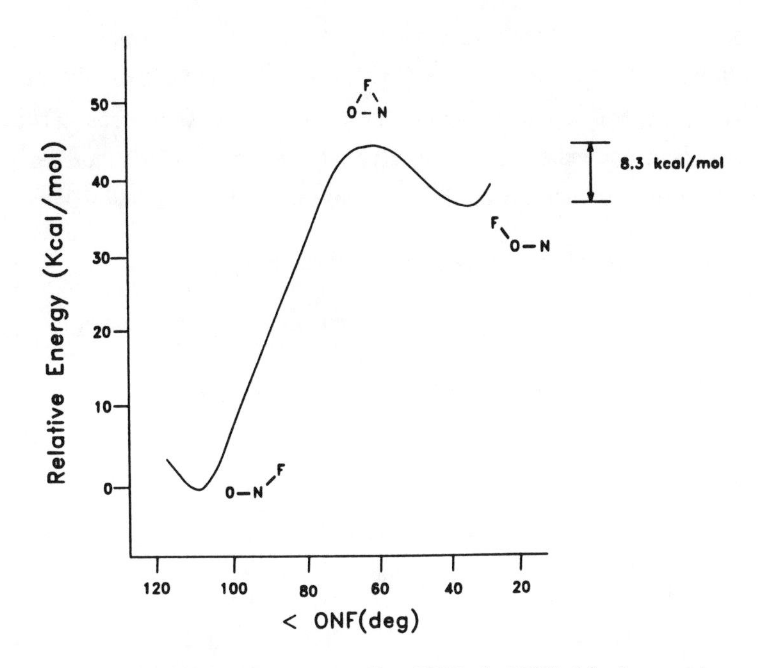

Figure 1.3 Potential energy curve for ONF -> FON (singlet state) as a function of ∠ONF. [Redrawn from reference 85.]

for the conversion (Figure 1.3). Computed rate constants for such a conversion as a function of temperature (Table 1.3) show that the predominant process (if any) should be the one resulting in the production of the F-NO isomer (*86*). In general, however, a dissociation of X-NO to NO and X is preferred over the XNO to XON isomerization, which for all intents and purposes is a migration of X from the nitrogen to the oxygen possibly via a cyclic intermediate (*89*).

1.3.2: Preparation of X-NO Compounds

The nitrosyl halides (X = Cl, Br, I) are preparable by the direct reaction of NO with the appropriate halogen (*92-94*) or via the reaction of alkyl nitrites with the electrophilic trimethylsilyl halides (*95*), i.e.,

$$NO + 1/2\ X_2 \rightleftharpoons XNO \qquad (1.11)$$

$$RONO + Me_3SiX \rightleftharpoons X\text{-}NO + Me_3SiOR \qquad (1.12)$$

Nitrosyl fluoride is a highly reactive gas which is also preparable via the use of transient intermediates such as F (*96,97*), F_3NO (*98*), CF_3 (*99*) or even HF (*100*) generated under appropriate experimental conditions. Nitrosyl chloride is a convenient synthetic reagent, and may be synthesized on a laboratory scale by employing the reactions outlined below *(101-104)*:

$$NaNO_2 + 2\ HCl \longrightarrow ClNO + H_2O + NaCl \qquad (1.13)$$

$$PCl_3 + HNO_3 + 2\ H_2O \longrightarrow ClNO + 2\ HCl + H_3PO_4 \qquad (1.14)$$

$$2\ NO_2 + KCl \longrightarrow ClNO + KNO_3 \qquad (1.15)$$

$$NO + NO_2Cl \longrightarrow ClNO + NO_2 \qquad (1.16)$$

Nitrosyl chloride may also be synthesized by the reaction of an aqueous mixture of hydrochloric and nitric acids

$$HNO_3 + 3\ HCl \rightleftharpoons ClNO + Cl_2 + 2\ H_2O \qquad (1.17)$$

and such a system has long been called "aqua regia" because of its ability to dissolve gold (*93*). The related trifluoronitrosomethane (*105,106*) and nitrosyl cyanide (*103,107-109*) are preparable via the following reactions:

$$CF_3COONO + NO \xrightarrow[\Delta]{} CF_3NO + CO_2 \qquad (1.18)$$

$$AgCN + ClNO \longrightarrow NCNO + AgCl \qquad (1.19)$$

while HNO and FONO have normally been generated in situ for spectroscopic study (*110,111*). For example, HNO is observable during the flash photolysis of compounds such as $MeNO_2$, $EtNO_2$, isoamyl nitrite, and mixtures of NO and NH_3 or mixtures of MeCHO or H_2CO and NO (*112*).

Some physical data for representative X-NO compounds are listed in Table 1.4.

Table 1.4 Physical Data for the Nitrosyl Halides and Related Compounds

Compound (X-NO)	ν_{NO} (cm^{-1})	X-N (Å)	N-O (Å)	X-N-O (deg)	Ref
F-NO	1852	1.517	1.13	109.9	*113-115*
Cl-NO	1800	1.975	1.14	113.3	*116,117*
Br-NO	1801	2.14	1.15	114	*117,118*
I-NO[a]	1809	-	-	-	*118*
H-NO	1562	1.063	1.206	109.1	*109,119*
FO-NO[a]	1716	1.50	1.20	110	*110*
NC-NO	1501	1.418	1.217	113.6	*103,120,121*

[a] matrix isolated

As indicated in the Table, the X-NO compounds possess bent ground-state geometries

$$X\diagup\ddot{N}=O$$

having a high degree of covalency of the X-N bond. In general, the X-NO compounds may be thought of as being formed either by (i) overlap of the lone p electron of X (s electron for H) with the singly occupied π^* orbital of NO, or (ii) interaction between the HOMO of the X^- anion and the LUMO of the NO^+ cation. Regardless of the view adopted, the changes in the N-O bond length and the NO stretching frequency are usually attributed to the electronegativity and nucleophilicity of the X group. Indeed, the net atomic charges on the nitrogen in various X-NO compounds (q_N) vary with the nucleophilicity of X, and actually correlate with Pearson's nucleophilicity parameter (*n*) (*122*). Thus, the more nucleophilic the anion X^-, the smaller is the net atomic charge on nitrogen. It is, therefore, not so surprising that q_N is largest and positive when the formal NO^+ fragment binds to hard species (e.g. H_2O), but tends to be zero or negative when bound to soft species (e.g. SCN^- or HS^-).

In general, the NO^+ cation prefers coordination to soft anions rather than to hard ones (*122*), the formation of nitrosyl thiocyanate from NO^+ and SCN^- being an illustrative example. The net atomic charges on the various atoms of the isolated NO^+ and SCN^- atoms are shown in Table 1.5.

Table 1.5 Net atomic Charges of NO^+ and SCN^-.[a]

$[NO]^+$		$[SCN]^-$	
q_N	+0.80	q_S	-0.25
q_O	+0.20	q_C	-0.07
		q_N	-0.68

[a] Data taken from reference 122.

Although the NO^+ cation might be expected to combine with the nitrogen of the SCN^- anion (due to its larger negative charge) in a charge-controlled reaction, it evidently prefers to combine with the ambident SCN^- anion at the sulfur atom in a typical soft-soft, orbital-controlled reaction resulting in a complete charge transfer from the nucleophile to the electrophile (*123*), if the calculated atomic charges of the NCS-NO molecule are any indication.

$$\overset{-0.17}{N} \equiv \overset{-0.06}{C} - \underset{+0.32}{S} - \overset{-0.02}{N} = \overset{-0.07}{O}$$

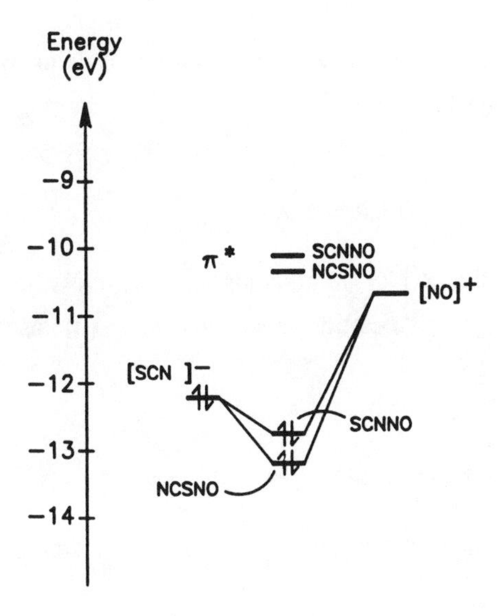

Figure 1.4 Change in the frontier orbital energy in the reaction of the nitrosyl cation with the thiocyanate anion. [Redrawn from reference 123.]

The changes in frontier-orbital (HOMO) energies accompanying the reaction of NO^+ with SCN^- at the sulfur and nitrogen atoms to produce ON-SCN and ON-NCS respectively are illustrated in Figure 1.4. As may be seen in the Figure, the energy of the bonding orbital is lowered in the case of *S*-nitrosation relative to *N*-nitrosation, which may reflect the fact that in general, such soft-soft interactions are counterbalanced by stabilizations due to *covalent* bonding (*123*).

Perhaps it would be wise to pause for a moment at this point and ask the question: What does this all mean? For one, understanding that the bonding in the X-NO compounds have only *degrees* of covalent bonding in the X-N link provides an appreciation of the different physical properties and even the reactivity patterns for these molecules. A comparison of ClNO, a typical nitrosyl halide, with nitrosyl thiocyanate, NCS-NO, illustrates this point. For ClNO, the net charge on the nitrosyl nitrogen is calculated to be +0.21, whereas for NCS-NO this computed charge is reduced to -0.02 (*123*). The fact that ClNO is indeed partly "ionic" in character is demonstrated by its use as an ionizing solvent, and by the fact that its solid phases consist of some $[Cl]^-[NO]^+$ units with ν_{NO} of 2090 cm^{-1} (*117*). Also, ClNO is a much better nitrosating reagent than is NCS-NO

(*124*), the majority of the syntheses of metal nitrosyls using X-NO reagents employing ClNO (see Chapter 2).

1.3.3: Reactions of X-NO Compounds

The reactions of the X-NO compounds may be classified into two main groups, namely (i) dissociation reactions and (ii) association (adduct forming) reactions.

1.3.3.1: Dissociation Reactions

The photodissociation reactions (eq 1.20) of various nitrosyl halides (*93,125-139*),

$$\text{X-NO} \longrightarrow \text{NO} + \text{X} \qquad (1.20)$$

nitrosyl cyanide (*140-148*), nitrosyl hydride (*91,111,149,150*) and methyl nitrite (*151*) have been extensively studied and, interestingly, the photodissociation of ClNO, i.e.,

$$\text{ClNO}_{(g)} \underset{-\Delta H}{\overset{h\nu}{\rightleftarrows}} \text{NO}_{(g)} + \tfrac{1}{2}\,\text{Cl}_{2\,(g)} \qquad (1.21)$$

has been proposed for photochemical syntheses as well as for the storage of solar energy (*139,152*). The mechanism of this photodissociation involves the following steps:

$$\text{ClNO} + h\nu \longrightarrow \text{NO} + \text{Cl}\cdot \qquad (1.22)$$

$$\text{Cl}\cdot + \text{ClNO} \longrightarrow \text{NO} + \text{Cl}_2 \qquad (1.23)$$

$$2\,\text{ClNO} + h\nu \longrightarrow 2\,\text{NO} + \text{Cl}_2 \qquad (1.24)$$

Nitrosyl chloride has a low dissociation energy and absorbs a significant amount of energy in the visible region of the solar spectrum (*125*). It also has a high quantum yield of 2.0 for NO formation (*153,154*) as well as a significant free energy change (4.9 kcal mol^{-1} of ClNO) and a high heat of reaction (9.0 kcal mol^{-1} of ClNO) (*139*). In order to keep the photodissociation process occurring, however, the photoproducts have to be separated from the ClNO reagent immediately as they are formed. Nevertheless, the heat of reaction can be utilized by recombining the photoproducts in a heat exchanger (*152*). Experiments utilizing fuel cells for this chemical storage of solar energy have been described (*152*), and NO laser emissions due to the vibro-rotational transitions of "excited" NO formed from the photodissociations of BrNO (*129,132*) and ClNO (*130,131*) have also been studied. However, discussions of the solar energy storage and these laser emissions are well beyond the scope of this book.

1.3.3.2: Association (Adduct Forming) Reactions

The X-NO (X = Cl, Br) compounds form 1:1 molecular complexes with numerous electron donors such as arenes (*155*) as well as with a series of Bronsted and Lewis acids such as HCl in inert matrices (*156-158*). Curiously, the IR NO-stretching frequencies for the 1:1 molecular complexes *shift to higher energies in both cases*, and do not depend on whether electron acceptors or electron donors are combined with the X-NO compounds. The shift of ν_{NO} to higher energies for the XNO:Lewis acid complex is easily understood since it would be expected that electron withdrawal from the π^* antibonding orbital of the N-O group in X-NO would increase the NO bond strength. For the XNO:Lewis base complexes, however, these shifts to higher energies are not easily explained. It has been suggested that this could be due to an increased polarization of X-NO (toward ionic character), or a change in bond angles and X-NO geometry upon complex formation coupled with a significant redistribution of charge within the X-NO fragment (*155*).

The X-NO compounds readily add to unsaturated organic molecules (*93,124,159*), e.g.,

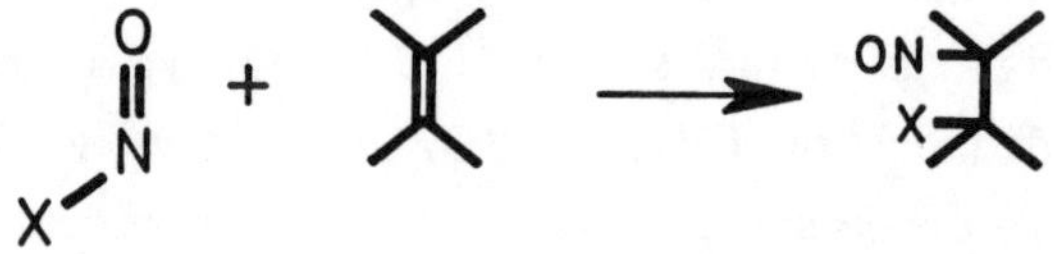

(1.25)

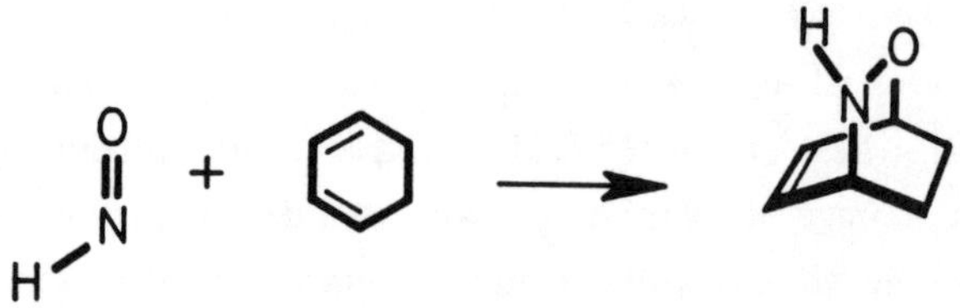

(1.26)

The reactions of *C*-nitroso compounds (R-NO) have been reviewed *(160)*, and will not be considered here. In a related vein, complexes of the form FNO.*n*HF (*n* = 3,4,7) have been isolated and their structures determined by X-ray crystallography *(161)*, but these also will not be discussed further.

By far the most important use of the nitrosyl halides is for the nitrosations of organic compounds and their derivatives *(124,162)*, e.g.

$$(Me_3Si)_3N + ClNO \longrightarrow (Me_3Si)_2NNO + Me_3SiCl \quad (1.27)$$

or for the nitrosations of metal complexes to form metal nitrosyl complexes (see Chapter 2). From the viewpoint of the environmental chemist, the gas phase reactions of the XNO compounds with atoms or small gaseous molecules or radicals *(92,112,163-169)* are very important, and these reactions will be considered in some detail in Chapter 8 of this book.

1.4: The Bonding of NO to Metal Centers

In principle, the NO molecule could bind to a metal center via the N or O ends of the molecule to produce either *nitrosyl* or *isonitrosyl* linkages, i.e.

M-NO	M-ON
nitrosyl	*isonitrosyl*

In practice, however, the binding of NO to metal centers is dominated by the nitrosyl form, and the reader's attention is drawn to the analogous description of the bonding in the nitrosyl halides discussed earlier in this chapter. Although the isolated M-ON isonitrosyl link has not yet been unequivocally demonstrated, there are two examples of the related M-NO-M' *nitrosyl-isonitrosyl* interactions that have been reported (Section 2.2.1.2).

It is expected that the M-ON link should occur with the more electropositive metals or hard metal centers. Indeed, theoretical work on Li(NO) and Na(NO) favors strong M-ON interactions and/or ground state cyclic structures (*170-175*). However, since there is insufficient experimental evidence for the existence of the M-ON bonding unit, it is excluded in subsequent discussions involving the bonding between metal centers and NO.

1.4.1: Oxidation States and the Valence Bond Approach

Historically, the bonding of the M-NO unit in metal nitrosyl complexes has been described by assigning formal oxidation states to the metal and the nitrosyl ligand (*176-180*). In such an assignment, the linear M-NO unit is deemed to possess a bound NO^+ ligand (which is isoelectronic with CO), and conversely, a bent M-NO unit is deemed to possess a bound NO^- ligand. In those terms, the bond formation between a metal and NO could involve, in principle, (i) a donation of an electron from NO to the metal (thus forming NO^+), or an acceptance of an electron from the metal (forming NO^-), and (ii) a donation of a lone pair of electrons from a σ-orbital (Figure 1.1) on the nitrosyl nitrogen to the metal. Coordination of nitric oxide as NO^+ to a metal center would thus involve a *net donation of 3 electrons* from NO to the metal, whereas coordination of NO as NO^- would involve a *net donation of only 1 electron* from NO to the metal. However, assigning oxidation states to M-NO links is undesirable, since the formal oxidation states in $Co(CO)_3NO$, $Fe(CO)_2(NO)_2$, $Mn(CO)(NO)_3$ and $Cr(NO)_4$ have the unrealistic values of -1,-2,-3 and -4, respectively !

In valence-bond terms, the resulting NO^+ group is considered to be *sp* hybridized at both the N and O atoms, and this linear bonding mode may then be represented by the following resonance forms:

$$\overset{-}{M}\leftarrow N\equiv\overset{+}{O}: \longleftrightarrow M=N=\ddot{O} \longleftrightarrow \overset{-}{M}=\overset{+}{N}=\ddot{O} \longleftrightarrow M\equiv\overset{+}{N}-\ddot{O}:^{-}$$

The bonding of NO as nitroside (NO^-) to electron-rich or electropositive metals has been linked to the bonding in the X-NO (X = halide, alkyl, etc) nitroso compounds. By definition, the attachment of NO in this fashion requires prior acceptance of an electron from the metal followed by a donation of an electron pair to the metal. In valence bond terms, this mode of bonding may be represented as:

$$M-\ddot{N}=\ddot{O}: \longleftrightarrow \overset{+}{M}=\ddot{N}-\ddot{O}:^{-}$$

both the N and O atoms being sp^2 hybridized. Interestingly, the requirement of NO bond orders of both 1 and 2 in the valence-bond pictures of the linear and bent MNO groups may be related to the fact that vibrational spectroscopy cannot be used as a reliable tool for distinguishing between these two types of groups.

1.4.2: The Molecular Orbital Approach

It is now generally accepted that the bonding of NO to a metal center involves a synergic interaction between the NO ligand orbitals and the metal orbitals as shown below:

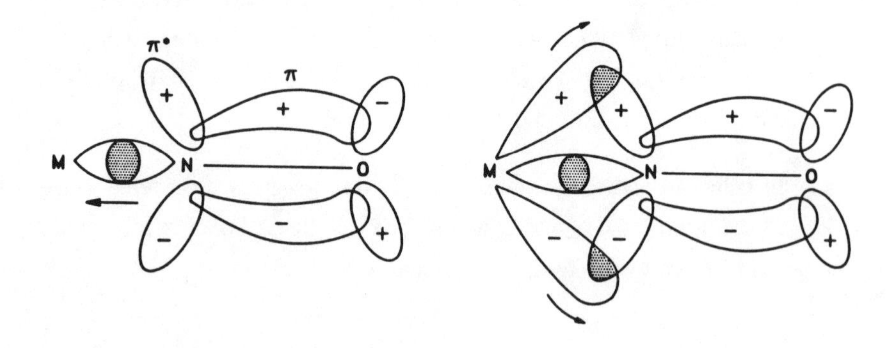

Thus, the M-NO bonding may be considered to be made up of two components (a) donation of electron density from a σ-type orbital on NO onto the metal, and (b) a donation of electron density from the occupied metal *d*-orbitals into the π^* antibonding orbitals of the NO ligand. This latter M to NO donation of electron density is normally referred to as "backdonation", a term which erroneously implies that it occurs as a consequence of nitrosyl-to-metal σ donation.

The bonding picture described above is analogous to that proposed for metal carbonyl bonding interactions *(181)*. However, there are significant differences in the natures of the electron distributions in the M-NO and M-CO links. One important difference stems from the fact that NO is more electronegative than CO *(182)*, the result being that the NO ligand is a better overall electron acceptor than CO *(183)*. Indeed, NO is normally considered to be a weaker σ-donor, but a better π-acceptor, than CO *(184-188)*. Another important difference is that in the M-NO link, the M-N bond is usually strong (c.f. the valence-bond picture), whereas the N-O bond is relatively weak. This contrasts with the M-CO link, where the M-C bond is relatively weak, and the C-O bond is strong *(183,189)*. An important chemical manifestation of this trend is in the reactivities of these groups: many metal nitrosyls lose their nitrosyl oxygens to form metal nitrides, whereas the displacement of intact CO from metal carbonyls is the common occurrence.

One final point in this comparison of NO and CO as ligands is that in the M-NO link there is a more equal electron-density distribution in the valence molecular orbitals of this group *(182,189)*, whereas in the M-CO link, more charge is transferred to the C atom than to the O atom *(189)*. The important feature to remember, however, is that the NO ligand is a strong π-acid capable of imparting unique stabilities to its complexes by increasing the HOMO-LUMO gaps of these complexes *(185,186,190)*.

1.4.3: Comparison of NO and NS as Ligands

It is appropriate, at this point, to compare the bonding of the NO ligand to the related NS ligand. It may be noted that NO is valence isoelectronic with

NS (sulfur nitride) *(191)*. However, NO and NS have one major difference in that NS has a relatively large dipole moment (1.86 D) compared to NO (0.16 D). Although the transient, monomeric NS radical has been thermally and photochemically generated *(192-194)*, it is unstable and rapidly dimerizes/polymerizes to $(NS)_x$ *(194,195)*. The NS^+ cation is also valence isoelectronic with NO^+, but unlike the latter, the positive charge in the cation is predominantly localized on the sulfur rather than the nitrogen atom *(196)*. Thus the NS^+ cation *(197-201)* is best referred to as the *thiazyl* rather than the *thionitrosonium* cation.

Notwithstanding the differences in free NO/NO^+ and NS/NS^+ mentioned above, the NS ligand binds through the nitrogen to metal centers to generate M-NS (*thionitrosyl*) complexes. Many such complexes are now known, *(194,202-210)*, and the binding properties of NO vs NS continue to be debated. It appears that whether NO is a better overall electron acceptor than NS or not depends largely on the charge density in the complex *(209,211)*. The data presented in Table 1.5 serve to illustrate this point well *(209)*.

Table 1.5 IR Data for some Isoelectronic $[CpM(CO)_2NX]^{n+}$ Complexes.[a]

	Nitrosyl (X = O)		Thionitrosyl (X = S)	
	ν_{NO} (cm^{-1})	ν_{CO} (cm^{-1})	ν_{NS} (cm^{-1})	ν_{CO} (cm^{-1})
$CpCr(CO)_2NX$	1706	2020	1165	2023
		1946		1963
$[CpMn(CO)_2NX]^+$	1878 sh	2114	1284	2108
	1838 vs	2075		2070
$[CpFe(CO)_2NX]^{2+}$	1982	2198	1388	2178
		2183		2165

[a] Data taken from reference 209.

In the neutral Cr complexes, the CO stretching frequencies for the thionitrosyl compound are slightly higher than those of the nitrosyl analogue, implying that the NS ligand is more electron-withdrawing than NO *(202,207)*. However, when the charge on the metal complex is increased, it appears that the NS ligand is able to provide more electron density to the metal center than the NO ligand. For instance, the slightly decreased CO stretching frequencies for the monocationic, manganese thionitrosyl complex (relative to its nitrosyl analogue) is consistent with the existence of a more electron-rich metal center in this compound. The magnitude of this shift increases to about 20 cm^{-1} in the dicationic Fe complexes, where the NS ligand appears to be a far superior overall electron-donating ligand than NO. It must also be noted that in going from Cr to Mn to Fe, the π-donor abilities of the metal decrease, thereby resulting in higher values for both the CO and NX stretching frequencies *(209)*.

In conclusion, it must be remembered that sulfur is much less electronegative than oxygen, and the nitrogen in NS thus bears a greater negative charge than does the nitrogen in NO. As a consequence of this, the NS ligand in a thionitrosyl compound may be able to vary its electron-accepting or -donating abilities depending on the electronic requirements of the metal center. It is not too surprising, therefore, that the NO vs NS dichotomy is still under active debate *(206,212-215)*. In closing this section, it may be noted that only a few polythionitrosyl complexes have been synthesized, and this may be due to the propensity of the NS ligands to couple in the metal's coordination sphere *(216)*.

1.4.4: The Enemark-Feltham Notation for Metal Nitrosyls

It is now generally accepted that the bonding in the $M(NO)_x$ moieties of metal nitrosyl complexes is largely of a covalent nature, and that the electron distribution in the M-NO unit is more or less evenly distributed in this group. As a result, it is unreasonable to assign formal oxidation states to the metal and the nitrosyl ligand in M-NO complexes (vide supra). Enemark and Feltham *(217)* have proposed the $\{M(NO)_x\}^n$ formalism for metal nitrosyls which presumes that the bonding extant in the $M(NO)_x$ fragments of metal nitrosyl complexes, $L_yM(NO)_x$, can be considered as the bonding extant in the hypothetical $M(NO)_x$ group that is perturbed by the presence of ancillary ligands (L_y). In this notation, the value of *n* is taken as the total number of electrons associated with the metal *d* and the π^* (NO) orbitals, or alternatively, the familiar number of *d* electrons on the metal in the "L_yM" group, plus the π^* electron from each of the NO ligands (which is roughly equivalent to considering all the NO ligands being bound as NO^+). For example, $Mn(NO)(CO)_4$ is classified as $\{Mn(NO)\}^8$. This is because (i) the "neutral" CO ligands do not change the effective *d*-electron count (of seven) of the metal in the "$Mn(CO)_4$" fragment, and (ii) adding one electron from the NO ligand to this *d*-electron count results in the value of *n* being 8. Similarly, CpNi(NO) is classified as $\{Ni(NO)\}^{10}$, since (i) the cyclopentadienyl monoanion *reduces* the *d*-electron count on Ni in the "CpNi" fragment from ten (in isolated Ni) to nine, and (ii) addition of one electron from the NO ligand brings this number back to ten. Other examples using this notation are shown below:

compound	Enemark-Feltham notation	
$Re(NO)(H)_2(PPh_3)_3$	$\{Re(NO)\}^6$	Re (d^{7-2}) + 1 π^* electron
$CpW(NO)_2PPh_3$	$\{W(NO)_2\}^7$	W (d^{6-1}) + 2 π^* electrons
$Ir(NO)Cl_2(PPh_3)_2$	$\{Ir(NO)\}^8$	Ir (d^{9-2}) + 1 π^* electron
$Co(NO)(CO)_3$	$\{Co(NO)\}^{10}$	Co (d^9) + 1 π^* electron

In these examples, CO and PPh_3 are considered to be neutral ligands, whereas Cp, H and Cl are considered to be monoanionic ligands. Thus, a simple way to arrive at the stated notation for each complex is to assign typical oxidation states to all the ligands except NO. Addition of the appropriate number of π^* electrons to the resultant metal *d*-electron count will then result in the correct value of *n*. By using such a formalism, a general analysis of the structures, bonding and reactivity patterns of metal nitrosyl complexes can be made, and this formalism is especially useful in predicting the $M(NO)_x$ geometries of a large number of metal nitrosyl complexes *(218-220)*, as will be discussed fully in the next Chapter.

1.5: References

1. Harvey, K. B.; Porter, G. B. *Introduction to Physical Inorganic Chemistry*; Addison-Wesley: Massachusetts, 1963; pp 157-159.
2. Lawlor, L. J.; Vasudevan, K.; Grein, F. *J. Am. Chem. Soc.* **1978**, *100*, 8062.
3. Jones, K. In *Comprehensive Inorganic Chemistry*; Pergamon: Oxford, 1973; Vol. 2, Chapter 19.
4. Wong, M. W.; Nobes, R. H.; Bouma, W. J.; Radom, L. *J. Chem. Phys.* **1989**, *91*, 2971.
5. Natalis, P.; Collin, J. E.; Delwiche, J.; Caprace, G.; Hubin, M. J. *J. Electron Spectrosc. Relat. Phenom.* **1979**, *17*, 421.
6. Allendorf, S. W.; Leahy, D. J.; Jacobs, D. C.; Zare, R. N. *J. Chem. Phys.* **1989**, 91, 2216.
7. Rudolf, H.; Mckoy, V. *J. Chem. Phys.* **1989**, 91, 2235
8. (a) Lee, K. Y.; Kuchynka, D. L.; Kochi, J. K. *Inorg. Chem.* **1990**, *29*, 4196. (b) Kim, E. K.; Kochi, J. K. *J. Am. Chem. Soc.* **1991**, *113*, 4962.
9. O'Keefe, A.; Derai, R.; Bowers, M. T. *Chem. Phys. Lett.* **1985**, *113*, 93.
10. Siegel, M. W.; Celotta, R. J.; Hall, J. L.; Levine, J.; Bennett, R. A. *Phys. Rev. A* **1972**, *6*, 607.
11. Teillet-Billy, D.; Fiquet-Fayard, F. *J. Phys.* **1977**, *10B*, L111.
12. Chalasinski, G.; Kukawska-Tarnawska, B. *J. Phys. Chem.* **1990**, *94*, 3450.
13. Coe, J. V.; Snodgrass, J. T.; Freidhoff, C. B.; McHugh, K. M.; Bowen, K. H. *J. Chem. Phys.* **1987**, *87*, 4302.
14. Sunder, W. A.; Wayda, A. L.; Distefano, D.; Falconer, W. E.; Griffiths, J. E. *J. Fluorine Chem.* **1979**, *14*, 299.
15. Eller, P. G.; Penneman, R. A. *J. Less-Common Met.* **1987**, *127*, 19.
16. Klavsut, G. N.; Markov, A. V.; Solov'ev, V. N. *Vesti Akad. Navuk BSSR, Ser. Fiz.-Energ. Navuk* **1989**, 41. CA111:237367y.
17. Sato, N.; Kigoshi, A. *J. Fluorine Chem.* **1982**, *20*, 365.
18. Sato, N. Kigoshi, A. *Thermochim. Acta* **1982**, *57*, 141.

19. Moravek, R. T.; Kauffman, G. B. *Inorg. Synth.* **1986**, *24*, 217.
20. Bougon, R.; Bui, H. T.; Lance, M.; Abazli, H. *Inorg. Chem.* **1984**, *23*, 3667.
21. Prinz, H.; Dehnicke, K.; Mueller, U. *Z. Anorg. Allg. Chem.* **1982**, *488*, 49.
22. Sato, N.; Kigoshi, A. *Thermochim. Acta* **1981**, *47*, 253.
23. Kigoshi, A. *Thermochim. Acta* **1979**, *29*, 147.
24. Thrasher, J. S.; Seppelt, K. *Z. Anorg. Allg. Chem.* **1985**, *529*, 85.
25. Faggiani, R.; Gillespie, R. J.; Kolis, J.; Malhotra, K. C. *J. Chem. Soc., Chem. Commun.* **1987**, 591.
26. Klavsut, G. N.; Markov, A. V.; Khod'ko, N. N. *Russ. J. Inorg. Chem.* **1989**, *34*, 2516.
27. Klavsut, G. N.; Markov, A. V.; Khod'ko, N. N. *Russ. J. Inorg. Chem.* **1989**, *34*, 1438.
28. Lee, K. C.; Aubke, F. *Inorg. Chem.* **1979**, *18*, 389.
29. Griffiths, J. E.; Sunder, W. A. *Spectrochim. Acta* **1979**, *35A*, 1329.
30. Dehnicke, K.; Prinz, H.; Massa, W.; Pebler, J.; Schmidt, R. *Z. Anorg. Allg. Chem.* **1983**, *499*, 20.
31. Christe, K. O.; Wilson, W. W.; Bougon, R. A. *Inorg. Chem.* **1986**, *25*, 2163.
32. Miki, E.; Tanaka, M.; Mizumachi, K.; Ishimori, T. *Bull. Chem. Soc. Jpn.* **1986**, *59*, 3275.
33. (a) Christe, K. O.; Wilson, W. W.; Wilson, R. D. *Inorg. Chem.* **1989**, *28*, 904. (b) Christe, K. O.; Curtis, E. C.; Dixon, D. A.; Mercier, H. P.; Sanders, J. C. P.; Schrobilgen, G. J. *J. Am. Chem. Soc.* **1991**, *113*, 3351.
34. Kroenke, P.; Wahren, M. *Isotopenpraxis* **1984**, *20*, 108. CA100:199394t.
35. Connelly, N. G.; Dragett, P. T.; Green, M.; Kuc, T. A. *J. Chem. Soc., Dalton Trans.* **1977**, 70.
36. Berthold, H. J.; Baethge, H. G.; Hoelscher, B. G.; Kienert, H. J.; Ludwig, W.; Molepo, J. M.; Wartchow, R. *Ber. Bunsenges. Phys. Chem.* **1983**, *87*, 245.
37. Callanan, J. E.; Granville, N. W.; Green, N. H.; Staveley, L. A. K.; Weir, R. D.; White, M. A. *J. Chem. Phys.* **1981**, *74*, 1911.
38. Finch, A.; Gates, P. N.; Page, T. H. *J. Inorg. Nucl. Chem.* **1980**, *42*, 292.
39. Cosby, P. C.; Helm, H. *J. Chem. Phys.* **1981**, *75*, 3882.

40. Bardsley, J. N. *Planet. Space Sci.* **1983**, *31*, 667.
41. Reents, W. D., Jr.; Freiser, B. S. *J. Am. Chem. Soc.* **1981**, *103*, 2791.
42. Morris, R. A.; Viggiano, A. A.; Dale, F.; Paulson, J. F. *J. Chem. Phys.* **1988**, *88*, 4772.
43. Heninger, M.; Fenistein, S.; Durup-Ferguson, M.; Ferguson, E. E.; Marx, R.; Mauclaire, G. *Chem. Phys. Lett.* **1986**, *131*, 439.
44. Wilcox, J. B.; Harbol, K. L.; Moran, T. F. *J. Phys. Chem.* **1981**, *85*,3415.
45. Dobler, W.; Federer, W.; Howorka, F.; Lindinger, W.; Durup-Ferguson, M.; Ferguson, E. *J. Chem. Phys.* **1983**, *79*, 1543.
46. Dotan, I.; Fehsenfeld, F. C.; Albritton, D. L. *J. Chem. Phys.* **1979**, *71*, 3280.
47. Aparina, E. V.; Balakai, A. A.; Kir'yakov, N. V.; Markin, M. I.; Tal'roze, V. L. *Dokl. Akad. Nauk SSSR* **1988**, *300*, 871. CA111:181038x.
48. Hiraoka, K.; Yamabe, S. *J. Chem. Phys.* **1989**, *90*, 3268.
49. Minh, T. N. *Chem. Phys. Lett.* **1985**, *117*, 571.
50. Ferguson, E. E.; Adams, N. G.; Smith, D.; Alge, E. *J. Chem. Phys.* **1984**, *80*, 6095.
51. Speller, C. V.; Fitaire, M.; Pointu, A. M. *J. Chem. Phys.* **1983**, *79*, 2190.
52. Eisele, F. L.; Ellis, H. W.; McDaniel, E. W. *J. Chem. Phys.* **1979**, *70*, 5924.
53. Dheandhanoo, S.; Johnson, R. *Planet. Space Sci.* **1983**, *31*, 933.
54. Mauclaire, G.; Heninger, M.; Fenistein, S.; Wronka, J.; Marx, R. *Int. J. Mass Spectrom. Ion Processes* **1987**, *80*, 99.
55. Richter, R.; Lindinger, W.; Ferguson, E. E. *J. Chem. Phys.* **1988**, *89*, 5692.
56. Castro, A.; Mosquera, M.; Prieto, M. F. R.; Santaballa, J. A.; Tato, J. V. *J. Chem. Soc., Perkin Trans. 2* **1988**, 1963.
57. Pullman, A.; Ranganathan, S. *Chem. Phys. Lett.* **1984**, *107*, 107.
58. Randeniya, L. K.; Zeng, X.; Smith, M. A. *Chem. Phys. Lett.* **1988**, *147*, 346.
59. Van Koppen, P. A. M.; Derai, R.; Kemper, P.; Liu, S.; Bowers, M. T. *Int. J. Mass Spectrom. Ion Processes* **1986**, *73*, 41.
60. Rakshit, A. B. *Int. J. Mass Spectrom. Ion Phys.* **1980**, *36*, 31.
61. Anchell, J. L.; Harriman, J. E. *J. Chem. Phys.* **1990**, *92*, 2943.

62. Dotan, I.; Barlow, S. E.; Ferguson, E. E. *Chem. Phys. Lett.* **1985**, *121*, 38.
63. Ferguson, E. E.; Smith, D.; Adams, N. G. *Int. J. Mass Spectrom. Ion Processes* **1984**, *57*, 243.
64. Kato, T. J. *Chem. Phys.* **1984**, *80*, 6105.
65. Dotan, I.; Fehsenfeld, F. C.; Albritton, D. L. *J. Chem. Phys.* **1979**, 71, 3289.
66. Dowek, D.; Dhuicq, D.; Barat, M. *Phys. Rev. A* **1983**, *28*, 2838.
67. Kim, E. K.; Kochi, J. K. *J. Org. Chem.* **1989**, *54*, 1692.
68. Brownstein, S.; Gabe, E.; Lee, F.; Tan, L. *J. Chem. Soc., Chem. Commun.* **1984**, 1566.
69. Brownstein, S.; Gabe, E.; Irish, B.; Lee, F. Louie, B.; Piotroski, A. *Can. J. Chem.* **1987**, *65*, 445.
70. Brownstein, S.; Gabe, E.; Lee, F.; Piotrowski, A. *Can. J. Chem.* **1986**, *64*, 1661.
71. Heo, G. S.; Hillman, P. E.; Bartsch, R. A. *J. Heterocycl. Chem.* **1982**, *19*, 1099.
72. Savoie, R.; Pigeon-Gosselin, M.; Rodrigue, A.; Chenevert, R. *Can. J. Chem.* **1983**, *61*, 1248.
73. Myers, C.; Frech, R. *Solid State Ionics* **1983**, *8*, 49.
74. Kummer, J. T. *Inorg. Synth.* **1979**, *19*, 51.
75. Ruan, J. Z.; Litt, M. H. *Macromolecules* **1988**, *21*, 882.
76. Collman, J. P.; McDevitt, J. T.; Leidner, C. R.; Yee, G. T.; Torrance, J. B.; Little, W. A. *J. Am. Chem. Soc.* **1987**, *109,* 4606.
77. Miyahara, M. *Chem. Pharm. Bull.* **1986**, *34*, 1950.
78. Vogel, F. L.; Forsman, W. C. US Patent 4461719 A, 1984. CA101:182182v.
79. Vogel, F. L.; Forsman, W. C. US Patent 4382882 A, 1983. CA99:31759c.
80. Inabe, T.; Kannewurf, C. R.; Lyding, J. W.; Moguel, M. K.; Marks, T. J. *Mol. Cryst. Liq. Cryst.* **1983**, *93*, 355.
81. Rubner, M.; Cukor, P.; Jopson, H.; Deits, W. *J. Electron. Mater.* **1982**, *11*, 261.
82. Rachdi, F.; Bernier, P.; Billaud, D.; Pron, A.; Przyluski, *J. Polymer* **1981**, *22*, 1605.

83. Billaud, D.; Pron, A.; Vogel, F. L.; Herold, A. *Mater. Res. Bull.* **1980**, *15*, 1627.
84. Iwaki, M.; Fujihana, T.; Okitaka, K. *Mater. Sci. Eng.* **1985**, *69*, 211.
85. (a) Curtiss, L. A.; Maroni, V. A. *J. Phys. Chem.* **1986**, *90*, 56. (b) Maity, D. K.; Bhattacharyya, S. P. *J. Am. Chem. Soc.* 1990, *112*, 3223.
86. Alberts, I. L.; Handy, N. C.; Palmieri, P. *Chem. Phys. Lett.* **1986**, *129*, 176.
87. Walch, S. P.; Rohlfing, C. M. *J. Chem. Phys.* **1989**, *91*, 2939.
88. Smolyar, A. E.; Zaretskii, N. P.; Charkin, O. P. *Zh. Neorg. Khim.* **1979**, *24*, 3160. CA92:65109m.
89. Bruna, P. J. *Chem. Phys.* **1980**, *49*, 39.
90. Bhatia, S. C.; Hall, J. H., Jr. *J. Chem. Phys.* **1985**, *82*, 1991.
91. Carter, S.; Mills, I. M.; Murrell, J. N. *J. Chem. Soc., Faraday Trans. 2* **1979**, *75*, 148.
92. Grimley, A. J.; Houston, P. L. *J. Chem. Phys.* **1980**, *72*, 1471.
93. Beckham, L. J.; Fessler, W. A.; Kise, M. A. *Chem. Rev.* **1951**, *48*, 319.
94. Forte, E.; Hippler, H.; Van der Bergh, H. *Int. J. Chem. Kinet.* **1981**, *13*, 1227.
95. Weiss, R.; Wagner, K. G. *Chem. Ber.* **1984**, *117*, 1973.
96. Foster, S. C.; Johns, J. W. C. *J. Mol. Spectrosc.* **1984**, *103*, 176.
97. Jacox, M. E. *J. Phys. Chem.* **1983**, *87*, 4940.
98. Kinkead, S. A.; Schreeve, J. M. *Inorg. Chem.* **1984**, *23*, 4174.
99. Sugawara, K.; Nakanaga, T.; Takeo, H.; Matsumura, C. *J. Phys. Chem.* **1989**, *93*, 1894.
100. Stepaniuk, N. J.; Lamb, B. J. US Patent 4908270 A, 1990. CA112:201683e.
101. Pass, G.; Sutcliffe, H. *Practical Inorganic Chemistry*; Chapman-Hall: London, 1968; pp 145-146.
102. Kalbandkeri, R. G.; Padma, D. K.; Murthy, A. R. V. *Z. Anorg. Allg. Chem.* **1979**, *450*, 103.
103. Bak, B.; Nielsen, O. J. *J. Labelled Compd. Radiopharm.* **1978**, *15(Suppl. Vol)*, 715.
104. Boughriet, A.; Coumare, A.; Fischer, J. C.; Wartel, M. *Int. J. Chem. Kinet.* **1988**, *20*, 775.
105. Umemoto, T.; Tsutsumi, H. *Bull. Chem. Soc. Jpn.* **1983**, 56, 631.
106. Jpn. Patent 57/70841 A2, 1982. CA97:144340m.

107. Jonkers, G.; Mooyman, R.; De Lange, C. A. *Chem. Phys.* **1981**, *57*, 97.
108. Nadler, I.; Noble, M.; Reisler, H.; Wittig, C. *J. Chem. Phys.* **1985**, *82*, 2608.
109. Dixon, R. N.; Johnson, P. *J. Mol. Spectrosc.* **1985**, *114*, 174.
110. Sorenson, S. A.; Noble, P. N. *J. Chem. Phys.* **1982**, *77*, 2483.
111. Tang, K. Y.; Fairchild, P. W.; Lee, E. K. C. *J. Phys. Chem.* **1979**, *83*, 569.
112. Cheskis, S. G.; Nadtochenko, V. A.; Sarkisov, O. M. *Int. J. Chem. Kinet.* **1981**, *13*, 1041.
113. Esposti, C. D.; Cazzoli, G.; Favero, P. G. *J. Mol. Spectrosc.* **1985**, *109*, 229.
114. Konarski, J. *Acta Phys. Pol.* **1986**, *A69*, 685.
115. Cazzoli, G.; Esposti, C. D.; Favero, P. G.; Palmieri, P. *Nuovo Cimento Soc. Ital. Fis.* **1984**, *3D*, 627.
116. Bai, Y. Y.; Ogai, A.; Qian, C. X. W.; Iwata, L.; Segal, G. A.; Reisler, H. *J. Chem. Phys.* **1989**, *90*, 3904.
117. Jones, L. H.; Swanson, B. I. *J. Phys. Chem.* **1991**, *95*, 86.
118. Feuerhahn, M.; Hilbig, W.; Minkwitz, R.; Engelhardt, U. *Spectrochim. Acta* **1978**, *34A*, 1065.
119. Hirota, E. *Nippon Kagaku Kaishi* **1986**, 1438.
120. Bell, S. *J. Chem. Soc., Faraday Trans. 2* **1981**, *77*, 321.
121. Bak, B.; Nicolaisen, F. M.; Nielsen, O. J.; Skaarup, S. *J. Mol. Struct.* **1979**, *51*, 17.
122. Joergensen, K. A.; El-Wassimy, M. T. M.; Lawesson, S. O. *Acta Chem. Scand.* **1983**, *B37*, 785.
123. Joergensen, K. A.; Lawesson, S. O. *J. Am. Chem. Soc.* **1984**, *106*, 4687.
124. Williams, D. L. H. *Nitrosation*; Cambridge Univ. Press; New York, 1988.
125. Picard-Bersellini, A.; Cheikh, M.; Broquier, M. *Chem. Phys.* **1989**, *133*, 461.
126. Bell, A. J.; Pardon, P. R.; Frey, J. G. *Mol. Phys.* **1989**, *67*, 465.
127. Qian, C. X. W.; Ogai, A.; Iwata, L.; Reisler, H. *J. Chem. Phys.* **1988**, *89*, 6547.
128. Bruno, A. E.; Bruhlmann, U.; Huber, J. R. *Chem. Phys.* **1988**, *120*, 155.
129. Del Barrio, J. I.; Tablas, F. M. G. *Opt. Commun.* **1986**, *58*, 245.
130. Bechara, J.; Cunningham, D. G.; Denvir, D. J.; Gillan, I.; Morrow, T.; McGrath, W. D. *Opt. Acta* **1986**, *33*, 473.

131. Del Barrio, J. I.; Rebato, J. R.; Tablas, F. M. G. *Chem. Phys. Lett.* **1985**, *122*, 59.
132. Del Barrio, J. I.; Rebato, J. R.; Tablas, F. M. G. *Opt. Commun.* **1985**, *55*, 280.
133. Hawkins, K. C.; Peacock, R.; Rutt, H. N. *J. Phys. D* **1985**, *18*, 191.
134. Wilson, M. W.; Rothschild, M.; Muller, D. F.; Rhodes, C. K. *J. Chem. Phys.* **1982**, *76*, 4452.
135. Moser, M. D.; Weitz, E.; Schatz, G. C. *J. Chem. Phys.* **1983**, *78*, 757.
136. Lahmani, F.; Lardeux, C.; Solgadi, D. *J. Chem. Phys.* **1982**, *77*, 275.
137. Moser, M. D.; Weitz, E. *Chem. Phys. Lett.* **1981**, *82*, 285.
138. Michael, J. V.; Lee, J. H. *J. Phys. Chem.* **1979**, *83*, 10.
139. Scharf, H. D.; Fleischhauer, J.; Leismann, H.; Ressler, I.; Schleker, W.; Weitz, R. *Angew. Chem., Int. Ed. Engl.* **1979**, *91*, 652.
140. Qian, C. X. W.; Ogai, A.; Reisler, H.; Wittig, C. *J. Chem. Phys.* **1989**, *90*, 209.
141. Klippenstein, S. J.; Khundkar, L. R.; Zewail, A. H.; Marcus, R. A. *J. Chem. Phys.* **1988**, *89*, 4761.
142. Nadler, I.; Pfab, J.; Radhakrishnan, G.; Reisler, H.; Wittig, C. *J. Chem. Phys.* **1983**, *79*, 2088.
143. Knee, J. L.; Khundkar, L. R.; Zewail, A. H. *J. Phys. Chem.* **1985**, *89*, 4659.
144. Qian, C. X. W.; Noble, M.; Nadler, I.; Reisler, H.; Wittig, C. *J. Chem. Phys.* **1985**, *83*, 5573.
145. Dupuis, M.; Lester, W. A., Jr. *J. Chem. Phys.* **1985**, *83*, 3990.
146. Khundkar, L. R.; Knee, J. L.; Zewail, A. H. *J. Chem. Phys.* **1987**, *87*, 77.
147. Pfab, J.; Haeger, J.; Krieger, W. *J. Chem. Phys.* **1983**, *78*, 266.
148. Gowenlock, B. G.; Radom, L. *Aust. J. Chem.* **1978**, *31*, 2349.
149. Dixon, R. N.; Noble, M. *Springer Ser. Chem. Phys.* **1979**, *6*, 81.
150. Petersen, J. C. *J. Mol. Spectrosc.* **1985**, *110*, 277.
151. Lahmani, F.; Lardeaux, C.; Solgadi, D. *Chem. Phys. Lett.* **1983**, *102*, 523.
152. Jain, R.; Schultz, J. S. *J. Membr. Sci.* **1986**, *26*, 313.
153. Pravilov, A. M.; Ryabov, S. E. *High Energy Chem.* **1982**, *16*, 331.
154. Vilesov, F. I.; Karpov, L. G.; Kozlov, A. S.; Pravilov, A. M.; Smirnova, L. G. *High Energy Chem.* **1978**, *12*, 468.

155. Bai, H.; Ault, B. S. *J. Phys. Chem.* **1990**, *94*, 606.
156. Lucas, D.; Allamandola, L. J.; Pimentel, G. C. *Croat. Chem. Acta* **1982**, *55*, 121.
157. Murata, S. *Kagaku Gijutsu Kenkyusho Hokoku* **1984**, *79*, 597. CA102:176330w.
158. Mueller, R. P.; Murata, S.; Huber, J. R. *Chem. Phys.* **1982**, *66*, 237.
159. Ensley, H. E.; Mahadevan, S. *Tetrahedron Lett.* **1989**, *30*, 3255.
160. Kirby, G. W. *Chem. Soc. Rev.* **1977**, *6*, 1.
161. Mootz, D.; Poll, W. *Z. Naturforsch., B.; Anorg. Chem., Org. Chem.* **1984**, *39B*, 1300.
162. Lavallee, F. A.; Russ, C. R. *Inorg. Nucl. Chem. Lett.* **1970**, *6*, 527.
163. Abbatt, J. P. D.; Toohey, D. W.; Fenter, F. F.; Stevens, P. S.; Brune, W. H.; Anderson, J. G. *J. Phys. Chem.* **1989**, *93*, 1022.
164. Wategaonkar, S. J.; Setser, D. W. *J. Chem. Phys.* **1989**, *90*, 251.
165. Sedlacek, A. J.; Wight, C. A. *J. Chem. Phys.* **1987**, *86*, 2787.
166. Finlayson-Pitts, B. J.; Ezell, M. J.; Wang, S. Z.; Grant, C. E. *J. Phys. Chem.* **1987**, *91*, 2377.
167. Napoleone, V.; Schelly, Z. A. *J. Phys. Chem.* **1980**, *84*, 17.
168. Wartel, M.; Boughriet, A.; Fischer, J. C. *Anal. Chim. Acta* **1979**, *110*, 211.
169. Finlayson-Pitts, B. J.; Livingston, F. E.; Berko, H. N. *J. Phys. Chem.* **1989**, *93*, 4397.
170. Smolyar, A. E.; Zaretskii, N. P.; Klimenko, N. M.; Charkin, O. P. *Zh. Neorg. Khim.* **1979**, *24*, 3165. CA92:82885d.
171. Harcourt, R. D.; Roso, W. *Int. J. Quantum Chem.* **1979**, *16*, 1033.
172. Girard-Dussau, N.; Dargelos, A.; Chaillet, M. *THEOCHEM* **1982**, *6*, 123.
173. Gerber, S.; Huber, H. *J. Phys. Chem.* **1989**, *93*, 545.
174. Andrews, W. L. S.; Pimentel, G. C. *J. Chem. Phys.* **1966**, *44*, 2361.
175. Brunet, J.-P.; Durand, G.; Chapuisat, X. *Chem. Phys.* **1985**, *97*, 271.
176. Moeller, T. *J. Chem. Educ.* **1946**, 441.
177. Moeller, T. *J. Chem. Educ.* **1946**, 542.
178. Raynor, J. B. *Inorg. Chim. Acta* **1972**, *6*, 347.
179. Griffith, W. P.; Lewis, J.; Wilkinson, G. *J. Inorg. Nucl. Chem.* **1958**, *7*, 38.
180. Lewis, J.; Irving, R. J.; Wilkinson, G. *J. Inorg. Nucl. Chem.* **1958**, *7*, 32.

181. Griffith, W. P. In *Comprehensive Inorganic Chemistry*; Pergamon: Oxford, 1973; Vol. 4, Chapter 46.
182. Stucky, G. D.; Rye, R. R.; Jennison, D. R.; Kelber, J. A. *J. Am. Chem. Soc.* **1982**, *104*, 5951.
183. Hedberg, L.; Hedberg, K.; Satija, S. K.; Swanson, B. I. *Inorg. Chem.* **1985**, *24*, 2766.
184. Nefedov, V. I. *Koord. Khim.* **1980**, *6*, 163.
185. Bursten, B. E.; Gatter, M. G. *J. Am. Chem. Soc.* **1984**, *106*, 2554.
186. Bursten, B. E.; Gatter, M. G.; Goldberg, K. I. *Polyhedron* **1990**, *9*, 2001.
187. Bursten, B. E.; Jensen, J. R.; Gordon, D. J.; Treichel, P. M.; Fenske, R. F. *J. Am. Chem. Soc.* **1981**, *103*, 5226.
188. Fenske, R. F.; Jensen, J. R. *J. Chem. Phys.* **1979**, *71*, 3374.
189. Avanzino, S. C.; Bakke, A. A.; Chen, H.-W.; Donahue, C. J.; Jolly, W. L.; Lee, T. H. Ricco, A. J. *Inorg. Chem.* **1980**, *19*, 1931.
190. Bursten, B. E.; Cayton, R. H. *Organometallics* **1987**, *6*, 2004.
191. Salahub, D. R.; Messmer, R. P. *J. Chem. Phys.* **1976**, *64*, 2039.
192. Anacona, J. R.; Davies, P. B. *Chem. Phys. Lett.* **1984**, *108*, 128.
193. Henshaw, T. L.; Ongstad, A. P.; Lawconnell, R. I. *J. Phys. Chem.* **1990**, *94*, 3602.
194. Herberhold, M. *Comments Inorg. Chem.* **1988**, *7*, 53.
195. Roesky, H. W.; Witt, M. *Comments Inorg. Chem.* **1981**, *1*, 183.
196. Karna, S. P.; Grein, F. *Chem. Phys.* **1986**, *109*, 35.
197. Mews, R. *Angew. Chem., Int. Ed. Engl.* **1976**, *15*, 691.
198. Apblett, A.; Chivers, T.; Fait, J. F. *Inorg. Chem.* **1990**, *29*, 1643.
199. Apblett, A.; Banister, A. J.; Biron, D.; Kendrick, A. G.; Passmore, J.; Schriver, M.; Stojanac, M. *Inorg. Chem.* **1986**, *25*, 4451.
200. Clegg, W.; Glemser, O.; Harms, K.; Hartman, G.; Mews, R.; Noltemeyer, M.; Sheldrick, G. M. *Acta Crystallogr., Sect. B* **1981**, *B37*, 548.
201. Apblett, A.; Chivers, T.; Fait, J. F. *J. Chem. Soc., Chem. Commun.* **1989**, 1596.
202. Greenhough, T. J.; Kolthammer, B. W. S.; Legzdins, P.; Trotter, J. *Inorg. Chem.* **1979**, *18*, 3548.
203. Lichtenberger, D. L.; Hubbard, J. L. *Inorg. Chem.* **1984**, *23*, 2718.
204. Glemser, O.; Mews, R. *Angew. Chem., Int. Ed. Engl.* **1980**, *19*, 883.
205. Mews, R.; Liu, C. *Angew. Chem., Int. Ed. Engl.* **1983**, *22*, 162.

206. Minelli, M.; Hubbard, J. L.; Lichtenberger, D. L.; Enemark, J. H. *Inorg. Chem.* **1984,** *23,* 2721.

207. Kolthammer, B. W. S.; Legzdins, P. *J. Am. Chem. Soc.* **1978,** *100,* 2247.

208. Beran, K.; Steinke, G.; Mews, R. *Chem. Ber.* **1989,** *122,* 1613.

209. Hartman, G.; Mews, R. *Angew. Chem.* **1985,** *97,* 218.

210. Roesky, H. W.; Pandey, K. K. *Adv. Inorg. Chem. Radiochem.* **1983,** *26,* 337.

211. Herberhold, M.; Smith, P. D.; Alt, H. G. *J. Organomet. Chem.* **1980,** *191,* 79.

212. Hubbard, J. L.; Lichtenberger, D. L. *Inorg. Chem.* **1980,** *19,* 3866.

213. Pandey, K. K.; Ahuja, S. R.; Goyal, M. *Indian J. Chem.* **1985,** *24A,* 1059.

214. Lichtenberger, D. L.; Hubbard, J. L. *Inorg. Chem.* **1985,** *24,* 3835.

215. Parisod, G.; Comisarow, M. B. *Adv. Mass Spectrom.* **1980,** *8A,* 212.

216. Kersting, M.; Hoffmann, R. *Inorg. Chem.* **1990,** *29,* 279.

217. Enemark, J. H.; Feltham, R. D. *Coord. Chem. Rev.* **1974,** *13,* 339.

218. Feltham, R. D.; Enemark, J. H. *Top. Stereochem.* **1981,** *12,* 155.

219. Michael, D.; Mingos, P.; Sherman, D. J. *Adv. Inorg. Chem.* **1989,** *34,* 293.

220. Mingos, D. M. P. *Inorg. Chem.* **1973,** *12,* 1209.

2

Synthesis and Characterization of Metal Nitrosyls

The chemical construction of the MNO group may be divided into two categories depending on whether the 'NO' group has to be introduced onto the metal center by chemical reagents capable of delivering 'NO' or whether a pre-formed nitrogen-containing ligand may be suitably derivatized by chemical means to produce the M-(NO) moiety *(1)*. In this chapter, the various methods that have been applied to the synthesis of metal nitrosyls are briefly discussed. Only procedures that result in the formation of the MNO group are presented. Other chemistry that does not result in the production of one or more MNO links (e.g., chemistry involving the transformation of other ancillary ligands) is not considered in this chapter. Common methods used to identify and characterize MNO group(s) are also presented later in this chapter. More detailed treatments on the synthesis and characterization of metal nitrosyl complexes are available in the literature *(2-4)*.

2.1: Synthesis of Nitrosyl Complexes

2.1.1: External Sources of NO

2.1.1.1: Gaseous Nitrogen Oxides

Nitrogen oxide is a paramagnetic molecule that can combine with metal-centered radicals. Consequently, gaseous nitric oxide (NO) or solutions of the dissolved gas *(5)* may be used to introduce NO onto a metal center. These reactions may either proceed without ligand substitution (i.e., simple adduct formation) or with ligand substitution (i.e. displacement of neutral or anionic ligands) in the metals' coordination spheres.

In the case of simple adduct formation without ligand substitution, the MNO link may be formed either *reversibly* or *irreversibly*. For example, reversible coordination of NO has been observed in a number of Fe-chelate complexes *(6-9)*.

$$Fe(edta) + 2\,NO \rightleftharpoons Fe(NO)_2(edta) \qquad (2.1)$$

$$Fe(acac) + NO \rightleftharpoons Fe(NO)(acac) \qquad (2.2)$$

A number of factors determine the nature and the extent of such adduct formation, and the reactions of NO with the $MnX_2(PR_3)$ complexes [R_3 = Pr_3, Bu_3, $PhMe_2$, $PhEt_2$] both in the solid state and in THF solutions exemplify this point *(10,11)*. When the halide is Cl or Br, reversible coordination of NO is observed (eq. 2.3).

$$MnX_2(PR_3) + NO \rightleftharpoons Mn(NO)X_2(PR_3) \qquad (2.3)$$

$$(X = Cl, Br)$$

However, when the halide is I, the product is unstable and rapidly decomposes via the oxidation of the phosphine ligand. Many cobalt macrocyclic complexes *(12-17)* are known to bind NO, and a variety of factors such as temperature and time determine the stability of the product complexes.

Although nitric oxide is a dangerous reagent to employ in metal atom reactions, some measure of success has been achieved in the condensation of metal vapors with NO gas to produce metal nitrosyl complexes *(18)*, e.g.

$$Co + BF_3 + 3\,PF_3 + NO \longrightarrow Co(NO)(PF_3)_3 + BF_3 \qquad (2.4)$$

$$Mn + BF_3 + PF_3 + 3NO \longrightarrow Mn(NO)_3PF_3 + BF_3 \qquad (2.5)$$

Boron trifluoride reduces the vapor pressure of NO in these experiments since it complexes weakly with NO in the condensed phase. Matrix isolated $Fe(NO)_x$ ($x \leq 4$) forms upon a similar condensation of Fe vapor and NO gas (*19*). In 100% NO, $Fe(NO)_4$ is the major product. On occasion, adsorption of NO onto surfaces (e.g. β-aluminas and zeolites) forms metal-nitrosyl units, but these units will be considered in detail in Chapter 5.

Nitric oxide may also displace ligands in a metal's coordination sphere. In such displacements, each molecule of NO displaces the equivalent of a three-electron ligand *(20,21)*, e.g.,

$$CpCr(CO)_3H + NO \longrightarrow CpCr(NO)(CO)_2 \quad (2.6)$$

$$Co(NO)(CO)_3 + 2NO \longrightarrow Co(NO)_3 + 3\ CO \quad (2.7)$$

$$Cr(CO)_6 + 4NO \longrightarrow Cr(NO)_4 + 6\ CO \quad (2.8)$$

Nitric oxide also displaces two-electron ligands in some monometallic compounds with the concomitant formation of the corresponding bimetallic complexes presumably via the dimerization of the resulting odd-electron species, e.g.,

$$CpCo(CO)_2 + NO \longrightarrow 1/2\ [CpCo(NO)]_2 + 2CO \quad (2.9)$$

or adds to polymetallic complexes with concomitant rupture of metal-metal bonds *(22,23)*, e.g.,

$$1/2\ M_2(OR)_6 + NO \longrightarrow [M(NO)(OR)_3]_n \quad (2.10)$$

$$(M = Mo, W;\ R = alkyl)$$

The term "reductive nitrosylation" has been used previously to describe the formation of an MNO link accompanied by the formal reduction of the metal *(24)*, e.g.,

$$WCl_6 \xrightarrow[MeCN]{NO} W(NO)Cl_3(MeCN)_2 \quad (2.11)$$

In this example (eq 2.11), the tungsten metal center is reduced from the +6 to the +2 oxidation state. However, since metal nitrosyls do not always lend themselves to classification via their oxidation states, this terminology is no longer commonly used.

Unfortunately, many of the reactions of NO gas with metal complexes are not always as straightforward as those shown above. For instance, salt formation may occur on occasion *(25)*,

$$OsCl_5 + 2\ NO \longrightarrow [NO][Os(NO)Cl_5] \qquad (2.12)$$

and it is also worth noting that a common route to the formation of metal oxo compounds involves the use of NO gas on a suitable precursor metal complex *(26)*.

Other nitrogen oxides have been employed for the syntheses of metal nitrosyl complexes. As with their use in nitrosations of organic compounds *(27)*, it is not clear what exactly the reacting species are. For example, dinitrogen trioxide (N_2O_3) has been employed in the nitrosation of metal complexes *(28-30)*, and a typical example of such a reaction is shown in eq 2.13. Thus, even though the reversible dissociation of N_2O_3 into NO and NO_2 is known to occur, it sometimes behaves as though it were actually $[NO]^+[NO_2]^-$ *(31-33)*.

$$Ir(CO)Cl(PPh_3)_2 + N_2O_3 \longrightarrow Ir(NO)(NO_2)(CO)Cl(PPh_3)_2 \qquad (2.13)$$

Equally perplexing is the fact that even though the related dinitrogen tetroxide (N_2O_4) undergoes normal homolytic dissociation into NO_2, it also dissociates into $[NO]^+[NO_3]^-$ in media of high dielectric constant, and it acts effectively as such *(34)* (eq 2.14).

$$Pt(PPh_3)_2(C_2H_4) + N_2O_4 \longrightarrow Pt(NO)(NO_3)(PPh_3)_2 + C_2H_4 \qquad (2.14)$$

The related trioxodinitrate dianion effectively causes replacement of CN^- by NO^- in the reaction summarized in eq 2.15 *(35,36)*,

$$[Ni(CN)_4]^{2} \xrightarrow{[N_2O_3]^{2-}} [Ni(NO)(CN)_3]^{2-} \qquad (2.15)$$

but this reaction is complicated and is also pH dependent.

2.1.1.2: Nitrosonium Salts

Many nitrosonium salts are now known, and their syntheses and properties have been detailed in Section 1.2. In general, these salts are hydrolyzed readily to produce nitrous acid (eq 2.16). Thus, they have to be handled under rigorously dry conditions *(37)*.

$$NO^+ + H_2O \rightleftharpoons HNO_2 + H^+ \quad (2.16)$$

In other protic solvents such as alcohols, the NO^+ cation is involved in a nitrosation reaction resulting in the production of alkyl nitrites (eq 2.17).

$$NO^+ + ROH \rightleftharpoons RONO + H^+ \quad (2.17)$$

However, the nitrosonium cation is stable in carefully purified solvents such as dichloromethane, nitromethane, acetonitrile, ethyl acetate, hexane, carbon tetrachloride and even aromatic solvents such as benzene, in which it forms a charge transfer complex (Section 1.2.2). Tetrahydrofuran is rapidly polymerized by NO^+ *(38,39)*, and it is not the solvent of choice for nitrosation reactions by this cation.

Nitrosonium salts are increasingly being employed for the generation of metal nitrosyls *(37)*. Indeed, some metal powders react with the NO^+ cation to generate nitrosyl species *(40)* (eq 2.18).

$$\text{Fe powder} \xrightarrow{NO^+} Fe_2(NO)_6^{2+} \quad (2.18)$$

The nitrosonium cation (NO^+) is formally a two-electron donor and is isoelectronic with CO. Consequently, it is not surprising that it forms simple 1:1 adducts with some coordinatively unsaturated metal complexes *(41)*, e.g.,

$$Ir(CO)ClL_2 + NO^+ \longrightarrow [Ir(NO)(CO)ClL_2]^+ \quad (2.19)$$

and also readily displaces CO (or other labile 2-electron ligands such as carbon disulfide) in a number of coordinatively saturated complexes to generate their nitrosyl analogues *(42-44)*, e.g.,

$$CpMn(CO)_3 + NO^+ \longrightarrow [CpMn(NO)(CO)_2]^+ + CO \quad (2.20)$$

$$Fe(CO)_2(PPh_3)_2(CS_2) + NO^+ \longrightarrow [Fe(CO)_2(PPh_3)_2NO]^+ + CS_2 \quad (2.21)$$

As was mentioned in Chapter 1, the nitrosonium cation may also function as a potent 1-electron oxidant *(45)*, and on occasion, simple oxidation of the organometallic reagent occurs. Whether oxidation or nitrosyl substitution occurs can been rationalized on the basis of initial NO^+ attack of the metal complex *(44)* to form an intermediate species with a bent MNO linkage. Displacement of a ligand L from the metal's

$$M-L \xrightarrow{NO^+} [M(L)(N{=}O)]^+ \begin{cases} \xrightarrow{-L} [M-NO]^+ & \text{substitution} \\ \xrightarrow{-NO} [M-L]^+ & \text{oxidation} \end{cases}$$

coordination sphere then results in an overall substitution reaction, whereas expulsion of the NO radical results in a net 1-electron oxidation of the initial metal complex.

Replacement of a halide or pseudohalide ligand by NO^+ is also known to occur in a number of reactions of the nitrosonium cation with metal complexes *(46)*, e.g.,

$$CpFe(dppm)(SPh) + NO^+ \longrightarrow [CpFe(NO)(dppm)]^{2+} + SPh^- \quad (2.22)$$

In alcoholic solvents, protonation (eq 2.24) instead of nitrosylation (eq 2.23) may occur at more basic metal centers *(47)*, due to the solvolysis equilibrium between NO^+ and H^+ (eq 2.17).

$$CpRh(CO)PPh_3 \xrightarrow{NO^+} [CpRh(NO)PPh_3]^+ \quad (2.23)$$

$$CpIr(CO)PPh_3 \xrightarrow{NO^+} [CpIr(CO)(PPh_3)H]^+ \quad (2.24)$$

Occasionally, however, the NO^+ reagent attacks a coordinated ligand instead of the metal center in a complex to form new C-N bonds, but these reactions will be treated separately in Chapter 7.

It is also worth mentioning that the related nitronium cation (NO_2^+) has been used to introduce the nitrosyl group onto metal centers (L = MeCN) *(48)*.

$$(MeCp)Mn(CO)_3 + NO_2^+ \longrightarrow [(MeCp)Mn(NO)(CO)_2]^+ + CO_2 \quad (2.25)$$

$$L_2Re(CO)_3Br + NO_2^+ \longrightarrow [L_2Re(NO)(CO)_2Br]^+ + CO_2 \quad (2.26)$$

Commercially available nitronium salts are usually contaminated with the nitrosonium cation *(49)*, and care must always be taken to prepare the nitronium reagent without any NO^+ impurity.

In the gas and liquid phases, nitric acid decomposes spontaneously to release NO_2 (or N_2O_4), which ionizes completely in nitric acid to give NO^+ and NO_3^-. Thus, it is not surprising that many reactions of nitric acid with metal complexes result in the formal addition of NO^+ *(50,51)*, e.g.

$$[M(CN)_6]^{4-} \xrightarrow{HNO_3} [M(NO)(CN)_5]^{2-} \quad (2.27)$$

(M = Fe, Ru, Os)

$$CpMn(CO)_3 \xrightarrow{HNO_3} [CpMn(NO)(CO)_2]^+ \quad (2.28)$$

Interestingly, silver nitrate appears to act as a nitrosylating agent for some Mo and W tricarbonyl compounds, although the mechanism remains unknown *(52)*.

2.1.1.3: Inorganic Nitroso Compounds, XNO

The fact that the nitrosyl halides exist in equilibrium with X_2 and NO (Section 1.3) helps to explain why they react with a variety of metal complexes to:

(i) add both X and NO to the metal center

(ii) add only NO (but not X) to the metal center, or

(iii) add only X (but not NO) to the metal center.

Nitrosyl chloride (ClNO) is the most widely used nitrosyl halide, and examples of all three possible modes of reactivity abound in the literature *(53)*. The reaction described by equation 2.29 *(54)* involves all 3 modes.

$$[CpCr(CO)_3]_2 \xrightarrow{ClNO} CpCr(NO)(CO)_2 + CpCr(NO)_2Cl + [CpCrCl_2]_2 \quad (2.29)$$

However, the addition of *both* X and NO to the metal center is a more common mode of reaction of XNO, as shown in eqs 2.30-2.33 *(30,55-57)*.

$$Ni(PPh_3)_4 \xrightarrow{ClNO} Ni(NO)Cl(PPh_3)_2 \quad (2.30)$$

$$Cp^*Ru(CO)_2Cl \xrightarrow{ClNO} Cp^*Ru(NO)Cl_2 \quad (2.31)$$

$$(arene)Cr(CO)_3 \xrightarrow{ClNO} [Cr(NO)_2Cl_2]_n \quad (2.32)$$

$$Ir(CO)Cl(PPh_3)_2 \xrightarrow{ClNO} Ir(NO)(CO)Cl_2(PPh_3)_2 \quad (2.33)$$

As mentioned previously, only the halide is transferred in some attempted nitrosylation reactions *(58,59)*, e.g.,

$$[Pt_2(P_2O_5H_2)_4]^{4-} \xrightarrow{ClNO} [Pt_2(P_2O_5H_2)_4Cl_2]^{4-} + 2NO \quad (2.34)$$

$$Cp^*W(CO)_3\text{-}W(O)_2Cp^* \xrightarrow{ClNO} Cp^*W(CO)Cl_2 + Cp^*W(O)_2Cl \quad (2.35)$$

In *n*-butanol, nitrosyl chloride equilibrates with hydrochloric acid via the reaction *(60)*

$$ClNO + n\text{-}BuOH \rightleftharpoons BuONO + HCl \quad (2.36)$$

Thus, protonation of metal complexes may well override nitrosyl formation. Also, the nitrosyl halides may effect oxidation of the metal complexes *(53)*, or may simply remove organic fragments from organometallic complexes *(61)*.

Related compounds such as $ONBr_3$, $S_2O_3NO^-$ and the unstable ONSCN are also known to transfer the NO group to inorganic and organic compounds *(28-30,62-64)*. The inefficiency of ONSCN to function as a good nitrosylating agent *(27,65)* is due in part to the fact that the ON-SCN molecule is more covalent in nature that are the nitrosyl halides *(66)*, and its lower reactivity correlates with the established trend that the more nucleophilic the anion (X^-), the lower the reactivity of the X-NO compound.

2.1.1.4: Organic Nitroso Compounds

Three major types of organic nitroso compounds have been employed for the nitrosylations of metal complexes; these are the *N*-, *C*-, and *S*-nitroso compounds (the organic RONO nitrites are considered in the next section).

(i) *N*-Nitroso Compounds

The majority of the nitrosylations employing organic *N*-nitroso compounds have involved the use of Diazald (*N*-methyl-*N*-nitroso-*p*-toluenesulfonamide), which can deliver the 'NO' group either as NO^+ (eq 2.37) or as NO (eq 2.38).

$$[CpCr(CO)_3]^- \xrightarrow{\text{Diazald}} CpCr(NO)(CO)_2 \qquad (2.37)$$

$$Cp^*Cr(CO)_3H \xrightarrow{\text{Diazald}} Cp^*Cr(NO)(CO)_2 \qquad (2.38)$$

Other *N*-nitroso compounds such as Et_2NNO and $NH_2C(=O)N(NO)Me$ have also been employed in similar reactions *(67)*. The susceptibility of N-N bond-cleavage in a variety of *N*-nitrosoamino compounds (R_2N-NO) resulting in the generation of NO in situ and the production of metal nitrosyls has been studied. In general, if R is electron-withdrawing, then nitrosylation occurs readily. However, if R is electron-donating, then nitrosylation appears not to proceed to any appreciable extent *(68)*.

(ii) *C*-Nitroso Compounds

The coordination and structural chemistry of *C*-nitroso compounds has been reviewed *(69-73)*. In addition to binding to metal centers, *C*-nitroso compounds may also be used to effect the nitrosylation of metal centers *(74)*, e.g.,

$$Pt(PPh_3)_4 + CF_3NO \longrightarrow (PPh_3)_2Pt(NO)(CF_3) \qquad (2.39)$$

(iii) *S*-Nitroso Compounds

S-Nitroso compounds have been used previously for effecting the nitrosation of organic compounds *(63,75)*. The RS-NO compounds are relatively unstable but are very reactive, and dissociate with the release of NO (eq 2.40).

$$2\ \text{RS-NO} \longrightarrow \text{RS-SR} + 2\text{NO} \qquad (2.40)$$

This dissociation is temperature-dependent (e.g., the dissociation of *S*-nitrosocysteine) *(76)*. Occasionally, $HgCl_2$ is added to catalyze these decompositions necessary for M-NO formation, but the nitrosylation reactions are complicated at times by the binding of the RS-SR byproduct to the metal center, e.g.,

$$\text{RhCl(PPh}_3)_3 \xrightarrow{\text{ON(Cys)}} \text{Rh(NO)Cl}_2(\text{PPh}_3)(\text{cys})_2 + \text{Rh(NO)Cl}_2(\text{PPh}_3)_2 \qquad (2.41)$$

Trityl thionitrite has also been used for the syntheses of thermally stable metal nitrosyls of the Groups 7 and 8 metals *(77)*.

2.1.1.5: Transfer of Coordinated NO

Some metal nitrosyls readily lose their coordinated nitrosyl groups (Chapter 7), and on occasion, this reactivity pattern may be exploited for the syntheses of other metal nitrosyl complexes, e.g.,

$$\text{M-NO} + \text{M}' \longrightarrow \text{M} + \text{M}'\text{-NO} \qquad (2.42)$$

Indeed, the $[Co(NO)_2X]_2$ complexes (X = halide) have been used extensively for the syntheses of other metal nitrosyls *(78-84)*, e.g.,

$$[\text{CpM(CO)}_3]_2 \xrightarrow{[\text{Co(NO)}_2\text{X}]_2} \text{CpM(NO)(CO)}_2 \qquad (2.43)$$

(M = Cr, Mo, W; X = Cl, Br, I)

$$\text{Mn}_2(\text{CO})_{10} \xrightarrow{[\text{Co(NO)}_2\text{Cl}]_2} \text{Mn(NO)}_3(\text{THF}) \qquad (2.44)$$

Other metal nitrosyls have also been used as nitrosylating agents *(85)*, e.g.,

$$NiCl_2(PPh_3)_2 \xrightarrow{Rh(NO)(PPh_3)_3} Ni(NO)Cl(PPh_3)_2 \qquad (2.45)$$

Nitrosyl porphyrin (P) complexes of Fe and Co have also been employed in such syntheses *(86)*, e.g.,

$$(P)Co(NO) + (P)Fe \longrightarrow (P)Co + (P)Fe(NO) \qquad (2.46)$$

$$[(P)Fe(NO)]^+ + (P)Co \longrightarrow [(P)Fe]^+ + (P)Co(NO) \qquad (2.47)$$

In a similar manner, some cobalt nitrosyls have been used to transfer their nitrosyl groups to hemoglobin and myoglobin *(87,88)*.

2.1.1.6: Nitrite (NO_2^-) Sources

(i) Inorganic Nitrites

The bis(triphenylphosphine)iminium nitrite reagent, PPN(NO_2), is a mild, versatile nitrosylating agent *(89)*, and has been used for the synthesis of monometallic and cluster complexes *(90-92)*.

$$Cr(CO)_6 + PPN[NO_2] \longrightarrow PPN[Cr(NO)(CO)_4] \qquad (2.48)$$

$$Ru(CO)_5 + PPN[NO_2] \longrightarrow PPN[Ru(NO)(CO)_3] \qquad (2.49)$$

Kinetic studies of reaction 2.48 indicate that this nitrosylation reaction is first-order in both reactants and is not inhibited by CO. A mechanism involving nitrite attack on a coordinated CO (with the subsequent loss of carbon dioxide) is consistent with the results obtained from these studies.

$$Cr(CO)_6 \xrightarrow{NO_2^-} [(CO)_5Cr{-}C({=}O){-}O{-}N{=}O]^- \xrightarrow{-CO} [(CO)_4Cr{-}C({=}O){-}O{-}N{=}O]^-$$

$$\longrightarrow [(CO)_4Cr(\kappa^2{-}C(O)ON(O))]^- \longrightarrow [(CO)_4Cr(CO_2)(N{=}O)]^- \xrightarrow{-CO_2} [Cr(CO)_4NO]^-$$

In acidic media, the nitrite ion equilibrates with the NO^+ cation (i.e., the reverse of eq 2.16),

$$NO_2^- + 2H^+ \rightleftharpoons H_2O + NO^+ \qquad (2.50)$$

and sometimes the NO_2^-/H^+ (nitrous acid) combination is necessary for the nitrosylation to take place (*93*), as shown in eqs 2.51-2.52.

$$[Ru(bipy)_2(H_2O)Cl]^+ + NO_2^- \longrightarrow \textit{no reaction} \qquad (2.51)$$

$$[Ru(bipy)_2(H_2O)Cl]^+ + NO_2^-/H^+ \longrightarrow [Ru(NO)(bipy)_2Cl]^{2+} \qquad (2.52)$$

(ii) Organic RONO

Trichloronitromethane (or its bromo analogue) reacts with various metal carbonyls or halides to form ionic nitrosonium salts *(94,95)*, e.g.

$$Fe_2(CO)_9 \xrightarrow{Cl_3CNO_2} [NO][FeOCl_2] \qquad (2.53)$$

$$TaCl_5 \xrightarrow{Cl_3CNO_2} [NO][TaOCl_4] \qquad (2.54)$$

or it may also react with metal compounds to form covalent nitrosyl complexes *(96-98)*.

$$OsCl_5 \xrightarrow{Cl_3CNO_2} Os(NO)Cl_3 \qquad (2.55)$$

Other organic nitrites such as acetyl nitrite or amyl nitrite may also be used in nitrosylation reactions, even though the identity of the "reactive" species may be ambiguous *(99,100)*. For example, acetyl nitrite, MeC(O)ONO, may also be represented as nitrosyl acetate, $[NO]O_2CMe$ (*101*), and indeed in protic solvents RONO complexes equilibrate with NO^+ (eq 2.17). They may also undergo slow hydrolysis in aqueous media *(101)*

$$AcONO + H_2O \rightleftharpoons AcOH + HNO_2 \qquad (2.56)$$

or may even undergo rapid NO exchange with alcohols *(102)*.

$$RONO + R'OH \rightleftharpoons ROH + R'ONO \qquad (2.57)$$

As with any other nitrosylating agent, RONO complexes may also react with bound organic groups. For example, *tert*-butylnitrite nitrosates the aromatic ring in some (arene)$Cr(CO)_3$ complexes rather than the metal centers to form nitrosyl complexes *(103)*.

2.1.1.7: Hydroxylamine

Hydroxylamine has been used widely to effect the nitrosylation of metal centers under a variety of basic, neutral, or even acidic conditions *(104-108)*. Many metal oxides, halides or cyanides react with NH_2OH (often in the presence of other reagents) to convert ultimately to their nitrosyl complexes *(109,110)*. For example,

$$\text{(P)FeCl} \xrightarrow[-40^{\circ}\text{C}]{\text{NH}_2\text{OH}} \text{(P)Fe(NH}_2\text{OH)} \xrightarrow[-20^{\circ}\text{C}]{} \text{(P)FeNO} \qquad (2.58)$$

reaction 2.58 describes the formal replacement of Cl^- by NO^- in a nitrosylation reaction of an Fe-Cl complex by hydroxylamine *(111)*. Many metal oxides may also be nitrosylated by hydroxylamine *(105)*, e.g.,

$$\text{Na}_2\text{WO}_4 + \text{NH}_2\text{OH.HCl/py} \longrightarrow \text{W(NO)}_2\text{Cl}_2\text{py}_2 \qquad (2.59)$$

Hydroxylamine also functions as a one-electron reducing agent, and occasionally, only a formal reduction of the metal complex occurs *(112)*. Interestingly, acetamidoxime (capable of generating NH_2OH in situ) has been employed as a source of "NO" in the reaction described by eq 2.60 (*113*).

$$\text{Mo(O)}_2\text{(acac)}_2 \xrightarrow{\text{MeC(NH}_2\text{)NOH}} \text{Mo(NO)(MeC(NH}_2\text{)NO)(acac)}_2 \qquad (2.60)$$

2.1.2: Reactions of Coordinated Ligands

2.1.2.1: Nitrite-Nitrosyl Conversion

The M-(NO_2) link (either as nitro, M-NO_2, or nitrito, M-ONO) may be deoxygenated by a number of processes to yield nitrosyl species *(114-118)*.

$$\text{Ni(NO}_2\text{)}_2\text{(PMe}_3\text{)}_2 + \text{CO} \longrightarrow \text{Ni(NO)(NO}_2\text{)(PMe}_3\text{)}_2 + \text{CO}_2 \qquad (2.61)$$

$$Ru(NO_2)_2L_2 \xrightarrow{2H^+} [Ru(NO)(NO_2)L_2]^{2+} \quad (2.62)$$

(L = azo-py)

$$[Fe(NO_2)(CO)_3(PPh_3)_2]^+ \xrightarrow[\Delta]{} [Fe(NO)(CO)_2(PPh_3)_2]^+ \quad (2.63)$$

$$[Ru(bpy)_2(CO)(NO_2)]^+ \xrightarrow{PCl_5} [Ru(NO)(bpy)_2Cl]^{2+} \quad (2.64)$$

Metal nitrates may be deoxygenated similarly *(119,120)*, e.g.

$$Ni(PEt_2Ph)_2(NO_3)_2 \xrightarrow[\Delta]{} Ni(NO)(NO_3)(OPEt_2Ph)_2 \quad (2.65)$$

$$[Ru_2(asp)_4(NO_3)].4H_2O \xrightarrow[\Delta]{} [Ru_2(NO)(asp)_4].4H_2O \quad (2.66)$$

Such deoxygenation processes of metal nitrite and nitrato complexes have been exploited synthetically for the oxidations of organic substrates, and these will be discussed fully in Chapter 7.

2.1.2.2: Oxidations of Coordinated Amines

In what are still poorly understood transformations, coordinated amines may be transformed to nitrosyls by oxidizing agents under appropriate reaction conditions *(121)*, e.g.,

$$[Ru(NH_3)_6]^{2+} \xrightarrow{Cl_2/NH_3(aq)} [Ru(NO)(NH_3)_5]^{3+} \quad (2.67)$$

The $[(O_{zeol})_3Ru(NO)(NH_3)_x]$ compounds (x = 1, 2) are also formed when $[Ru(NH_3)_6]^{3+}$ complexes in X- and Y-zeolites are heated to 450 K in flowing O_2 *(122,123)*.

2.1.2.3: Protonation of Aminato-type Ligands

Addition of acids to aminato-type linkages may, on occasion, form nitrosyl complexes *(124)*.

$$Rh(PPh_3)_2[ONN(p\text{-tol})O] \longrightarrow Rh(NO)(PPh_3)_2Cl_2 \quad (2.68)$$

The related $RuL_2(NO)Cl_3$ complex is also obtainable via this route.

2.1.2.4: Thionitrosyl-Nitrosyl Interchange

Recent claims concerning a thionitrosyl-nitrosyl interchange have been reported *(125,126)*, e.g.

$$Rh(NS)(CO)Cl_2L_2 \xrightarrow{ClNO} Rh(NO)(CO)Cl_2L_2 \qquad (2.69)$$

$$Ru(NS)Cl_3L_2 \xrightarrow{ClNO} Ru(NO)Cl_3L_2 \qquad (2.70)$$

$$(L = PPh_3, AsPh_3)$$

It is not known for certain whether the metal-nitrogen bond remains intact in such interchanges, but this work certainly warrants further investigation.

2.1.2.5: Reactions of Azides

In what appears to be an unusual reaction, hypofluorous acid acts an oxygenating agent in the transformation of a coordinated azide to a coordinated nitrosyl group *(127)*.

$$[Cr(H_2O)_5(N_3)]^{2+} \xrightarrow{HOF} [Cr(NO)(H_2O)_5]^{2+} \qquad (2.71)$$

2.2: Characterization of the MNO Group

The NO group can exhibit a number of bonding modes to metal centers. These modes may be broadly classified into two main groups, namely:

terminal - in which the NO group binds to only one metal center via the N atom (this mode may even occur in polymetallic complexes)

bridging - in which the NO group is attached to more than one metal center either via the N atom alone, or via both the N and O atoms of the same NO group.

Terminal NO ligands may adopt either linear or bent MNO geometries. In practice, however, there are only a few metal nitrosyl complexes displaying MNO bond angles of exactly 180°. Most so-called linear MNO linkages are actually slightly bent (with deviations of up to 10°), and it is not uncommon to encounter the phrase "essentially linear" in many descriptions of such linkages. *Bridging* NO ligands are encountered in bimetallic, trimetallic and other polymetallic compounds. In the cases where both N and O atoms are involved in the bridge, the term "isonitrosyl" has been used to describe the MON part of the interaction.

Single-crystal X-ray crystallography remains as the most reliable technique for the unambiguous determination of the coordination mode(s) of the NO ligand(s) in the solid state. However, IR spectroscopy is the most widely used methodology for the determination of electron-density changes of the MNO link of a metal nitrosyl complex in a reaction solution, whereas nitrogen NMR spectroscopy is becoming increasingly popular in the prediction of MNO bond geometries of metal nitrosyl complexes both in solution and in the solid state. Other techniques such as Mössbauer spectroscopy, NQR and ESR have also been employed *(4,128,129)*, but these techniques do not unambiguously identify the MNO link, and as such are not covered in this section. Gas-phase electron diffraction studies have also been used for the determination of the structures of some metal nitrosyls *(130,131)*.

2.2.1: X-ray crystallography

In beginning this section, it is worth noting that caution must be exercised whenever the MNO bond angles of a metal nitrosyl complex are extracted from published data and discussed. The O-atoms of nitrosyl groups usually exhibit relatively large thermal motions, and even though the use of low temperature X-ray crystallography may generate more accurate data on the location of the N and the O atoms, some problems are still encountered with disorder in crystals of some MNO complexes. This is especially problematic with complexes possessing bent nitrosyl groups, where in some cases two-fold **(A)**, three-fold **(B)** and four-fold **(C)** crystallographic disorder may be present.

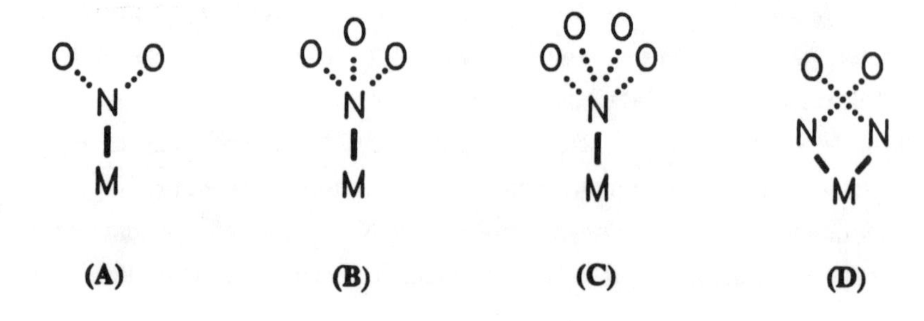

(A) **(B)** **(C)** **(D)**

On occasion, the N-atom position of the MNO group may also be disordered (**D**), giving rise to cases where more than one model may be required to interpret the X-ray data. Excellent descriptions of such problems have been presented elsewhere *(2)*.

As pointed out in Section 1.4, the bonding in the "$M(NO)_x$" moieties of metal nitrosyl complexes is best considered to be the bonding in the hypothetical $M(NO)_x$ group perturbed by the presence of ancillary ligands *(2)*. Thus, metal nitrosyls are conveniently classified as $\{M(NO)_x\}^n$ species, where n may be defined as (i) the total number of electrons associated with the metal d and/or π^*(NO) orbitals, or (ii) the familiar number of d electrons on the metal if the nitrosyl ligand is considered to be bonded formally as NO^+.

2.2.1.1: The Linear η^1-NO Link

Mononitrosyl Complexes

The majority of mononitrosyl complexes $\{M(NO)\}^n$ characterized so far are six-coordinate, and for complexes with $n = 0 - 6$, these electrons reside in the d_{xy} ($2B_2$) and d_{xz},d_{yz} ($8E$) orbitals which range from bonding to non-bonding.

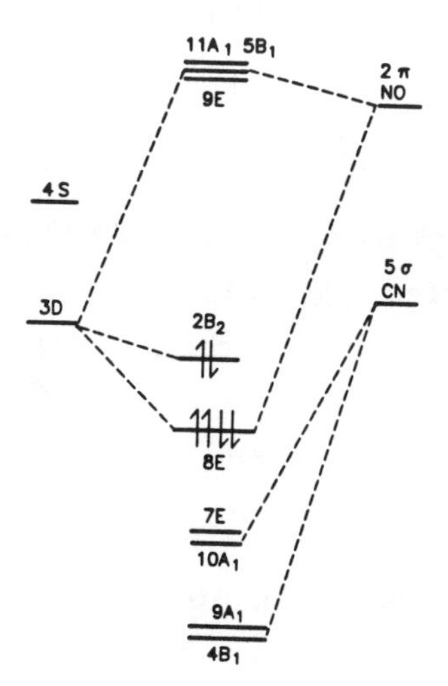

Figure 2.1. Molecular-orbital diagram of the nitroprusside ion. [Adapted from reference 132.]

Thus, no distortions in the MNO group are expected to result from the occupation of these orbitals (*2*). Consequently, it is clear that for mononitrosyl complexes $\{M(NO)\}^n$ with $n \leq 6$, the MNO link will be essentially linear. A typical MO diagram for a six-coordinate mononitrosyl complex (nitroprusside) is shown in Figure 2.1 *(132)*. As may be deduced from Figure 2.1, the added electrons (i.e. $n \geq 7$) will occupy an antibonding π-type orbital (such as *9E*) involving M, N and O. Consequently, bending of the MNO group relieves the electronic strain in the complex. Interestingly, the variation of the observed M-N-O bond angles in six-coordinate mononitrosyl complexes has been compared with the related variation of the O-N-O bond angles of the $[NO_2]^{+1,0,-1}$ group *(3)*.

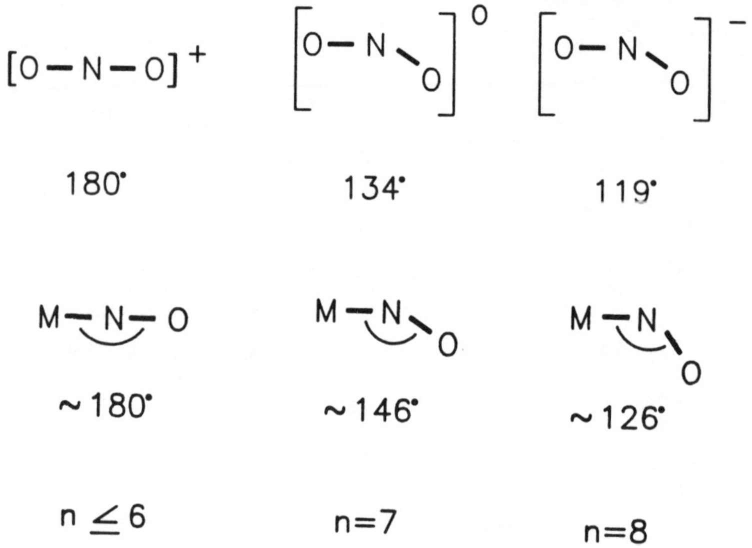

$[O-N-O]^+$ $[O-N\searrow O]^0$ $[O-N\searrow O]^-$

180° 134° 119°

M–N–O M–N↘O M–N↘O

~180° ~146° ~126°

$n \leq 6$ $n=7$ $n=8$

Representative examples of the variation of MNO bond angles with *n* in six-coordinate complexes are listed in Table 2.1. As may be seen from the data in Table 2.1, there are dramatic changes in the MNO bond angles as *n* changes from 6 to 7 and 8 in these six-coordinate complexes. Indeed, the average bond angles for $\{MNO\}^7$ and $\{MNO\}^8$ complexes have been calculated to be $145 \pm 10°$ and $125 \pm 10°$ respectively *(2)*.

In six-coordinate mononitrosyl complexes, the *trans* M-L bonds appear to be "long and weak" when the ν_{NO}s of the complexes are less than *ca.* 1800 cm^{-1}.

Table 2.1 Structural Data for Some Six-Coordinate Mononitrosyl Complexes

Complex	*n*	M-N(Å)	N-O(Å)	MNO (deg)	Ref
$CpW(NO)(CH_2SiMe_3)_2$	4	1.757(8)	1.226(10)	169.5(6)	*133*
$Cr(NO)(NO_2)_2(py)_3$	5	1.68(1)	1.15(1)	180	*134*
$[CpCr(NO)]_2(\mu\text{-}OMe)_2$	5	1.689(8)	1.199(10)	166.3(7)	*135*
$CpRe(NO)(PPh_3)(CHO)$	6	1.777(8)	1.190(1)	178.0(8)	*136*
$CpMo(NO)(\eta^4\text{-diene})^a$	6	1.767(2)	1.213(3)	172.2(2)	*137*
$[Fe(NO)(das)_2(NCS)]BF_4$	7	1.729(9)	1.246(13)	158.6(9)	*138*
Fe(NO)(1-MeIm)(TPP)	7	1.743(4)	1.133(11)	140.1(8)	*139,140*
$[Co(NO)(NH_3)_5]Cl_2$	8	1.871(6)	1.154(7)	119(1)	*141*
$[Co(NO)(das)_2(NCS)]NCS$	8	1.87(1)	1.18(2)	132.3(14)	*142*

[a] diene = *trans*-2,5-dimethyl-2,4-hexadiene

On the other hand, these *trans* M-L bonds appear to be "short and strong" when the ν_{NO}s of the complexes are greater than *ca.* 1800 cm^{-1} (Section 3.4.1).

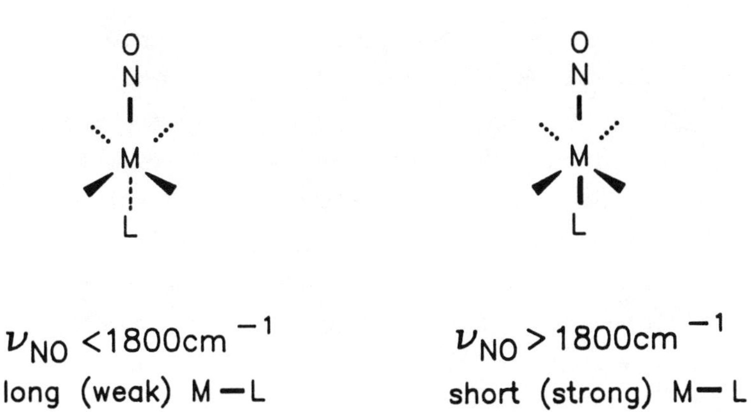

Five-coordinate metal nitrosyl complexes may either adopt tetragonal pyramidal (TP) structures, or trigonal bipyramidal (TBP) structures, although sometimes the distinction between these two structures is not unambiguous *(143,144)*. Generally, TP complexes may be viewed as six-coordinate complexes possessing one vacant site, and indeed, the coordination geometries adopted by their nitrosyl ligands (Table 2.2) follow similar patterns as for the six-coordinate complexes.

Table 2.2 Structural Data for Some Five-Coordinate Mononitrosyl Complexes

Complex	*n*	MNO (deg)	Ref
TP Complexes			
Mn(NO)TPP	6	177.8(3)	*148*
Fe(NO)TPP	7	149.2(6)	*149*
$Co(NO)(S_2CNMe_2)_2$	8	135.1(15)	*150*
TBP Complexes			
$Mn(NO)(CO)_4$	8	180(0)	*151*
$Co(NO)Cl_2(PPh_2Me)_2$		164.5(6)	*152*

There are exceptions to this general rule. For example, the $[Fe(NO)(CN)_4]^{2-}$ complex is formally $\{FeNO\}^7$ but contains a linear FeNO group *(145)*: In this complex, the odd electron resides largely on the CN^- ligands and not the FeNO group *(146,147)*. Five-coordinate TBP complexes may have their nitrosyl ligands in either the apical or equatorial positions. The structurally characterized examples are all of the $\{MNO\}^8$ type, and the majority possess equatorial nitrosyl ligands. The use of bulky and/or chelating ancillary ligands may force the NO group into the axial position. Nevertheless, the MNO groups in these complexes are essentially linear.

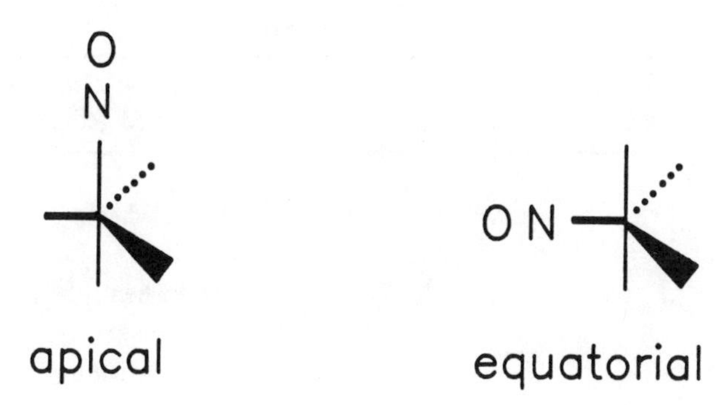

Almost all of the structurally characterized four-coordinate metal nitrosyl complexes are of the $\{MNO\}^{10}$ type, and possess pseudotetrahedral coordination geometries with linear MNO groups *(2)*. A notable exception is $[Ir(NO)Cl(PPh_3)]_2(\mu\text{-}O)$ which is formally $\{MNO\}^8$ and square planar at each metal center with linear IrNO groups *(153)*.

Approximately trigonal planar geometries exist for three-coordinate $\{MNO\}^{10}$ complexes. An example is bimetallic $[Ni(NO)]_2(\mu\text{-}Me_2pz)_2$, which contains linear NiNO groups *(154)*.

A number of "mononitrosyl" M(NO) groups may coexist in polymetallic complexes *(90)*. For instance, the two NO groups in the $[Rh(NO)Br\{(Ph_2PO)_2H\}]_2$ dimer are terminally bonded to both metals, and each MNO link is bent (RhNO *ca.* 129°) *(155)*. A concise summary of the predicted geometries of mononitrosyl complexes is presented in Table 2.3.

Dinitrosyl Complexes

In principle, metal dinitrosyl complexes may exist in either the *trans*- or *cis*-dinitrosyl forms:

ON — M — NO

trans

M(NO)(NO)

cis

Table 2.3: Predicted Geometries for Mononitrosyl Complexes

Coord. No	Structure	Predicted geometry
6	octahedral	linear when $n \leq 6$
		bent when $n \geq 7$
5	TP	linear when $n \leq 6$
		linear or bent when $n = 7$
		bent when $n = 8$
	TBP	linear when $n = 8$
4	tetrahedral	linear when $n = 10$
	distorted tetrahedron	slightly bent when $n = 10$

Given the strong π-acid nature of the NO ligand, it is not surprising that the majority of dinitrosyl complexes possess *cis*-symmetry. Molecular orbital calculations predict that (i) complexes with the $\{M(NO)_2\}^4$ configuration could adopt either linear *trans* or bent *cis* structures although the *trans* arrangement will minimize the repulsion of the NO groups, (ii) the addition of electrons up to $n = 8$ will favor the *cis* structure, and (iii) the *cis* $\{M(NO)_2\}^{10}$ species will possess a populated 1_{b1} orbital that is localized entirely on the NO groups and is antibonding with respect to the two N and the two O-atoms. Thus the population of this orbital might cause an increase in the NMN bond angle, thereby causing the O-atoms to move further apart *(3)*. In any event, it is common to find a slight bending of the MNO group in metal dinitrosyl complexes. Representative MNO and NMN bond angles for some dinitrosyl complexes are shown in Table 2.4.

Table 2.4. Representative Bond Angles (deg) for Some Dinitrosyl Complexes

complex	*n*	M-N-O	N-M-N	Ref
$CpCr(NO)_2Cl$	6	170.0(3)	93.9(1)	*156*
		168.8(3)		
$[Mn(NO)_2(P(OMe)_2Ph)_3]BF_4$	8	168(1)	116.5(5)	*157*
		170(1)		
$Fe(NO)_2Cl(PPh_3)$	9	166.4(5)	115.6(3)	*158*
		165.5(5)		
$Fe(NO)_2(dppe)$	10	178.0(3)	125.4(2)	*159*
		176.0(6)		

It has been noted that for $\{M(NO)_2\}^{10}$ complexes with NMN bond angles of less than *ca.* 130°, the two O atoms bend toward each other to give an "*attracto*" conformation,

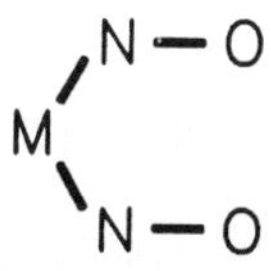

"*attracto*"
O-M-O < N-M-N

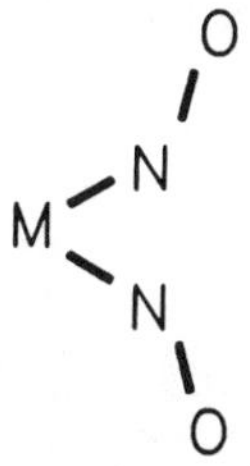

"*repulso*"
O-M-O > N-M-N

whereas with complexes with N-M-N angles greater than *ca.* 130°, the O atoms move further apart to result in a "*repulso*" conformation. In general, it has been observed (with some exceptions, of course) that the *attracto* conformation is characteristic of the first row (3d) transition-metal dinitrosyls with good π-accepting ligands, whereas the *repulso* conformation occurs with the second- (4d) and third-row (5d) transition metal dinitrosyls with poor π-acceptors *(160)*. Again, some caution must be used in assigning such conformations, since occasionally the differences in NMN and OMO angles are often very small (of the order of a few degrees), and are sometimes not statistically significant *(159,161)*.

It must also be noted that not all metal dinitrosyls possess two linear NO groups. For example, the $[Ru(NO)_2Cl(PPh_3)_2]^+$ and $[Os(NO)_2(OH)(PPh_3)_2]^+$ complexes possess *both* linear *and* bent NO groups in the solid state *(162,163)*.

Trinitrosyl Complexes

Only one polynitrosyl complex, namely $Mn(NO)_3PPh_3$, has been crystallographically characterized. Its solid state molecular structure reveals that it possesses tetrahedral geometry (non-crystallographic C_3 symmetry) with linear NO groups, with the MNO bond angles averaging 177.2(7)° *(164)*.

2.2.1.2: Bridging Nitrosyls

(i) The η^1-μ-NO Bridging Ligand:

The simplest cases are the ones in which one NO group bridges two metal centers (μ) via the N atom (η^1).

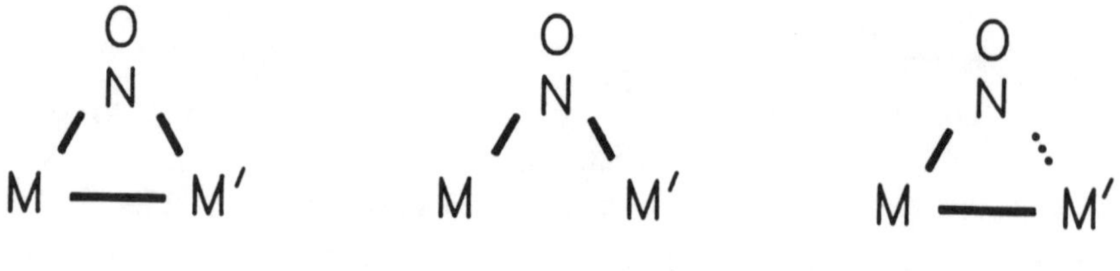

O N M — M′ **(A)**

O N M M′ **(B)**

O N M — M′ **(C)**

These μ-NO bridges may be symmetrically (**A**, **B**) or asymmetrically (**C**) disposed, and may be held together by one or more metal-metal bonds (**A**,

C), or may be held together solely by the NO bridge (**B**). By far the most common example of η^1-μ-NO bimetallic complexes possess structure (**A**), although a growing number of structures (**B**) are being found. There are a few examples of the asymmetric bridge (**C**), where metal-metal bonds may or may not be present *(165)*. Typical structurally characterized NO-bridging complexes are listed in Table 2.5.

Table 2.5: Structural Data for Some NO-Bridged Complexes

compound	M-M' (Å)	M-N (Å)	M-N-M (deg)	ref
Structure (A)				
$[CpCr(NO)]_2(\mu\text{-}NO)_2$ (Cr-Cr)	2.615(1)	1.960(3)	86.7(1)	*166*
$(CpFe)_2(\mu\text{-}NO)_2$ (Fe=Fe)	2.326(4)	1.768(9)	82.3(4)	*167*
Structure (B)				
$[Cp^*CoCl]_2(\mu\text{-}NO)(\mu\text{-}Cl)$	3.108	1.80 1.84(1)	117.2(7)	*168*
$Pt_4(\mu\text{-}NO)_2(O_2CMe)_6$	3.311(2)	1.912(15)	119.9(13)	*169*
$[(PPh_3)Cl_2Ir(\mu\text{-}NO)(\mu\text{-}dppn)PdCl]PF_6$	3.327(2)	2.030(15) 1.953(17)	113.3(7)	*170*
Structure C				
$[Pt_2(\mu\text{-}NO)(\mu\text{-}dppm)_2Cl_2]BF_4$ (no Pt-Pt bond)	3.186(2)	1.83(2) 2.02(2)	112(1)	*171*
$PPN[Fe_2W(\mu\text{-}CO)(\mu\text{-}NO)(\mu_3\text{-}Ctol)(CO)_6Cp]$ (W-Fe)	2.702(2)	1.911(11) 2.286(12)	94.9(3)	*172*

Although the largest number of η^1-μ-NO complexes are *homobimetallic*, i.e. M = M', there is presently a growing number of *heterobimetallic* η^1-μ-NO complexes being synthesized *(170,173-175)*. Complete listings of structural parameters for such complexes may be found elsewhere *(2,90,170)*.

(ii) The η^1-μ_3-NO Bridging Ligand

Triply bridging NO groups (bridged via the N atom) have structural parameters similar to those of the doubly-bridging type (η^1-μ) discussed above (*2,90*) and are discussed in Section 5.2.8.

(iii) The η^2-μ_4-NO Bridging Ligand

Only two complexes possessing isonitrosyl (η^2, M-N-O-M') linkages have been structurally characterized in the solid state, and these are the tetranuclear compounds of rhenium *(176)* and Mo/Co *(177)*.

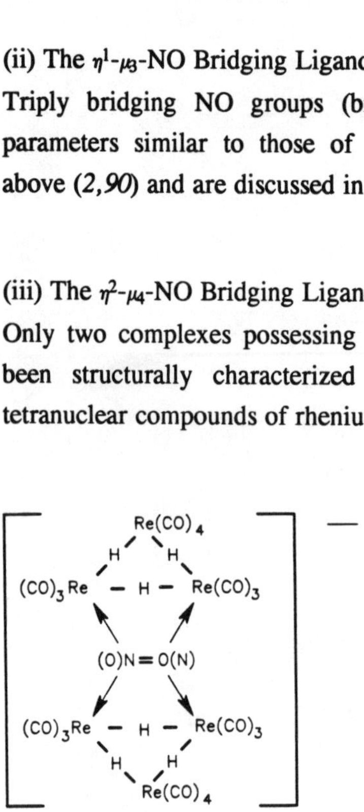

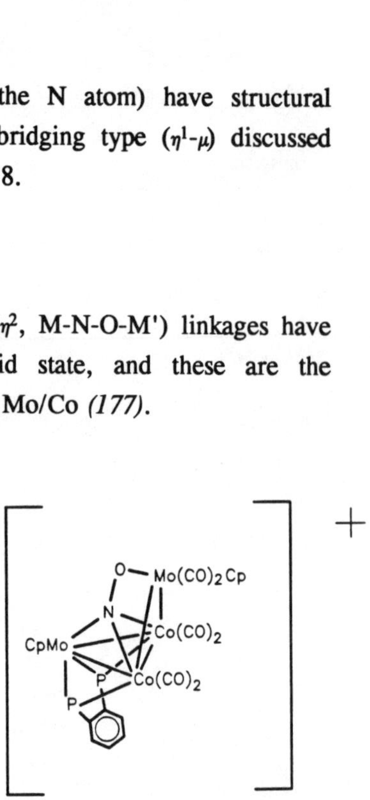

Given the tendency for nitrosyl groups to interact with metallic and non-metallic Lewis acids (Chapter 7), it is somewhat surprising that more complexes of this type have not been structurally characterized.

2.2.2: Nitrogen NMR Spectroscopy

Both ^{14}N and ^{15}N nuclei are NMR active even though they both have low NMR sensitivities *(4,178-180)* (Table 2.6). ^{15}N has a very low natural abundance and possesses very long relaxation times. ^{14}N is quadropolar, but has a high natural abundance and possesses small relaxation times.

Fortunately, the chemical shifts of ^{14}N and ^{15}N are interchangeable, due to the fact that primary isotope effects are very small and are generally of the order of experimental error *(4)*. Typical ranges of these shifts are listed in Table 2.7 and displayed in Fig 2.2.

Table 2.6. NMR Parameters for Some Common Nuclei.

Nucleus	spin (*I*)	Nat. Abund (%)	Nuc. Quad. Mom. (x 10^3)
^{1}H	1/2	99.985	-
^{2}H	1	0.015	2.73
^{13}C	1/2	1.11	-
^{31}P	1/2	100	-
^{14}N	1	99.63	16.0
^{15}N	1/2	0.37	-
^{17}O	5/2	0.037	-26.1

N chemical shifts are dominated by the paramagnetic term and as may be noted in Figure 2.2, a very wide range of N chemical shifts (well over 1000 ppm) are exhibited by metal nitrosyl complexes. For linear mononitrosyl complexes, the ^{15}N shielding observed in the NMR spectra tends to increase across the series of transition metals, and increase down a particular group *(181-184)*. That the N chemical shifts of bent (or bridging) nitrosyls are more deshielded than those of the linear nitrosyls appears to be a general trend *(185)*.

Table 2.7. ^{15}NMR Data for Some MNO Complexes Relative to Liquid Nitromethane.

compound	medium	δ (^{15}N), ppm
$NOBF_4$	SO_2	-7.5
linear nitrosyls		
$CpCr(NO)(CO)_2$	$CHCl_3$	49.0
$CpMo(NO)(CO)_2$	$CHCl_3$	37.6
$CpW(NO)(CO)_2$	$CHCl_3$	16.5
$Ru(NO)Cl_3(PMePh_2)_2$	CD_2Cl_2	-30.9
bent nitrosyls		
$Co(NO)(ketox)_2$	$CHCl_3$	740.3
$[Rh(NO)Cl(CO)(PPr_3)_2]^+$	CD_2Cl_2	368.3
dinitrosyl complexes		
$CpCr(NO)_2Cl$	$CHCl_3$	184.6
$CpMo(NO)_2Cl$	$CHCl_3$	185.4
$CpW(NO)_2Cl$	$CHCl_3$	172.9

Interestingly, the availability of inexpensive isotope labels has allowed the use of ^{15}N NMR spectroscopy to become more popular in the study of metal-nitrosyl complexes than ^{14}N NMR spectroscopy. The use of nitrogen NMR spectroscopy in the study of fluxional processes involving the NO groups of MNO complexes in solution and the use of solid-state CP-MAS ^{15}N NMR spectroscopy as an invaluable tool for the determination of the geometries of NO ligands in a metal nitrosyl complex are currently being developed *(186,187)*.

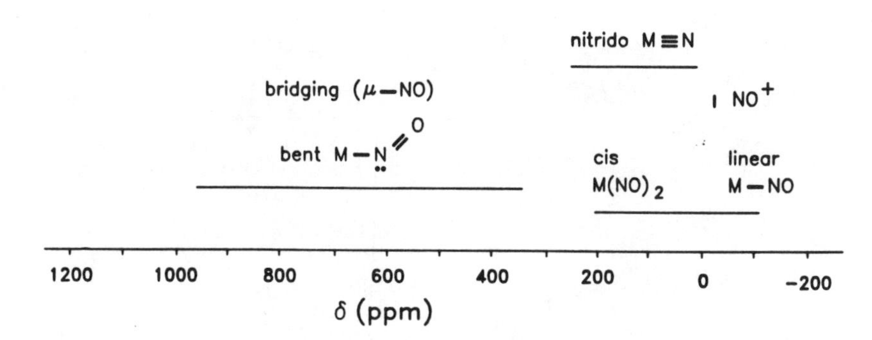

Figure 2.2 Typical ranges of N chemical shifts displayed by metal nitrosyl complexes (referenced to liquid nitromethane).

2.2.3: IR Spectroscopy

Although the MNO group is expected to exhibit ν_{NO}, ν_{MN} and δ_{MNO}, only ν_{NO} has been identified in most cases in the vibrational spectra of metal nitrosyl complexes. There is a striking resemblance in the patterns of IR-stretching vibrations of MNO complexes and MCO complexes. Indeed, the assignment of ν_{NO} in polynitrosyl complexes follows the well-established patterns described for metal carbonyl compounds (*188,189*). However, ν_{NO} bands are generally a little broader than those due to ν_{CO}, and in mixed carbonyl-nitrosyl complexes this feature usually helps in the assignments of ν_{NO} (Figure 2.3). Also, unlike ν_{CO} where the ^{13}C-O satellites are easily observable (at about $\Delta\nu = 45$ cm^{-1}) even in non-enriched samples, this is not the case with ν_{NO}.

As mentioned in Section 1.4, a synergic electronic interaction exists in the MNO bond consisting of σ-donation of electron density from the nitrosyl ligand to the metal followed by backdonation of electron density from the filled metal d orbitals into the π^* antibonding orbital of the NO group. In simplistic terms, therefore, an increased backdonation of electron density from M to NO should result in the population of this π^* orbital,

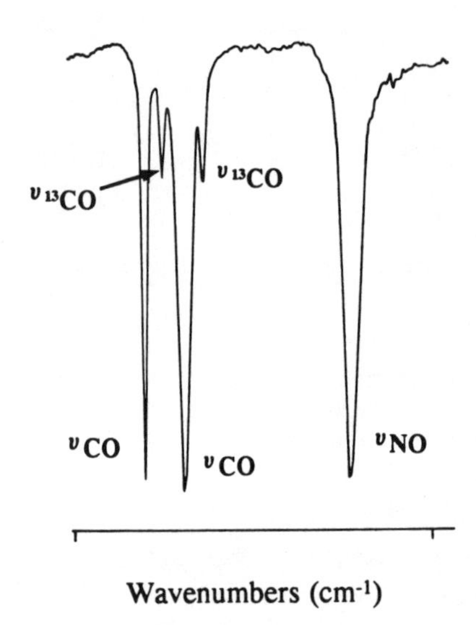

Figure 2.3 The IR spectrum of $CpCr(NO)(CO)_2$ in the 2200 - 1500 cm^{-1} region.

and hence a *weakening* of the N-O bond (and a *lowering* of the ν_{NO}) should occur. Conversely, a net donation of electron density from the NO group to the metal center should result in a *stronger* N-O bond, and hence a *higher* ν_{NO}. Such a simplistic view is useful when a single metal nitrosyl complex with a linear nitrosyl group is under investigation, as in the monitoring of the progress of a chemical reaction. Thus, *within the same complex*, changes in ν_{NO} are often a good indication of the electron density changes at the metal center. Consider, for instance, the reaction of the formally 16-electron complex, $CpW(NO)(CH_2SiMe_3)_2$ with trimethylphosphine:

$$CpW(NO)(CH_2SiMe_3)_2 + PMe_3 \longrightarrow CpW(NO)(CH_2SiMe_3)_2(PMe_3) \quad (2.72)$$
$$\nu_{NO} = 1541\ cm^{-1} \qquad\qquad \nu_{NO} = 1520\ cm^{-1}$$

The lowering of the ν_{NO} ($\Delta\nu$ = 21 cm^{-1}) (*190*) is consistent with the formation of a more electron-rich metal center in the 18-electron product complex, a process that results in an increased backdonation of electron density into the NO π^* antibonding orbital.

Changes in the electron density at the metal center may not necessarily affect the M-NO bond when there is *restricted* π-overlap in the

MNO bond (for example, in bent nitrosyl complexes). For example, the ν_{NO}s of both complexes in eq 2.73 are identical, and the extra electron density provided by the added CO ligand is not transferred onto the NO ligand (*191*).

$$Rh(NO)Cl_2(PPh_3)_2 + CO \longrightarrow Rh(NO)Cl_2(PPh_3)_2(CO) \qquad (2.73)$$
$$\nu_{NO} = 1630\ cm^{-1} \qquad\qquad \nu_{NO} = 1630\ cm^{-1}$$

When considering a *number* of metal nitrosyl complexes, it is not always possible to find a simple relationship between ν_{NO} and the electronic population or charge of the NO group, since the observed ν_{NO} is influenced by other factors such as the electronic effects of other ligands, the nature of the metal, the structure of the complex, and even the overall charge of the complex *(192,193)*.

Mononitrosyl Complexes

When nitric oxide (ν_{NO} 1870 cm^{-1}) binds to a metal center, its ν_{NO} may either *increase* or *decrease*, and it is not always possible to predict the nature of the shift, although in the majority of cases the ν_{NO} is *decreased* relative to that of free NO. Typical IR stretching frequencies for some mononitrosyl complexes are shown in Table 2.8.

As evidenced by the data in Table 2.8, there is *no correlation* between the ν_{NO} and the MNO bond angles in metal nitrosyl complexes, and it is incorrect to attempt to assign MNO geometries (linear, bent, bridging) based on the observed ν_{NO} in the IR spectrum of the complex. Some empirical rules have been proposed for distinguishing between linear and bent NO groups based on the observed nitrosyl stretching frequency, and these rules allow for the "adjustment" of the ν_{NO} based on the position of the metal in the periodic table, the charge on the complex, and the types of ancillary ligands present in the complex *(193)*. In any event, the regions of observed ν_{NO}s for linear, bent and even bridging MNO groups overlap over a wide range, as shown in Figure 2.4.

Table 2.8. IR Stretching Frequencies and MNO Bond Angles for Some Mononitrosyl Complexes

Compound	ν_{NO} (cm^{-1})	M-N-O (deg)	Ref
Linear MNO			
$[Fe(NO)(CN)_5]^{2-}$	1944	178.3(1)	*194*
$[Ru(NO)Cl_5]^{2-}$	1887	176.7(5)	*195*
$Mn(NO)(CO)_4$	1759	180	*196*
$Cp^*W(NO)I_2$	1630	163.9(46)	*197*
$Ir(NO)(PPh_3)_3$	1600	180	*198*
Bent MNO			
$[Ir(NO)(I)(CO)(PPh_3)_2]^+$	1720	125(3)	*199*
$Ni(NO)(N_3)(PPh_3)_2$	1710	152.7(7)	*200*
$Co(NO)(S_2CNMe_2)_2$	1630	134.5(17)[a]	*201*
		135.7(14)	
$Ir(NO)Me(I)(PPh_3)_2$	1525	120(2)	*202*
$Co(NO)(salox)_2$	1672	116(2)	*203*

[a] disordered

In general, only one ν_{NO} is expected for a mononitrosyl complex of general formula $L_nM(NO)$. However, there are some instances where more than one ν_{NO} may be observed, namely,

(i) when the complex exists in linear-bent isomeric forms (either static or fluxional), then each isomer may exhibit a unique ν_{NO}.

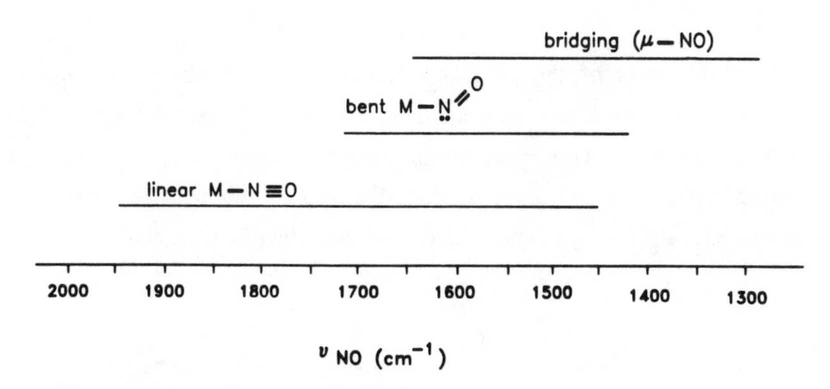

Figure 2.4 Typical ranges of nitrosyl IR-stretching frequencies for metal nitrosyl complexes.

Examples of such complexes include $[(allyl)Ir(NO)(PPh_3)_2]^+$ *(204)*, $Co(NO)Cl_2(PR_3)_2$ (R_3 = Ph_2Me, Me_3) *(205,206)* and $Cp_2V(NO)I$ *(207,208)*. Interestingly, it has been observed that the grinding of the linear MNO isomer of the cobalt complex (PR_3 = PPh_2Me) partially transforms it into its bent MNO isomer: thus two ν_{NO}s are observed in the Nujol mull spectrum of the complex *(205)*.

(ii) when the ν_{NO} is sensitive to the spin states of the metal, then different ν_{NO}s will be observed corresponding to each kind of spin state. For example, Fe(NO)(salphen) exhibits two ν_{NO}s: one for the high-spin state at room temperature (Fe: S = 3/2, ν_{NO} = 1724 cm^{-1}), and the other for the low-spin state at liquid-nitrogen temperatures (Fe: S = 1/2, ν_{NO} = 1643 cm^{-1}) *(209)*.

(iii) sometimes, a "splitting" of the ν_{NO} is observed. This situation arises not only because of solid state effects, but may also be due to the choice of solvent medium for recording the IR spectrum of the complex. For example, some nitrosyl complexes of cobalt, manganese and iron show a splitting of ν_{NO} in some solvents *(188,210,211)*.

(iv) when different conformations of a metal nitrosyl complex are present, each conformation may exhibit a slightly different ν_{NO}.

Dinitrosyl Complexes

For metal nitrosyls of the general formula $L_nM(NO)_2$, only one ν_{NO} is expected if the ON-M-NO bond angle is 180° (i.e., if the NO groups are *trans* to each other). This arises because the symmetric stretching vibration of the $M(NO)_2$ group does not alter the dipole moment, whereas the antisymmetric stretching vibration does and thus absorbs IR radiation.

O— N— M —N —O   O— N— M —N —O

sym   *antisym*

For non-linear $M(NO)_2$ groups, *both* the symmetric and antisymmetric stretching vibrations are IR active:

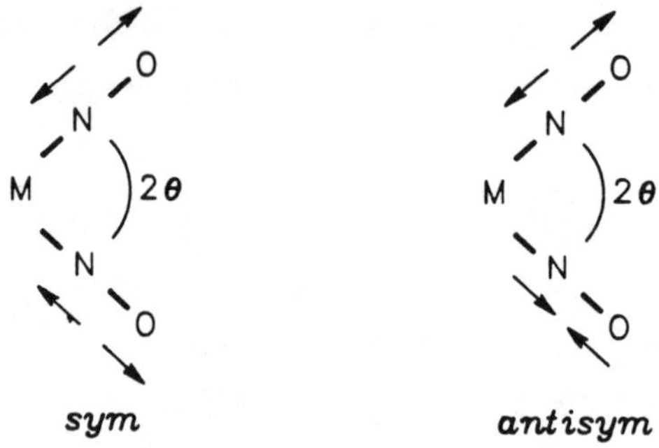

sym   *antisym*

and thus, two ν_{NO}s are observed in this case. The symmetric stretching vibration occurs at *higher* energy than the antisymmetric stretching vibration, and the two IR bands will be of equal intensity if the ON-M-NO bond angle is 90°. However, if the angle is not 90°, then the relative intensities of the two IR bands may then be expressed by *(212-215)*:

$$\frac{I_{sym}}{I_{as}} = \cot^2 \theta$$

Thus, as the angle between the two NO groups (2θ) increases, the intensity of ν_s gets smaller (Fig 2.5).

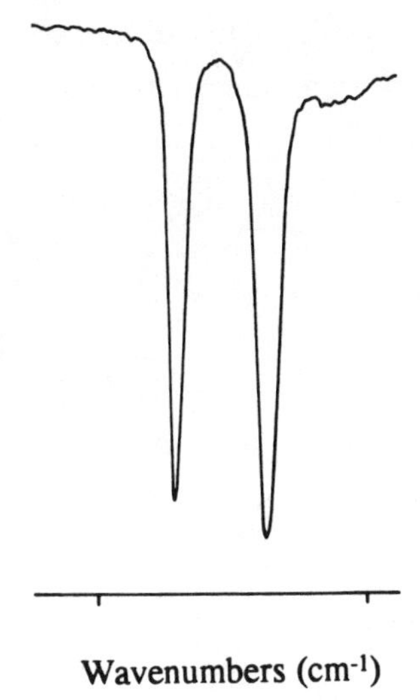

Figure 2.5. The IR spectrum of $CpCr(NO)_2Cl$ in the 2000 - 1500 cm^{-1} region (2θ = 93.9°)

Polynitrosyl Complexes

A number of metal trinitrosyls of the form $Mn(NO)_3(L)$ (L = CO, phosphines, arsines, etc.) have been synthesized *(21)*, and they exhibit two nitrosyl-stretching frequencies, thereby implying a tetrahedral arrangement of the ligands about the metal center. The solid-state molecular structure of $Mn(NO)_3(PPh_3)$ confirms this tetrahedral arrangement of C_3 symmetry *(164)*. As expected, the general class of $Fe(NO)_3L$ compounds (L = halide or phosphine) also display two nitrosyl-stretching frequencies *(40,53,216,217)*.

The tetranitrosyl complex, $Cr(NO)_4$, exhibits a single ν_{NO} at 1716 cm^{-1} *(21,218)*, consistent with the presence of a single type of linear nitrosyl ligand in this formally 18-electron compound. Thus, tetrahedral geometry has been proposed for this compound. On the other hand, matrix-isolated $Fe(NO)_4$ is pseudotetrahedral *(19)* and contains one bent and three linear ligands, consistent with its formulation as an 18-electron compound.

Although IR spectroscopy appears to be the most widely used technique for the characterization of metal-nitrosyl complexes, it must be noted that it alone cannot be used to unambiguously assign the mode of linkage of the NO group to the metal center.

2.3: References

1. Caulton, K. G. *Coord. Chem. Rev.* **1975**, *14*, 317.
2. Feltham, R. D.; Enemark, J. H. *Top. Stereochem.* **1981**, *12*, 155.
3. Enemark, J. H.; Feltham, R. D. *Coord. Chem. Rev.* **1974**, *13*, 339.
4. Michael, D.; Mingos, P.; Sherman, D. J. *Adv. Inorg. Chem.* **1989**, *34*, 293.
5. Young, C. L. *Solubility Data Ser.* **1981**, *8*, 336.
6. Ogura, K.; Watanabe, M. *Electrochim. Acta* **1982**, *27*, 111.
7. Littlejohn, D.; Chang, S. G. *J. Phys. Chem.* **1982**, *86*, 537.
8. Hishinuma, Y.; Kaji, R.; Akimoto, H.; Nakajima, F.; Mori, T.; Kamo, T.; Arikawa, Y.; Nozawa, S. *Bull. Chem. Soc. Jpn.* **1979**, *52*, 2863.
9. Zang, V.; van Eldik, R. *Inorg. Chem.* **1990**, *29*, 4462.
10. Barratt, D. S.; McAuliffe, C. A. *J. Chem. Soc., Chem. Commun.* **1984**, 594.
11. Barratt, D. S.; McAuliffe, C. A. *J. Chem. Soc., Dalton Trans.* **1987**, 2497.
12. Miki, E.; Saito, K.; Mizumachi, K.; Ishimori, T. *Bull. Chem. Soc. Jpn.* **1983**, *56*, 3515.
13. Kirsenko, V. N.; Lampeka, Y. D.; Yatsimirskii, K. B. *Theor. Experim. Chem.* **1983**, *19*, 680.
14. Jezowska-Trzebiatowska, B.; Gerega, K.; Formicka-Kozlowska, G. *Bull. Acad. Pol. Sci., Ser. Sci. Chi.* **1982/83**, *30*, 81.
15. Jezowska-Trzebiatowska, B.; Gerega, K.; Formicka-Kozlowska, G. *Inorg. Nucl. Chem. Lett.* **1980**, *16(9-12)*, 563.
16. Jezowska-Trzebiatowska, B.; Gerega, K.; Formicka-Kozlowska, G. *Inorg. Chim. Acta* **1980**, *40*, 187.

17. Jezowska-Trzebiatowska, B.; Gerega, K.; Vogt, A. *Inorg. Chim. Acta* **1978**, *31*, 183.
18. Middleton, R.; Hull, J. R.; Simpson, S. R.; Tomlinson, C. H.; Timms, P. L. *J. Chem. Soc., Dalton Trans.* **1973**, 120.
19. Doeff, M. M.; Pearson, R. G.; Barrett, P. H. *Inorg. Chim. Acta* **1986**, *117*, 151.
20. Sabherwal, I. H.; Burg, A. B. *J. Chem. Soc., Chem. Commun.* **1970**, 1001.
21. Satija, S. K.; Swanson, B. I. *Inorg. Synth.* **1976**, *16*, 1.
22. Chisholm, M. H.; Huffman, J. C.; Kelly, R. L. *Inorg. Chem.* **1980**, *19*, 2762.
23. Chisholm, M. H.; Cotton, F. A.; Extine, M. W.; Kelly, R. L. *Inorg. Chem.* **1979**, *18*, 116.
24. Bencze, L.; Kohan, J. *Inorg. Chim. Acta* **1982**, *65*, L17.
25. Weber, R.; Dehnicke, K. *Z. Naturforsch., B: Anorg. Chem., Org. Chem.* **1984**, *39B*, 262.
26. Bottomley, F.; Sutin, L. *Adv. Organomet. Chem.* **1988**, *28*, 339.
27. Williams, D. L. H. *Nitrosation*; Cambridge University: New York, 1988.
28. Jain, K. C.; Pandey, K. K.; Agarwala, U. C. *Indian J. Chem.* **1981**, *20A*, 1124.
29. Pandey, K. K.; Tiwari, R. D.; Agarwala, U. C. *Indian J. Chem., Sect. A* **1981**, *20A*, 370.
30. Pandey, K. K.; Agarwala, U. C. *Indian J. Chem., Sect. A* **1981**, *20A*, 240.
31. Jain, K. C.; Pandey, K. K.; Parashad, R.; Singh, T.; Agarwala, U. C. *Indian J. Chem., Sect. A* **1980**, *19A*, 1089.
32. Pandy, K. K.; Agarwala, U. C. *Synth. React. Inorg. Met.-Org. Chem.* **1981**, *11*, 25.
33. Pandey, K. K.; Agarwala, U. C. *Z. Anorg. Allg. Chem.* **1979**, *457*, 235.
34. Kubota, M.; Koerntgen, C. A.; McDonald, G. W. *Inorg. Chim. Acta* **1978**, *30*, 119.
35. Bonner, F. T.; Hughes, M. N. *Comments Inorg. Chem.* **1988**, *7*, 215.
36. Bonner, F. T.; Akhtar, M. J. *Inorg. Chem.* **1981**, *20*, 3155.
37. Mocella, M. T.; Okamoto, M. S.; Barefield, E. K. *Syn. React. Inorg. Metal-Org. Chem.* **1974**, *4*, 69.
38. Eckstein, Y.; Dreyfuss, P. *J. Polym. Sci., Polym. Chem. Ed.* **1979**, *17*, 4115.

39. Seung, S. L. N.; Dreyfuss, P.; Fetters, L. *J. Polym. Bull. (Berlin)* **1985**, *13(4)*, 337.
40. Herberhold, M.; Klein, R. *Angew. Chem., Int. Ed. Engl.* **1978**, *17*, 454.
41. Hodgson, D. J.; Payne, N. C.; McGinnety, J. A.; Pearson, R. G.; Ibers, J. A. *J. Am. Chem. Soc.* **1968**, *90*, 4486.
42. Connelly, N. G.; Dahl, L. F. *J. Chem. Soc., Chem. Commun.* **1970**, 880.
43. Touchard, D.; Le Bozec, H.; Dixneuf, P. *Inorg. Chim. Acta* **1979**, *33*, L141.
44. Connelly, N. G.; Demidowicz, Z.; Kelly, R. L. *J. Chem. Soc., Dalton Trans.* **1975**, 2335.
45. Lee, K. Y.; Kuchynka, D. L.; Kochi, J. K. *Inorg. Chem.* **1990**, *29*, 4196.
46. Treichel, P. M.; Molzahn, D. C.; Wagner, K. P. *J. Organomet. Chem.* **1979**, *174*, 191.
47. Connelly, N. G.; Davies, J. D. *J. Organomet. Chem.* **1972**, *38*, 385.
48. Kolobova, N. E.; Lobanova, I. A.; Zdanovich, V. I. *Izv. Akad. Nauk SSSR, Ser. Khim.* **1980**, 1651. CA93(25):239580w.
49. Elsenbaumer, R. L. *J. Org. Chem.* **1988**, *53*, 437.
50. Baran, E. J.; Muller, A. *Z. Anorg. Allg. Chem.* **1969**, *370*, 283.
51. Piper, T. S.; Cotton, F. A.; Wilkinson, G. *J. Inorg. Nucl. Chem.* **1955**, *1*, 165.
52. Shiu, K.-B.; Chang, C.-J. *J. Organomet. Chem.* **1990**, *395*, C47.
53. Legzdins, P.; Malito, J. T. *Inorg. Chem.* **1975**, *14*, 1875.
54. Kolthammer, B. W. S.; Legzdins, P.; Malito, J. T. *Inorg. Chem.* **1977**, *16*, 3173.
55. Hoyano, J. K.; Legzdins, P.; Malito, J. T. *Inorg. Synth.* **1978**, *18*, 126.
56. Chang, J.; Seidler, M. D.; Bergman, R. G. *J. Am. Chem. Soc.* **1989**, *111*, 3258.
57. Herberhold, M.; Haumaier, L. *J. Organomet. Chem.* **1978**, *160*, 101.
58. Hedden, D.; Roundhill, D. M.; Walkinshaw, M. D. *Inorg. Chem.* **1985**, *24*, 3146.
59. Alt, H. G.; Engelhardt, H. E.; Hayen, H. I.; Rogers, R. D. *J. Organomet. Chem.* **1989**, *366*, 287.
60. Napoleone, V.; Schelly, Z. A. *J. Phys. Chem.* **1980**, *84*, 17.
61. Ashok, R. F. N.; Gupta, M.; Arulsamy, K. S.; Agarwala, U. C. *Inorg. Chim. Acta* **1985**, *98*, 169.

62. Pandey, K. K.; Datta, S.; Agarwala, U. *Transition Met. Chem. (Weinheim)* **1979**, *4(6)*, 337.

63. Iglesias, E.; Williams, D. L. H. *J. Chem. Soc., Perkin Trans. 2* **1989**, *(4)*, 343.

64. Abia, L.; Castro, A.; Iglesias, E.; Leis, J. R.; Pena, M. E. *J. Chem. Res., Synop.* **1989**, *(4)*, 106.

65. Joergensen, K. A.; Lawesson, S. O. *J. Am. Chem. Soc.* **1984**, *106*, 4687.

66. Joergensen, K. A.; El-Wassimy, M. T. M.; Lawesson, S. O. *Acta Chem. Scand., Ser. B* **1983**, *B37*, 785.

67. Kan, C. T.; Hitchcock, P. B.; Richards, R. L. *J. Chem. Soc., Dalton Trans.* **1982**, 79.

68. Khan, M. I.; Agarwala, U. C. *Bull. Chem. Soc. Jpn.* **1986**, *59*, 1285.

69. Cameron, M.; Gowenlock, B. G.; Vasapollo, G. *Chem. Soc. Rev.* **1990**, *19*, 355.

70. Cameron, M.; Gowenlock, B. G.; Vasapollo, G. *J. Organomet. Chem.* **1991**, *325*, 325.

71. Cameron, M.; Gowenlock, B. G.; Vasapollo, G. *J. Organomet. Chem.* **1989**, *378*, 493.

72. Kirby, G. W. *Chem. Soc. Rev.* **1977**, *6*, 1.

73. Fantucci, P.; Pizzoti, M.; Porta, F. *Inorg. Chem.* **1991**, *30*, 2277.

74. Green, M.; Osborn, R. B. L.; Rest, A. J.; Stone, F. G. A. *J. Chem. Soc. A* **1968**, 2525.

75. Williams, D. L. H. *Chem. Soc. Rev.* **1985**, *14*, 171.

76. Pandey, D. S.; Saini, S. K.; Agarwala, U. C. *Bull. Chem. Soc. Jpn.* **1987**, *60*, 3031.

77. Pandey, D. S.; Agarwala, U. C. *Inorg. Chim. Acta* **1989**, *159*, 197.

78. Naeumann, F.; Rehder, D. *Z. Naturforsch., B: Anorg. Chem., Org. Chem.* **1984**, *39B*, 1647.

79. Schiemann, J.; Weiss, E.; Naeumann, F.; Rehder, D. *J. Organomet. Chem.* **1982**, *232*, 219.

80. Naeumann, F.; Rehder, D. *J. Organomet. Chem.* **1980**, 204, 411.

81. Naeumann, F.; Rehder, D.; Pank, V. *Inorg. Chim. Acta* **1984**, *84*, 117.

82. Brauel, A.; Rehder, D. *Z. Naturforsch., B.: Chem. Sci.* **1987**, *42*, 605.

83. Rehder, D.; Ihmels, K.; Wenke, D.; Oltmanns, P. *Inorg. Chim. Acta* **1985**, *100*, L11.

84. Oltmanns, P.; Rehder, D. *J. Organomet. Chem.* **1985**, *281*, 263.

85. Sacco, A.; Vasapollo, G.; Giannoccaro, P. *Inorg. Chim. Acta* **1979**, *32*, 171.

86. Mu, X. H.; Kadish, K. M. *Inorg. Chem.* **1990**, *29*, 1031.

87. Doyle, M. P.; Van Doornik, F. J.; Funckes, C. L. *Inorg. Chim. Acta* **1980**, *46*, L111.

88. Doyle, M. P.; Pickering, R. A.; Dykstra, R. L.; Cook, B. R. *J. Am. Chem. Soc.* **1982**, *104*, 3392.

89. Stevens, R. E.; Yanta, T. J.; Gladfelter, W. L. *J. Am. Chem. Soc.* **1981**, *103*, 4981.

90. Gladfelter, W. L. *Adv. Organomet. Chem.* **1985**, *24*, 41.

91. Mantell, D. R.; Gladfelter, W. L. *J. Organomet. Chem.* **1988**, *347*, 333.

92. Stevens, R. E.; Fjare, D. E.; Gladfelter, W. L. *J. Organomet. Chem.* **1988**, *347*, 373.

93. Godwin, J. B.; Meyer, T. J. *Inorg. Chem.* **1971**, *10*, 471.

94. Dehnicke, K.; Prinz, H.; Massa, W.; Pebler, J.; Schmidt, R. *Z. Anorg. Allg. Chem.* **1983**, *499*, 20.

95. Prinz, H.; Dehnicke, K.; Mueller, U. *Z. Anorg. Allg. Chem.* **1982**, *488*, 49.

96. Dehnicke, K.; Prinz, H. *Chem. -Ztg.* **1983**, *107 (9)*, 247.

97. Schumacher, C.; Weller, F.; Dehnicke, K. *Z. Anorg. Allg. Chem.* **1982**, *495*, 135.

98. Dehnicke, K.; Loessberg, R. *Chem. -Ztg.* **1981**, *105 (10)*, 305.

99. Crookes, M. J.; Roy, P.; Williams, D. L. H. *J. Chem. Soc., Perkin Trans. 2* **1989**, *(8)*, 1015.

100. Kyte, A. B.; Jones-Parry, R.; Whittaker, D. *J. Chem. Soc., Chem. Commun.* **1982**, 74.

101. Casado, J.; Castro, A.; Mosquera, M.; Prieto, M. F. R.; Tato, J. V. *Monatsh. Chem.* **1984**, *115*, 669.

102. Doyle, M. P.; Terpstra, J. W.; Pickering, R. A.; LePoire, D. M. *J. Org. Chem.* **1983**, *48*, 3379.

103. Senechal, D.; Senechal-Tocquer, M. C.; Gentric, D.; Le Bihan, J. Y.; Caro, B.; Gruselle, M.; Jaouen, G. *J. Chem. Soc., Chem. Commun.* **1987**, 632.

104. Bhattacharyya, R.; Saha, A. M.; Ghosh, P. N.; Mukherjee, M.; Mukherjee, A. K. *J. Chem. Soc., Dalton Trans.* **1991**, 501.

105. Sarkar, S.; Mohammad, I. B.; Subramanian, P. *Chem. Lett.* **1985**, 1633.

106. Bhattacharyya, R. G. *Proc. Indian Natl. Sci. Acad.* **1986**, *52A*, 973.

107. Sarkar, S.; Subramanian, P. *Inorg. Chim. Acta* **1979**, *35*, L357.

108. Bhattacharyya, R.; Saha, A. M. *J. Chem. Soc., Dalton Trans.* **1984**, 2085.

109. Bhattacharyya, R. G.; Bhattacharjee, G. P.; Ghosh, N. *Polyhedron* **1983**, *2*, 543.

110. Bhattacharyya, R.; Ghosh, N.; Battacharjee, G. P. *J. Chem. Soc., Dalton Trans.* **1989**, 1963.

111. Feng, D. W.; Ryan, M. D. *Inorg. Chem.* **1987**, *26*, 2480.

112. Aoyagi, K.; Mukaida, M.; Kakihana, H.; Shimizu, K. *J. Chem. Soc., Dalton Trans.* **1985**, 1733.

113. Chilou, V.; Gouzerh, P.; Jeannin, Y.; Robert, F. *Inorg. Chim. Acta* **1987**, *133*, 205.

114. Kriege-Simondsen, J.; Elbaze, G.; Dartiguenave, M.; Feltham, R. D.; Dartiguenave, Y. *Inorg. Chem.* **1982**, *21*, 230.

115. Bhaduri, S.; Sarma, K. R.; Narayan, B. A. *Transition Met. Chem. (Weinheim)* **1981**, *6*, 206.

116. Deb, A. K.; Paul, P. C.; Goswami, S. *J. Chem. Soc., Dalton Trans.* **1988**, 2051.

117. Baker, P. K.; Broadley, K.; Connelly, N. G. *J. Chem. Soc., Chem. Commun.* **1980**, 775.

118. Kukushkin, V. Y.; Tkachuk, V. M.; Popov, A. M.; Egorova, M. B. *Zh. Neorg. Khim.* **1989**, *34*, 2444.

119. Kukushkin, Y. N.; Ageeva, E. D.; Aleshin, V. E. *Zh. Obshch. Khim.* **1980**, *50*, 1877. CA93(16):160501n.

120. Carvill, A.; Higgins, P.; McCann, G. M.; Ryan, H.; Shiels, A. *J. Chem. Soc., Dalton Trans.* **1989**, 2435.

121. Schreiner, A. F.; Lin, S. W.; Hauser, P. J.; Hopcus, E. A.; Hamm, D. J.; Gunter, J. D. *Inorg. Chem.* **1972**, *11*, 880.

122. Pearce, J. R.; Gustafson, B. L.; Lunsford, J. H. *Inorg. Chem.* **1981**, *20*, 2957.

123. Verdonck, J. J.; Schoonheydt, R. A.; Jacobs, P. A. *J. Phys. Chem.* **1981**, *85*, 2393.

124. Ahmed, M.; Edwards, A. J.; Jones, C. J.; McCleverty, J. A.; Rothin, A. S.; Tate, J. P. *J. Chem. Soc., Dalton Trans.* **1988**, 257.

125. Pandey, K. K.: Jain, K. C.; Agarwala, U. C. *Inorg. Chim. Acta* **1981**, *48*, 23.

126. Udupa, K. N.; Jain, K. C.; Khan, M. I.; Agarwala, U. C. *Inorg. Chim. Acta* **1983**, *74,* 191.
127. Appelman, E. H.; Thompson, R. C.; Engelkemeir, A. G. *Inorg. Chem.* **1979**, *18,* 909.
128. Murgich, J.; Ambrosetti, R. *J. Magn. Reson.* **1987**, *74,* 344.
129. Kokulich, S. G.; Rund, J. V.; Pauley, D. J.; Bumgarner, R. E. *J. Am. Chem. Soc.* **1988**, *110,* 7356.
130. Hedberg, L.; Hedberg, K.; Satija, S. K.; Swanson, B. I. *Inorg. Chem.* **1985**, *24,* 2766.
131. Hedberg, K.; Hedberg, L.; Hagen, K.; Ryan, R. R.; Jones, L. H. *Inorg. Chem.* **1985**, *24,* 2771.
132. Fenske, R. F.; DeKock, R. L. *Inorg. Chem.* **1972**, *11,* 437.
133. Legzdins, P.; Rettig, S. J.; Sanchez, L. *J. Am. Chem. Soc.* **1985**, *107,* 1411.
134. Lukehart, C. M.; Troup, J. M. *Inorg. Chim. Acta* **1977**, *22,* 81.
135. Hardy, A. D. U.; Sim, G. A. *Acta Crystallogr., Sect. B* **1979**, *B35*, 1463.
136. Wong, W.-K.; Tam, W.; Strouse, C. E.; Gladysz, J. A. *J. Chem. Soc., Chem. Commun.* **1979**, 530.
137. Hartmann, G.; Mews, R. *Angew. Chem., Int. Ed. Engl.* **1985**, *24*, 202.
138. Enemark, J. H.; Feltham, R. D.; Huie, B. T.; Johnson, P. L.; Swedo, K. B. *J. Am. Chem. Soc.* **1977**, *99*, 3285.
139. Piciulo, P. L.; Rupprecht, G.; Scheidt, w. R. *J. Am. Chem. Soc.* **1974**, *96,* 5293.
140. Scheidt, W. R.; Piciulo, P. L. *J. Am. Chem. Soc.* **1976**, *98,* 1913.
141. Pratt, C. S.; Coyle, B. A.; Ibers, J. A. *J. Chem. Soc. A* **1971**, 2146.
142. Enemark, J. H.; Feltham, R. D.; Riker-Nappier, J.; Bizot, K. F. *Inorg. Chem.* **1975**, *14,* 624..
143. Hoffmann, R.; Chen, M. M. L.; Elian, M.; Rossi, A. R.; Mingos, D. M. P. *Inorg. Chem.* **1974**, *13,* 11.
144. Pierpont, C. G.; Eisenberg, R. *J. Am. Chem. Soc.* **1971**, *93,* 4905.
145. Schmidt, J.; Kühr, H.; Dorn, W. L.; Kopf, J. *Inorg. Nucl. Chem. Lett.* **1974**, *10*, 55.
146. Bottomley, F.; Grein, F. *J. Chem. Soc., Dalton Trans.* **1980**, 1359.
147. Hawkins, T. W.; Hall, M. B. *Inorg. Chem.* **1980**, *19,* 1735.
148. Scheidt, W. R.; Hatano, K.; Rupprecht, G. A.; Piciulo, P. L. *Inorg. Chem.* **1979**, *18,* 292.

149. Scheidt, W. R.; Frisse, M. E. *J. Am. Chem. Soc.* **1975**, *97*, 17.

150. Enemark, J. H.; Feltham, R. D. *J. Chem. Soc., Dalton Trans.* **1972**, 718.

151. Frenz, B. A.; Enemark, J. H.; Ibers, J. A. *Inorg. Chem.* **1969**, *8*, 1288.

152. Brock, C. P.; Collman, J. P.; Dolcetti, G.; Farnham, P. H.; Ibers, J. A.; Lester, J. E.; Reed, C. A. *Inorg. Chem.* **1973**, *12*, 1304.

153. Cheng, P. T.; Nyburg, S. C. *Inorg. Chem.* **1975**, *14*, 327.

154. Chong, K. S.; Rettig, S. J.; Storr, A.; Trotter, J. *Can. J. Chem.* **1979**, *57*, 3090.

155. English, R. B.; Steyn, M. M. V. *S. Afr. J. Chem.* **1984**, *37*, 177.

156. Greenhough, T. J.; Kolthammer, B. W. S.; Legzdins, P.; Trotter, J. *Acta Crystallogr., Sect. B* **1980**, *B36*, 795.

157. Laing, M.; Reiman, R. H.; Singleton, E. *Inorg. Chem.* **1979**, *18*, 2666.

158. Kopf, J.; Schmidt, J. *Z. Naturforsch., B: Anorg. Chem., Org. Chem.* **1975**, *30*, 149.

159. Wah, H. L. K.; Postel, M.; Pierrot, M. *Inorg. Chim. Acta* **1989**, *165*, 215.

160. Martin, R. L.; Taylor, D. *Inorg. Chem.* **1976**, *15*, 2970.

161. Le Borgne, G.; Mordenti, L.; Riess, J. G.; Roustan, J. L. *Nouv. J. Chim.* **1986**, *10*, 97.

162. Pierpont, C. G.; Eisenberg, R. *Inorg. Chem.* **1972**, *11*, 1088.

163. Clark, G. R.; Waters, J. M.; Whittle, K. R. *J. Chem. Soc., Dalton Trans.* **1975**, 463.

164. Wilson, R. D.; Bau, R. *J. Organomet. Chem.* **1980**, *191*, 123.

165. Calderón, J. L.; Cotton, F. A.; DeBoer, B. G.; Martinez, N. *J. Chem. Soc., Chem. Commun.* **1971**, 1476.

166. Calderón, J. L.; Fontana, S.; Frauendorfer, E.; Day, V. W. *J. Organomet. Chem.* **1974**, *64*, C10.

167. Calderón, J. L.; Fontana, S.; Frauendorfer, E.; Day, V. W.; Iske, S. D. A. *J. Organomet. Chem.* **1974**, *64*, C16.

168. Strauss, R. C.; Keller, E.; Brintzinger, H. H. *J. Organomet. Chem.* **1988**, *340*, 249.

169. de Meester, P.; Skapski, A. C. *J. Chem. Soc., Dalton Trans.* **1973**, 1194.

170. Tiripicchio, A.; Camellini, M. T.; Neve, F.; Ghedini, M. *J. Chem. Soc., Dalton Trans.* **1990**, 1651.

171. Ghedini, M.; Neve, F.; Mealli, C.; Tiripicchio, A.; Ugozzoli, F. *Inorg. Chim. Acta* **1990**, *178*, 5.

172. Delgado, E.; Jeffery, J. C.; Simmons, N. D.; Stone, F. G. A. *J. Chem. Soc., Dalton Trans.* **1986**, 869.
173. Tiripicchio, A.; Lanfredi, A. M. M.; Ghedini, M.; Neve, F. *J. Chem. Soc., Chem. Commun.* **1983**, 97.
174. Weiner, W. P.; Hollander, F. J.; Bergman, R. G. *J. Am. Chem. Soc.* **1984**, *106,* 7462.
175. Neve, F.; Ghedini, M. *Inorg. Chim. Acta* **1990**, *175*, 111.
176. Beringhelli, T.; Ciani, G.; D'Alfonso, G.; Molinari, H.; Sironi, A.; Freni, M. *J. Chem. Soc., Chem. Commun.* **1984**, 1327.
177. Kyba, E. P.; Kerby, M. C.; Kashyap, R. P.; Mountzouris, J. A.; Davis, R. E. *J. Am. Chem. Soc.* **1990**, *112,* 905.
178. Von Philipsborn, W.; Muller, R. *Angew. Chem., Int. Ed. Engl.* **1986**, *25*, 383.
179. Mason, J. *Chem. Rev.* **1981**, *81*, 205.
180. Levy, G. C.; Lichter, R. L. *Nitrogen-15 Nuclear Magnetic Resonanace Spectroscopy*; Wiley-interscience: New York, 1979.
181. Evans, D. H.; Mingos, D. M. P.; Mason, J.; Richards, A. *J. Organomet. Chem.* **1983**, *249*, 293.
182. Bell, L. K.; Mason, J.; Mingos, D. M. P.; Tew, D. G. *Inorg. Chem.* **1983**, *22,* 3497.
183. Bultitude, J.; Larkworthy, L. F.; Mason, J.; Povey, D. C.; Sandell, B. *Inorg. Chem.* **1984**, *23,* 3629.
184. Duffin, P. A.; Larkworthy, L. F.; Mason, J.; Stephens, A. N.; Thompson, R. M. *Inorg. Chem.* **1987**, *26,* 2034.
185. Bell, L. K.; Mingos, D. M. P.; Tew, D. G.; Larkworthy, L. F.; Sandell, B.; Povey, D. C.; Mason, J. *J. Chem. Soc., Chem. Commun.* **1983**, 125.
186. Mason, J.; Mingos, D. M. P.; Sherman, D.; Wardle, R. W. M. *J. Chem. Soc., Chem. Commun.* **1984**, 1223.
187. Mason, J.; Mingos, D. M. P.; Schaefer, J.; Sherman, D.; Stejskal, E. O. *J. Chem. Soc., Chem. Commun.* **1985**, 444.
188. Adams, D. M. *Metal-Ligand and Related Vibrations*; Edward Arnold: London, U.K., 1967; pp 84-111 and 268-271.
189. Kettle, S. F. A. *Inorg. Chem.* **1965**, *4,* 1661.
190. Legzdins, P.; Rettig, S. J.; Sanchez, L. *Organometallics* **1988**, *7,* 2394.
191. Collman, J. P.; Farnham, P.; Dolcetti, G. *J. Am. Chem. Soc.* **1971**, *93,* 1788.

192. Nakamoto, K. *Infrared and Raman Spectra of Inorganic and Coordination Compounds*, 4th. ed.; Wiley-Interscience: New York, 1986; pp 291-310.
193. Haymore, B. L.; Ibers, J. A. *Inorg. Chem.* **1975**, *14*, 3060.
194. Manoharan, P. T.; Hamilton, W. C. *Inorg. Chem.* **1963**, *2*, 1043.
195. Veal, J. T.; Hodgson, D. J. *Inorg. Chem.* **1972**, *11*, 1420.
196. Frenz, B. A.; Enemark, J. H.; Ibers, J. A. *Inorg. Chem.* **1969**, *8*, 1288.
197. Dryden, N. H.; Legzdins, P.; Einstein, F. W. B.; Jones, R. H. *Can. J. Chem.* **1988**, *66*, 2100.
198. Albano, V. G.; Bellon, P.; Sansoni, M. *J. Chem. Soc. (A) 1971*, 2420.
199. Hodgson, D. J.; Ibers, J. A. *Inorg. Chem.* **1969**, *8*, 1282.
200. Enemark, J. H. *Inorg. Chem.* **1971**, *10*, 1952.
201. Enemark, J. H.; Feltham, R. D. *J. Chem. Soc., Dalton Trans.* **1972**, 718.
202. Mingos, D. M. P.; Robinson, W. T.; Ibers, J. A. *Inorg. Chem.* **1971**, *10*, 1043.
203. Gallagher, M.; Ladd, M. F. C.; Larkworthy, L. F.; Povey, D. C. Salib, K. A. R. *J. Crystallogr. Spectros. Res.* **1986**, *16*, 967.
204. Schoonover, M. W.; Baker, E. C.; Eisenberg, R. *J. Am. Chem. Soc.* **1979**, *101*, 1880.
205. Brock, C. P.; Collman, J. P.; Dolcetti, G.; Farnham, P. H.; Ibers, J. A.; Lester, J. E.; Reed, C. A. *Inorg. Chem.* **1973**, *12*, 1304.
206. Alnaji, O.; Peres, Y.; Dartiguenave, M.; Dahan, F.; Dartiguenave, Y. *Inorg. Chim. Acta* **1986**, *114*, 151.
207. Bottomley, F.; Darkwa, J.; White, P. S. *J. Chem. Soc. Dalton Trans.* **1985**, 1435.
208. Bottomley, F.; Darkwa, J.; White, P. S. *J. Chem. Soc., Chem. Commun.* **1982**, 1039.
209. Fitzsimmons, B. W.; Larkworthy, L. F.; Rogers, K. A. *Inorg. Chim. Acta* **1980**, *44*, L53.
210. Horrocks, W. D., Jr.; Mann, R. H. *Spectrochim. Acta* **1965**, *21*, 399.
211. Fairey, M. B.; Irving, R. J. *Spectrochim. Acta* **1964**, *20*, 1757.
212. Crabtree, R. H. *The Organometallic Chemistry of the Transition Metals*; Wiley-Interscience: New York, 1988; pp 233-235.
213. Lukehart, C. M. *Fundamental Transition Metal Organometallic Chemistry*; Brooks/Cole: Montery, CA 1985; pp 176-182.
214. Poletti, A.; Foffani, A.; Cataliotti, R. *Spectrochim. Acta* **1970**, *26A*, 1063.
215. Beck, W.; Melnikoff, A.; Stahl, R. *Chem. Ber.* **1966**, *99*, 3721.

216. Hieber, W.; Beck, W. Z. *Naturforsch., B: Anorg. Chem., Org. Chem.* **1958**, *13B*, 194.

217. Hieber, W.; Jahn, A. Z. *Anorg. Allg. Chem.* **1959**, *301*, 301.

218. Satija, S. K.; Swanson, B. I.; Crichton, O.; Rest, A. J. *Inorg. Chem.* **1978**, *17*, 1737.

3

Coordination Compounds

In this chapter, coordination complexes containing nitrosyl ligands are surveyed according to their general types and metal group number. First, the simple metal nitrosyls and metal carbonylnitrosyls are discussed. This is followed by a discussion of metal nitrosyl hydrides, halides and their derivatives. Later, metal nitrosyl compounds containing other nitrogen-, oxygen-, and sulfur-containing ligands are presented and briefly discussed, since these compounds have some interesting chemistry of their own. Earlier reviews on the subject of coordination nitrosyl compounds are available, and the reader is advised to consult these for a historical perspective of the field *(1-3)*.

3.1: Binary Metal Nitrosyls

The $Cr(NO)_4$ complex is a red-black, volatile solid that melts at 38-39 °C *(4)*. Its single ν_{NO} of 1716 cm^{-1} is consistent with a tetrahedral structure for this compound with linear NO ligands. It is obtained by the photolysis of chromium hexacarbonyl in the presence of NO via the intermediate species, $Cr(NO)(CO)_x$ ($x = 4,5$) *(5)*, and it has been the subject of theoretical and spectroscopic studies *(6-10)*. The only other reported tetranitrosyl is $Fe(NO)_4$, which probably contains one bent and three linear NO ligands in a pseudotetrahedral arrangement around the metal *(11,12)*. If this is so, then this compound is isostructural with $Fe(NO)_3Cl$. Nevertheless, other species such as $Fe(NO)_3$ are formed during the matrix isolation of the tetranitrosyl *(11)*.

The red-black trinitrosyl $Co(NO)_3$ is formed from the reaction of $Co_2(CO)_8$ with NO to produce $Co(NO)(CO)_3$ as the intermediate species, which then reacts further with NO to form the final product *(13)*.

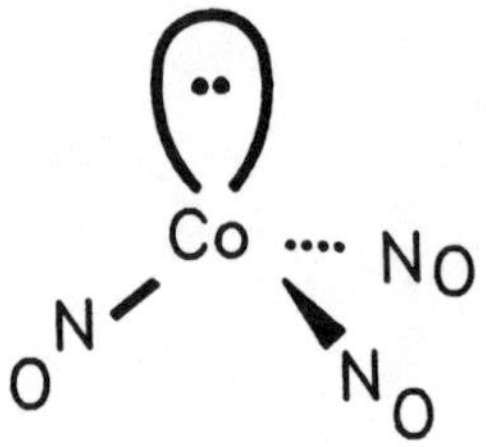

It possesses pyramidal C_{3v} symmetry (ν_{NO} 1860 and 1795 cm^{-1}) and is very basic, even more so than ammonia or trimethylamine. Consequently, it forms adducts (via the Co atom) with Lewis acids such as BF_3, $(BH_3)_2$ and $B(Me)_3$, the latter of which is stable in solution.

A compound of formulation $Mn(NO)_3$ has been observed spectroscopically during the *uv* photolysis of $Mn(NO)_3(CO)$ in Ar matrices *(14)*. In N_2 matrices, the $Mn(NO)_3(N_2)$ complex is formed instead. The dimeric dicationic $[Fe_2(NO)_6]^{2+}$ compound is obtained during the reaction of $Fe(NO)_2(CO)_2$ with the nitrosonium cation *(15)*.

3.2: Carbonylnitrosyls and Their Derivatives

The structures of the following isoelectronic molecules have been investigated by electron diffraction *(8)*.

$Cr(NO)_4$ $Mn(NO)_3(CO)$ $Fe(NO)_2(CO)_2$ $Co(NO)(CO)_3$ $Ni(CO)_4$

All these molecules are tetrahedral and contain linear MNO and MCO groups. Furthermore, these isoelectronic molecules are related to adjacent members in this series by the transfer of a single proton between metal and ligand nuclei *(16)*. Selected structural parameters for these complexes are collected in Table 3.1.

Table 3.1 Structural Parameters for the Isoelectronic $M(NO)_{4-n}(CO)_n$ Complexes (M = Cr, Mn, Fe, Co, Ni)

	$Cr(NO)_4$	$Mn(NO)_3(CO)$	$Fe(NO)_2(CO)_2$	$Co(NO)(CO)_3$	$Ni(CO)_4$
C-O (Å)		1.145	1.140	1.140	1.141
N-O (Å)	1.171	1.167	1.171	1.167	
M-C (Å)		1.949	1.883	1.842	1.838
M-N (Å)	1.763	1.717	1.688	1.671	
M-C Bond Order		1.26	1.51	1.67	1.65
M-N Bond Order	1.89	2.13	2.26	2.42	

Data taken from references 8 and 17.

As may be seen in the Table, all the C-O and N-O bond lengths in these complexes are longer than those of free CO (1.128 Å) and NO (1.15 Å), respectively. Also noteworthy is the observation that in these molecules, the M-N bond orders are substantially greater that the M-C bond orders, an observation that is consistent with the fact that the NO ligand is a much better π-acceptor than CO.

As mentioned in the previous section, only a few simple metal nitrosyls of the form $M(NO)_x$ are known. There are, however, many complexes of the form $M(NO)_x(L)_y$ in which the "$M(NO)_x$" moiety is bound (and stabilized) by neutral Lewis bases such as carbon monoxide and phosphines. In this section, the various classes of complexes of this type that have been synthesized to date are presented. Examples of some first-row transition metal carbonylnitrosyls are listed in Table 3.2.

Table 3.2 Some First Row Carbonylnitrosyls

Compound	ν_{NO} (cm^{-1})	Ref
$V(NO)(CO)_5$	1695	*18*
$[Cr(NO)(CO)_4]^-$	1608	*19*
$Mn(NO)(CO)_4$	1759	*20*
$[Mn(NO)(CO)_3]^{2-}$	1480	*21*
$Mn(NO)_3(CO)$	1733	*4*
$Fe(NO)_2(CO)_2$	1805,1765	*15*
$Co(NO)(CO)_3$	1807	*20,22*

3.2.1 Vanadium and Tantalum

The vanadium complex $V(NO)(CO)_5$ is thermally unstable, and is prepared by the reaction of $[V(CO)_6]^-$ with NO^+ at low temperature *(18)*. The carbonyl ligands may be replaced by Lewis bases (L) such as phosphines, arsines and amines to generate the corresponding $V(NO)(CO)_{5-n}(L)_n$ compounds *(23-26)*. In all cases, facile monosubstitution produces complexes with *trans*-$V(NO)(CO)_4(L)$ geometries. The remarkable reactivity of $V(NO)(CO)_5$ is attributed to the *trans* effect of the NO ligand *(27)*, and indeed, its thermal instability results from facile CO dissociation at room temperature. The dinitrosyl complexes $[V(NO)_2L_4]^+$ are preparable by various routes *(28,29)*. No direct evidence for the existence of $Ta(NO)(CO)_5$ has yet been obtained during the reaction of $[Ta(CO)_6]^-$ with NO^+. However, the addition of trimethylphosphine to the reaction mixture produces small amounts of $Ta(NO)(CO)_3(PMe_3)_2$ and $Ta(NO)(CO)_4(PMe_3)$ *(18)*.

3.2.2: Chromium, Molybdenum and Tungsten

No neutral complexes of the form $M(NO)_x(CO)_y$ have been reported for the Group 6 metals. However, the trimetallic $[W(NO)(CO)_3]_3(\mu\text{-}CN)_3$ complex is known, and contains no metal-metal bonds but rather W-NC-W bridges *(30)*. Several cationic complexes having the general formula $[M(NO)L_5]^{n+}$ (n = 1, 2 ; L = CNR, NCR, CO, PR_3) are known *(31-39)*. The coordinatively unsaturated $[W(NO)L_4]^+$ cations (L = CO, phosphines) are electrophilic enough to bind fluoride-containing anions to the W center via W-F interactions *(40-43)*. The η^1-acrolein adduct $[W(NO)(CO)_3(PMe_3)(\eta^1\text{-acrolein})]^+$ has been crystallographically characterized *(44)*, and it is known that the $[W(NO)L_4]^+$ precursor complexes catalyze Diels-Alder reactions *(44,45)*. The dinitrosyl cations $[M(NO)_2L_4]^{2+}$ are well-known complexes, and they are also strongly electrophilic *(35,38,46-49)*.

Interestingly, reaction of $Cr(CO)_6$ with nitrite results in the formation of anionic $[Cr(NO)(CO)_4]^-$ *(19,50)*, which reacts further with the Group 6 $M(CO)_5(THF)$ complexes to generate the homo- and heterobimetallic complexes $[CrM(NO)(CO)_9]^-$ (M = Cr, Mo, W) *(19)*.

3.2.3 Manganese and Technetium

The $Mn(NO)(CO)_4$ compound possesses trigonal bipyramidal geometry with an axial NO group *(51)*. A carbonyl ligand in this complex may be displaced by other ligands to produce the corresponding $Mn(NO)(CO)_3L$ species *(52,53)*. The related $Mn(NO)(CO)_3(CNR)$ complex is obtained, however, from the alkylation of $[Mn(NO)(CO)_3CN]^-$ *(54)*. Chemical reduction of $Mn(NO)(CO)_3L$ (L = CO, PPh_3) results in the formation of the $[Mn(NO)(CO)_3]^{2-}$ dianion, which reacts with $Mn(NO)(CO)_4$ or $Fe(CO)_5$ to produce $[Mn(NO)_2(CO)_6]^{2-}$ and $[MnFe(NO)(CO)_7]^{2-}$, respectively *(21)*. The $[Mn(NO)(CO)_4]^-$ monoanion has not been isolated, but has been detected by ESR spectroscopy *(55)*. Curiously, the cationic derivatives of manganese and technetium carbonylnitrosyls appear to be restricted to compounds such as $[Mn(NO)_2\{P(OMe)_2Ph\}_3]^+$ *(56)* and $[Tc(NO)(CNCMe_3)_5]^{2+}$ *(57)*.

The trinitrosyl, $Mn(NO)_3(CO)$, is preparable via the photolysis of $Mn_2(CO)_{10}$ in the presence of NO gas *(4)*

$$Mn_2(CO)_{10} \longrightarrow 2\ Mn(NO)(CO)_4 + 2\ CO \qquad (3.1)$$

$$Mn(NO)(CO)_4 + 2\ NO \longrightarrow Mn(NO)_3(CO) + 3\ CO \qquad (3.2)$$

The carbonyl ligand in $Mn(NO)_3(CO)$ is labile, and indeed, the analogous $Mn(NO)_3L$ species (L = phosphines, arsines, stibines, amines) are well-known compounds *(4,25,58)*. The crystal structure of $Mn(NO)_3(PPh_3)$ reveals that this complex possesses tetrahedral geometry with essentially linear MnNO groups *(59)*.

3.2.4: Iron, Ruthenium and Osmium

The solid-state molecular structure of the PPN salt of the $[Fe(NO)(CO)_3]^-$ anion reveals the presence of discrete cations and anions, and a tetrahedral arrangement of CO and NO ligands around the metal center in the anion *(60)*. In solution, however, the Na^+ salt of this anion exists as tight ion pairs due to a Na^+-nitrosyl interaction *(60,61)*, as detected by IR spectroscopy (Table 3.3).

Table 3.3 IR Stretching Frequencies (cm^{-1}) and Relative Abundances of Tight Ion Pairs of $Na[Fe(NO)(CO)_3]$ in Various Solvents at 25 °C.[a]

Solvent	$[Fe(CO)_3(NO^-Na^+)$		$Na^+[Fe(NO)(CO)_3]$		% TIP
	ν_{CO}	ν_{NO}	ν_{CO}	ν_{NO}	
Et_2O	2001, 1900	1599			100
THF	1900, 1885	1615	1978, 1875	1646	48
DME	1988, 1882	1618	1978, 1877	1647	20

[a] $PPN[Fe(NO)(CO)_3]$ exhibits IR bands at 1978, 1876 and 1645 cm^{-1} in THF. Data taken from reference 60.

Oxidation of the $[Fe(NO)(CO)_3]^-$ anion by the ferricinium cation at - 80 °C results in the generation of the transient $Fe(NO)(CO)_3$ radical which is detectable by ESR spectroscopy *(62)*. Other anionic derivatives of the form $[M(NO)L_3]^-$ are known for all three Group 8 metals *(63-65)*.

The molecular structure of the cationic mononitrosyls $[Fe(NO)(CO)(PPh_3)_2L]^+$ (L = CO, MeCN) reveal that they possess trigonal bipyramidal structures with apical phosphine ligands *(66,67)*. Other $[Fe(NO)L_4]^+$ compounds have been prepared *(63,68,69)* and presumably possess similar geometries. Electrochemical reduction of the $[M(NO)(dppe)_2]^+$ complexes (M = Fe, Ru, Os) results in the production of the corresponding neutral $M(NO)(dppe)_2$ radicals, which in the case of Fe, is prone to dissociate one end of the phosphorous-containing ligand *(64)*. In any event, all three congeners are reducible to their monoanions *(64)*. The related $[Fe(NO)_2(CO)_2]^-$ anion has been detected by ESR spectroscopy in γ-irradiated single crystals of $Fe(NO)_2(CO)_2$ *(70)*.

As expected, Lewis bases displace the carbonyl ligands in $Fe(NO)_2(CO)_2$ to afford the $Fe(NO)_2(CO)_{2-n}(L)_n$ derivatives *(71-78)*, and indeed, many complexes of the form $Fe(NO)_2L_2$ have been reported in the literature *(52,72,79,80-83)*. A related compound, namely $Fe(NO)(N_2Ar)(CO)(P^*)$ (P* = chiral aminophosphine) has been obtained by the reaction of $[Fe(NO)(CO)_3]^-$ with aryldiazonium salts in the presence of the phosphine *(84,85)*.

Phosphido-bridged complexes of general formula "$Fe(NO)_2L_2$" may be synthesized via the reactions outlined in eqs 3.3 - 3.5 *(83,86-89)*

$$Fe(NO)_2(PPh_2H)_2 + 2\ BuLi \longrightarrow Li_2[Fe(NO)_2(PPh_3)_2] + 2\ BuH \quad (3.3)$$

$$Li_2[Fe(NO)_2(PPh_3)_2] + Cp_2MCl_2 \longrightarrow Cp_2M(\mu\text{-}PPh_2)_2Fe(NO)_2 + 2\ LiCl \quad (3.4)$$

$$(M = Zr, Ti)$$

$$Fe(NO)_2(CO)_2 + Sn(PPh_2)_4 \longrightarrow Fe(NO)_2(\mu\text{-}PPh_2)_2Sn(\mu\text{-}PPh_2)_2Fe(NO)_2 \quad (3.5)$$

The bis-phosphine complexes $M(NO)_2(PPh_3)_2$ for Ru and Os are also known *(90-93)*. Interestingly, the Ru complex forms a 1:1 adduct with sulfur dioxide to generate $Ru(NO)_2(PPh_3)_2(SO_2)$, whose spectroscopic properties are consistent with it possessing one linear and one bent NO ligand *(94)*. The electrochemical oxidation of the pseudotetrahedral $Ru(NO)_2(PPh_3)_2$ compound proceeds via an irreversible two-electron process that involves a molecular rearrangement to its planar D_{2h} isomer *(95)*.

The neutral trinitrosyl radical $Fe(NO)_3(CO)$ has been generated during the low-temperature photolysis of $Fe(NO)_2(CO)_2$ in the presence of NO *(96)*. The stable cationic derivatives, $[Fe(NO)_3L]^+$ (L = PR_3), are obtained by the reaction of dimeric $[Fe_2(NO)_6]^{2+}$ with phosphines *(15)*.

3.2.5: Cobalt, Rhodium and Iridium

Replacement of the carbonyl ligands in $Co(NO)(CO)_3$ affords the $Co(NO)(CO)_{3-x}(L)_x$ complexes (x = 1,2) *(52,76,77,80-82,97-99)*. Indeed, the $M(NO)(PPh_3)_3$ compounds for all three metals are known *(93,100-103)*. Photolysis of $Co(NO)(CO)_3$ in the gas phase results in the sequential loss of two carbonyl ligands and a nitrosyl ligand to afford predominantly Co(CO), which reacts with $Co(NO)(CO)_3$ to form binuclear $Co_2(NO)(CO)_4$ *(104)*. Interestingly, $Co(NO)(PPh_3)_3$ reacts with sulfur dioxide to produce tetrahedral $Co(NO)(PPh_3)_2(\eta^1\text{-}SO_2)$ which contains a linear NO group and a sulfur-bound SO_2 ligand. In contrast, the Rh analogue contains a bent NO ligand and an $\eta^2\text{-}SO_2$ group *(105)*. Complexes of the form $Co(NO)(CNR)_3$ (R = alkyl, aryl) have also been prepared *(106,107)*. The cationic $[Co(NO)L_4]^{n+}$ complexes (n = 1, 2) are preparable by the reaction of their $[CoL_4]^+$ precursors with NO *(108,109)* or by the treatment of cobalt trinitrate with NO in the presence of the appropriate base *(110)*. Surprisingly, treatment of $Co(NO)(CO)_2(PPh_3)$ with excess $(Me_3CN)_2S$ affords $[Co(NO)(\mu_3\text{-}NCMe_3)]_4$ which has a cubane-type structure *(111)*.

The $M(NO)(PPh_3)_3$ complexes for Rh and Ir undergo facile reactions with perhalocarboxylic acids to afford $M(NO)(O_2CR)_2(PPh_3)_2$ derivatives which may undergo further reactions to produce other $M(NO)(X_2)(PPh_3)_2$ compounds *(112,113)*. Furthermore, the $M(NO)(PPh_3)_3$ compounds undergo phosphine exchange reactions with other Lewis bases *(114)*.

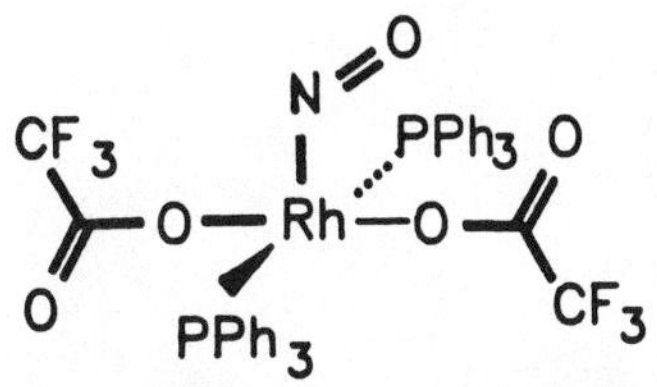

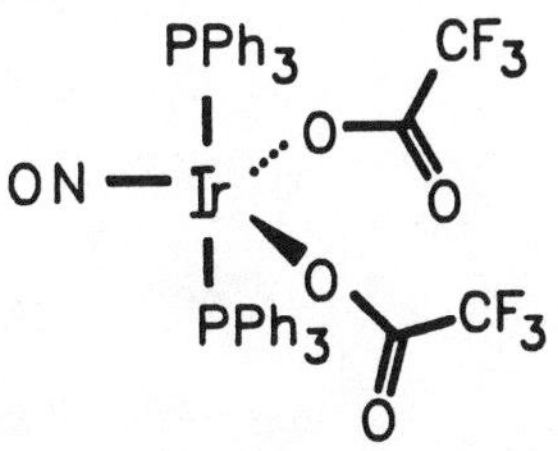

The cationic $[Ir(NO)(PPh_3)_2(MeCN)_3]^{2+}$ is synthesized from its neutral diiodo precursor via halide abstraction, and serves as a convenient precursor to various $[Ir(NO)(PPh_3)_2L_n]^{2+}$ complexes *(115)*.

The neutral $Co(NO)_2(CO)_2$ dinitrosyl radical has been obtained during the photolysis of a solution of $Co(NO)(CO)_3$ containing dissolved NO *(116)*. It is believed that the $Co(NO)_2(CO)_2$ radical possesses planar (D_{2h} symmetry) geometry with the extra electron delocalized onto the NO ligands *(116)*. The cationic $[Co(NO)_2L_2]^+$ compounds are more readily available, however, and many are also isolable *(78,117-121)*. The electrochemical reduction of $[Co(NO)_2\{P(OEt)_3\}_2]BPh_4$ in acetonitrile proceeds via a reversible one-electron reduction to afford, presumably, the neutral $Co(NO)_2\{P(OEt)_3\}_2$ complex. This presumption is based on the fact that exposure of the reduced solution containing this complex to air results in the quantitative regeneration of the starting cationic dinitrosyl complex *(120)*. Interestingly, the $[M(NO)_2L_2]^+$ complexes for Rh and Ir are effective reagents for the conversion of NO and CO into N_2O and CO_2 via the formation of *N-N*-bonded *cis*-N_2O_2 complexes *(122)*. Although the crystal structures of the Rh and Ir complexes are nearly identical, the molecular structure of the Rh complex is best described as intermediate between tetrahedral and square planar *(123)*.

3.2.6: Nickel

A number of $[Ni(NO)L_3]^+$ complexes have been prepared, where L = phosphine or phosphite *(69,78,124)*. The crystal structure of $[Ni(NO)(PMe_3)_3]^+$ reveals that the coordination geometry around the Ni atom is a slightly distorted tetrahedron with a NiNO bond angle of 175.4° *(125)*. The bimetallic $[Ni_2(\mu\text{-}NO)(CNMe)_2(dppm)_2]PF_6$ is preparable by the substitution of a μ-CNMe group in the neutral precursor complex by NO^+ *(126)*.

3.3: Nitrosyl Hydrides

A listing of some neutral nitrosyl hydrides is presented in Table 3.4.

Table 3.4 Some Neutral Nitrosyl Hydrides

Compound	ν_{M-H} (cm^{-1})	ν_{NO} (cm^{-1})	δ_H (ppm)	Ref
$HCr(NO)(CO)_4$		1740	- 4.85	*19*
$HW(NO)(CO)_2(PPh_3)_2$	1681	1628	0.05	*127*
$HMn(NO)_2(PPh_3)_2$		1684, 1640		*53*
$HRe(NO)_2(PPh_3)_2$	1780	1620, 1580		*128*
$H_2Re(NO)(PPh_3)_3$	1950, 1800	1640	- 0.90 - 2.1	*129*
$HFe(NO)(PF_3)_3$	1900	1839	5.87	*130*
$HFe(NO)(CO)(PPh_3)_2$		1680	- 5.0	*131*
$HRu(NO)(PPh_3)_3$	1965	1640	- 6.6	*132-134*
$HOs(NO)(PPh_3)_3$	2050	1635	- 7.3	*134a*
$H_2Os(NO)Cl(PPr_3)_2$	2130, 2050	1713	- 2.16 - 10.03	*134b*
$HRu_6C(CO)_{15}(NO)$		1774	- 21.38	*135*

All the Group 6 non-organometallic nitrosyl hydrides are of the form *trans*-$[HM(NO)L_4]$, i.e.,

The chromium hydride, $HCr(NO)(CO)_4$, is synthesized by the protonation of $[Cr(NO)(CO)_4]^-$ *(19)*, whereas the tungsten hydride $HW(NO)(CO)_2(PPh_3)_2$ is prepared by the reaction of $[W(NO)(CO)_3(PPh_3)_2]^+$ with H^- *(127)*. Other nitrosyl hydrides of Mo and W are also preparable by other routes *(32, 136,137)*, and the M-H bond in the W congeners is generally prone to undergo insertion by molecules such as carbon dioxide, carbon disulfide, aryldiazonium salts and aryl azides *(127)*.

The crystal structure of $H_2Re(NO)(PPh_3)_3$ reveals that this compound possesses a distorted octahedral geometry in which the equatorial plane is occupied by the NO, one PPh_3 and the two mutually-*cis* hydride ligands *(129)*. Other six-coordinate rhenium hydrides of the general formula $HRe(NO)L_3(X)$ have also been reported in the literature *(138-140)*.

Iron nitrosyl hydrides $HFe(NO)L_3$ are known *(63,131,134,141,142)* and, interestingly, whereas $HFe(NO)(PPh_3)_3$ adopts a *cis* H-Fe-NO conformation *(131)*, its Ru congener adopts the corresponding *trans* conformation *(143)* in what is essentially a distorted trigonal bipyramidal arrangement.

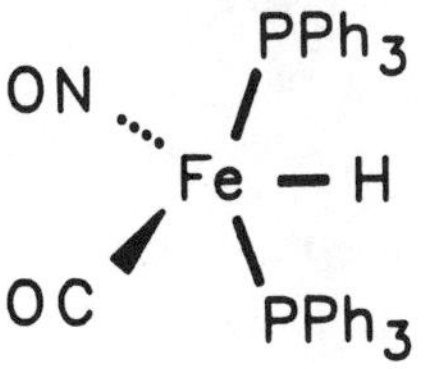

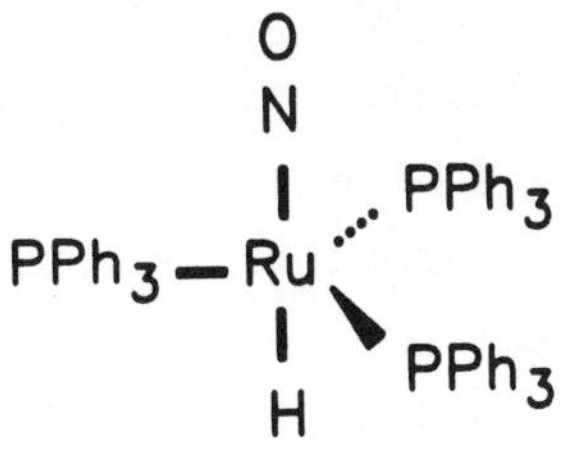

Protonation of the Os congener $HOs(NO)(PPh_3)_3$ yields the cationic dihydride $[H_2Os(NO)(PPh_3)_3]^+$ *(142)*. The neutral dihydride $H_2Os(NO)(PPh_3)_2(CS_2Me)$ is obtained by the reduction of $[(X)Os(NO)(PPh_3)_2(CSMe)]^+$ with borohydride *(144)*.

Three isomers of the pentacoordinate cation in the $[HIr(NO)(PPh_3)_3]ClO_4$ salt have been isolated, but only two have been structurally characterized by X-ray methods. Both forms exhibit distorted trigonal bipyramidal geometries around the central metal. However, the black isomer contains the NO and H ligands in axial positions *(145)*, whereas the brown isomer contains two phosphine ligands in the axial positions *(146)*. The cobalt hydride $[HCo(NO)\{P(OEt)_2Ph\}_4]^+$ has not been isolated, but has been observed when NO gas is reacted with the $[HCo\{P(OEt)_2Ph\}_4]^+$ complex *(148)*. The neutral iridium hydrides $HIr(NO)(PPh_3)_2X$ (X = halide) are also known *(147)*.

Several bimetallic tungsten hydrides of the form (μ-H)$W_2(NO)(CO)_7L_2$ (L = CO, CNR, THF, phosphine, $(C_5H_4PPh_2)_2Fe$) have been synthesized and structurally characterized *(148-150)*. A complex of composition $Cs_2[Mo_2(NO)_2Cl_5H]$ has been reported, and may contain a bridging hydride ligand *(151)*. Many hydride complexes of metal nitrosyl clusters are known, but these appear to be restricted to complexes of Ru *(135,152-156)* and Os *(156-158)*.

Dihydrogen complexes

UV photolysis of $Fe(NO)_2(CO)_2$ and $Co(NO)(CO)_3$ in liquid xenon at -104 °C and under 10 - 20 atm of H_2 yields $Fe(NO)_2(CO)(H_2)$ and $Co(NO)(CO)_2(H_2)$ respectively. The coordinated molecular dihydrogen ligands are identifiable by IR spectroscopy by their unique $\nu_{H\text{-}H}$ at *ca.* 3000 cm^{-1} (cf. $\nu_{D\text{-}D}$ at *ca.* 2200 cm^{-1}) *(159)*. When the pressure of H_2 is released, regeneration of the parent nitrosyl compounds occurs. This latter observation is entirely consistent with a recent molecular-orbital calculation which does not predict the $Co(NO)(CO)_2(H_2)$ complex to be stable with respect to H_2 loss *(160)*.

3.4: Nitrosyl Halides

3.4.1: Pentahalo- and Pentacyanonitrosyls

Although many pentahalonitrosyl complexes of the form $[M(NO)X_5]^{2-}$ have been synthesized for the Group 6 *(161)*, Group 7 *(162,163)*, Group 8 *(164-169)*, Group 9 *(170)* and Group 10 *(171,172)* elements, only a few have been structurally characterized by X-ray crystallography *(161,164,166,173-175)*. Representative examples of these compounds are listed in Table 3.5, together with their NO-stretching frequencies and their M-X bond distances.

Table 3.5 Pentahalonitrosyl Complexes

Compound	ν_{NO} (cm^{-1})	M-X_{tr} (Å)	M-X_{eq} (Å)	Ref
$(AsPh_4)_2[Mo(NO)Cl_5]$	1670	2.467	2.39 (av)	*176*
$K_2[Os(NO)F_5].H_2O$	1824	1.955	1.98 (av)	*164,173,177*
$(Hphen)_2[Os(NO)Br_5]_2$ $.H_2O$	1812	2.478	2.53 (av)	*178*
$(NH_4)_2[Ru(NO)Cl_5]$	1916	2.357	2.38 (av)	*179*
$K[Ir(NO)Cl_5].H_2O$	2008	2.286	2.34 (av)	*170*

As with other six-coordinate mononitrosyl complexes, it is generally observed that when the ν_{NO} of the complex is less than *ca.* 1800 cm^{-1}, the *trans* M-X_{ax} bond (with respect to the NO ligand in the other axial position) is longer than the equatorial M-X bond. Conversely, when the ν_{NO} of the complex is greater than *ca.* 1800 cm^{-1}, the *trans* M-X_{ax} bond is shorter than the M-X_{eq} bond. For a given metal, the increase in donor properties of F<Cl<Br<I generally implies an increase in M-X_{ax} bond lengths relative to M-X_{eq}. For example, in the $[Os(NO)X_5]^{2-}$ complexes, Os-F_{ax} is shorter than Os-F_{eq}, whereas Os-I_{ax} is longer than Os-I_{eq} *(173)*. The same trend is

also observed for the Ru congeners $[Ru(NO)X_5]^{2-}$, where the Ru-X_{ax} bond is shorter than Ru-X_{eq} bond in the chloride and bromide complexes, but longer in the iodide analogue *(166,175)*. Exceptions to this general rule occur due to unusual inter- and intramolecular contacts and in cases where the magnitude of ν_{NO} is particularly sensitive to the nature of the counterions in the complex. Consistent with the general rule, however, is the observation that $Cs[Re(NO)Cl_5]$ (ν_{NO} 1720 cm^{-1}) possesses a labile *trans* chloride ligand *(163)*. Indeed, this *trans* chloride ligand is readily substituted by solvent molecules to produce *trans*-$[Re(NO)Cl_4(solv)]^-$ *(162)*. Curiously, the previously formulated aquo-compounds $M_2[Ru(NO)X_5].2H_2O$ (M = Rb, Cs; X = Cl, Br) have been determined by X-ray crystallography to be actually *trans*-$M_2[Ru(NO)X_4(H_2O)]X.H_2O$ *(180)*.

The *trans*-fluoro ligands in many $[M(NO)F_5]^{2-}$ compounds (M = Ru, Os; ν_{NO} > 1800 cm^{-1}) do not participate in substitution reactions, the substitutions only occurring in the equatorial positions *(181,182)*. The $K[Ir(NO)X_5].H_2O$ compounds (X = Cl, Br) have been characterized by X-ray crystallography and both of these compounds exhibit the expected *trans*-shortening of the M-X_{ax} bond (X = Cl; ν_{NO} = 2008 cm^{-1}) *(170)*.

The one-electron oxidations of $[M(NO)X_5]^{2-}$ for Ru and Os produce the corresponding low-spin $[M(NO)X_5]^-$ monoanions *(169,183,184)*. The related nitrosonium-nitrosyl complex $NO[Os(NO)Cl_5]$ is also obtained by the reaction of NO with $OsCl_5$ *(185)*. A trichlorotin analogue of the osmium pentahalonitrosyl complex, namely $[Os(NO)Cl_3(SnCl_3)_2]^{2-}$, has been obtained and characterized by X-ray crystallography *(186)*.

The pentacyanonitrosyl complexes $[M(NO)(CN)_5]^{n-}$ are related to their pentahalonitrosyl analogues in structure. Several of these pentacyanonitrosyls are now known *(187)*, and have been the subjects of molecular-orbital calculations *(188,189)*. Representative examples of these compounds are listed in Table 3.6.

The pentacyanonitrosyl complexes of Group 5 appear to be restricted to vanadium *(190,199)* whereas those of Group 6 are limited to chromium and molybdenum *(190,192,200,201)*. Many manganese pentacyanonitrosyl complexes have been isolated, and many of these are polymeric *(201-203)*.

Table 3.6 Pentacyanonitrosyl Complexes

Compound	ν_{NO} (cm^{-1})	Ref
$K_3[V(NO)(CN)_5].2H_2O$	1530	*190*
$K_3[Cr(NO)(CN)_5].H_2O$	1637	*191*
$(PPh_4)_3[Mo(NO)(CN)_5]$	1585	*192*
$K_4[Mo(NO)(CN)_5]$	1455	*192*
$K_3[Mn(NO)(CN)_5].2H_2O$	1700	*190*
$K_3[Re(NO)(CN)_5].3H_2O$	1640	*193*
$K_2[Mn(NO)(CN)_5]$	1885	*194,195*
$Na_2[Fe(NO)(CN)_5].2H_2O$	1942	*196*
$K_3[Co(NO)(CN)_5].2H_2O$	2180	*190*
$Na_2[Ru(NO)(CN)_5].2H_2O$	1926	*197,198*

The $[Mn(NO)(CN)_5]^{2-}$ anion is readily reduced to the trianion *(204)*, which is also isolable *(202,205,206)*. The vast majority of studies on the pentacyanonitrosyls have, however, been effected with the Group 8 metals.

The $[Fe(NO)(CN)_5]^{2-}$ anion (commonly referred to as nitroprusside *(207)*) is by far the most studied of the pentacyanonitrosyls. Indeed, many X-ray studies of this anion have been carried out, and the counterions employed to obtain crystals have included Ba *(208)*, Zn *(209)*, Tl *(210)*, Ca *(211)*, Sr *(212)*, Cd *(213)*, Mn *(213)*, K *(214)*, and, of course, Na *(196,214,215)*. A single-crystal neutron diffraction study of Na-nitroprusside has also been carried out *(216)*, but the results do not significantly differ from those obtained by X-ray diffraction studies. The nitroprusside complexes containing cations such as NH_4^+ *(217)* and VO_2^+ (*218*) have also been obtained. Indeed, the nitroprusside anion has been the object of numerous spectroscopic studies *(201,219-226)*. The bonding in nitroprusside is not straightforward, since it occasionally involves a cyanide bridge.

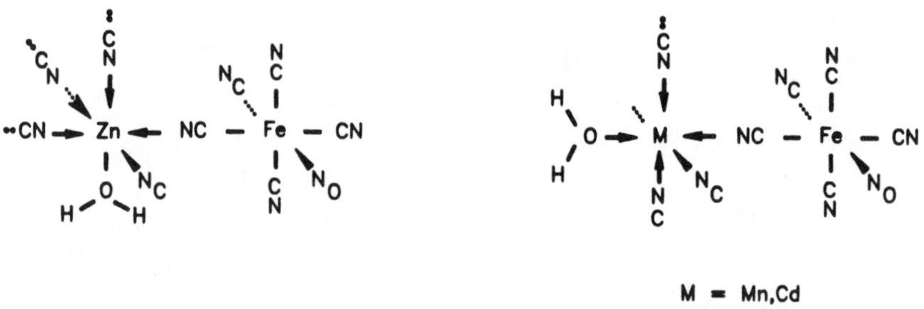

M = Mn,Cd

For example, in the zinc complex, the Zn^{2+} counterion is octahedrally coordinated by five cyanide ligands and one water molecule *(209)*. Related structures are also observed for the Mn^{2+} and Cd^{2+} complexes *(213)*, and for the $[(NH_3)_5Cr\text{-}NC\text{-}Fe(NO)(CN)_4]^+$ bimetallic compound *(227,228)*. It has been pointed out *(189)* that the bond-distance data in some of these pentacyanonitrosyl complexes may not be reflective of the *trans* influence of the NO ligand, since only small energies are required for bond-distance changes of up to 0.15 Å.

Irradiation of ground-state $Na_2[Fe(NO)(CN)_5].2H_2O$ with 400 - 540 nm light produces a long-living metastable state ($\tau > 10^7$ sec) which is formed by charge-transfer from the d_{xy} orbital to the antibonding π^* NO orbital *(229,230)*. In a related study, the NO-stretching frequency in excited-state nitroprusside was found to be 1835 cm^{-1}, which is 105 cm^{-1} lower than that in the ground state molecule, and which was inferred to reflect the bending of the NO ligand *(231)*. A similar linear-to-bent NO conformation change was also inferred in a similar study involving $K_3[Mn(NO)(CN)_5]$ *(232)*.

Although the reduction behavior of nitroprussides has been the subject of much controversy over the past few years, it is now firmly established that upon one-electron reduction, one of the cyanide groups dissociates *(233)*. Thus, it has been shown (by ESR) that the primary reduction product of nitroprusside in aqueous solution is $[Fe(NO)(CN)_4]^{2-}$

(via the release of CN^-) *(234)*. In aprotic solvents, however, the low-temperature (77 K) reduction of nitroprusside does indeed generate the trianion, which upon annealing produces $[Fe(NO)(CN)_4]^{2-}$ and free cyanide *(235)*.

3.4.2: Vanadium

The polymeric trinitrosyl $[V(NO)_3Cl_2]_n$ reacts with coordinating solvents with concomitant displacement of two nitrosyl ligands to form $V(NO)Cl_2L_2$ (L = THF), or reacts with solvents in the presence of reducing agents to form $[V(NO)_2ClL_2]_n$ (L = MeCN) *(29,237)*. The related cyano complex $V(NO)_2L_2(CN)_2$ is also known *(236)*.

3.4.3: Chromium, Molybdenum and Tungsten

The mononitrosyl complexes $[M(NO)Cl_3]_n$ (M = Mo, W) react with Lewis bases to form the corresponding $M(NO)Cl_3L_2$ compounds *(238)*. The low-temperature reaction of $MCl_3(PMe_3)_3$ (M = Mo, W) with NO affords the monomeric seven-coordinate $M(NO)Cl_3(PMe_3)_3$ complexes *(239)*. The related $[Mo(NO)Cl_3(POCl_3)]$ is actually dimeric with Cl bridges *(176)*. Other halonitrosyl compounds of Mo include the anions $[Mo(NO)Cl_4]^-$ and $[Mo(NO)Cl_5]^{2-}$, the latter of which has been discussed in the previous section *(161,176)*. A large number of complexes $M(NO)XL_4$ (M = Mo, W; L = CO, MeCN, PR_3) have been prepared *(127,240-243)*, and it has been determined that the substitution of CO in the W complexes $W(NO)X(CO)_4$ (X = Cl, Br, I) by triphenylphosphine proceeds via a dissociative mechanism *(244)*. However, the stronger nucleophile $P(n\text{-}Bu)_3$ replaces CO in these complexes by both dissociative and associative mechanisms. Related complexes containing transition metal-main group metal-metal bonds, namely $M(NO)L_4M'R_3$ (M = Cr, Mo, W; M' = Sn, Ge, Pb, Si) are also known *(245-248)*.

The simplest dinitrosyl halides of the Group 6 metals are polymeric $[M(NO)_2X_2]_n$ *(38,249)*. As expected, these polymeric complexes are cleaved by Lewis bases to produce the $M(NO)_2X_2L_2$ derivatives, some of which (M = Mo) can also be obtained by reaction of dimeric $Mo_2X_4L_4$ with

NO *(38,250-255,257)*. The polymeric $[M(NO)_2X_2]_n$ complexes of Mo and W also react with halides to produce the corresponding anionic $[M(NO)_2X_3]_2^{2-}$ (M = Mo, W) and $[M(NO)_2X_4]^{2-}$ (M = W) compounds *(256)*. The electrochemical reduction of $Mo(NO)_2Cl_2L_2$ (L = MeCN, L_2 = bipy) results in the formation of their corresponding 19-electron monoanions *(46)*.

The trinitrosyl halides *fac*-$W(NO)_3Cl_3$ *(249)* and *fac*-$[W(NO)_3Br_2]_2$ *(258)* (i.e., *fac* with respect to the three nitrosyl ligands) react with coordinating solvents with concomitant denitrosylation to generate $W(NO)_2X_2L_2$ (L = solvent).

3.4.4: Manganese, Technetium and Rhenium

The $MnX_2(PR_3)$ complexes bind NO to form $Mn(NO)X_2(PR_3)$ (X = Cl, Br, I; R = alkyl, aryl) *(259)*. The related technetium compound $Tc(NO)Br_2(CNCMe_3)_3$ has been characterized by X-ray crystallography *(57)*, and the paramagnetic $Tc(NO)Cl_3(PPh_3)_2$ complex has also been isolated *(260)*. Other Group 7 mononitrosyl compounds of the form $M(NO)X_3L_2$, $[M(NO)X_4]^-$ and $[M(NO)X_5]^-$ have also been prepared *(261-266)*.

The manganese compound $Mn(NO)_2\{P(OMe)_2Ph\}_2Cl$ possesses trigonal bipyramidal geometry with a planar $\{Mn(NO)_2Cl\}$ group, and it represents one of the few examples of a trigonal bipyramidal dinitrosyl compound that contains two equatorial NO groups *(267)*.

Descriptions of many nitrosyl halides of rhenium can be found in the literature *(268,269)*. The reported $Re(NO)Cl_3$ is hydrated *(270)* and eventually undergoes hydrolysis to a material formulated as $Re(NO)(OH)_3(H_2O)$ *(271)*. The five-coordinate paramagnetic complex $Re(NO)Cl_2(PPh_3)_2$ forms adducts with Lewis bases to generate the six-coordinate $Re(NO)Cl_2(PPh_3)_2L$ compounds *(269)*. The dinitrosyl compound $Re(NO)_2Cl_3$ is associated via chloride bridges and reacts with coordinating solvents such as MeCN to form $Re(NO)_2Cl_3(solv)$ *(272,273)*, or with Lewis bases such as $SbPh_3$ to form $Re(NO)_2Cl_2L_2$ *(274)*. Addition of chloride to

$Re(NO)_2Cl_3$ yields, as expected, $[Re(NO)_2Cl_4]^-$ *(272)*. Alternatively, addition of fluoride to the neutral compound produces $[Re(NO)_2Cl_2F_2]^-$ *(275)*. The tetrachloro anionic complex undergoes halide exchange when exposed to excess BBr_3, ultimately converting to the tetrabromide complex $[Re(NO)_2Br_4]^-$, which has been crystallographically characterized *(276)*. Lastly, nitrogen-bearing monodentate ligands cleave the $[Re(NO)(CO)_2Cl_2]$ dimers to generate the monomeric $Re(NO)(CO)_2Cl_2L$ complexes *(277,278)*.

3.4.5: Iron, Ruthenium and Osmium

Mononitrosyl trihalo complexes of the Group 8 metals include $[Fe(NO)Cl_3]^-$ *(279)*, $[Ru(NO)Cl_3]_n$ *(280)* and $Os(NO)Cl_3$ *(281)*. Lewis base adducts of the form $M(NO)X_3L_2$ and $[M(NO)X_4L]^-$ have been prepared for Ru and Os *(93,178,282-290)*. The insoluble $[Ru(NO)I_2]_n$ compound also reacts with Lewis bases to form the corresponding monomers *(291)*. Interestingly, the coordinatively unsaturated, four-coordinate $Ru(NO)Cl(PPh_3)_2$ complex oxidatively adds various RX molecules to an extent that is greater that its isoelectronic carbonyl analogues $M(CO)Cl(PPh_3)_2$ (M = Rh, Ir) *(292,293)*. The related $Os(NO)Cl(PPr_3)_2$ complex has been prepared recently *(134b)*.

The dimeric $[Ru(NO)Cl_4]_2^{2-}$ dianion is preparable by the reaction of the neutral dichloride with chloride *(280)*, whereas the monomeric $[Os(NO)Cl_4]^-$ monoanion is prepared from $(NO)_2[Os(NO)Cl_5]$ by treatment with chloride *(294)*. Other mononitrosyl complexes of the form $[M(NO)XL_4]^{0/2+}$ and $[M(NO)X_4L_4]^+$ are also known for the Group 8 metals *(287,295-297)*, as are the $X_nM'Fe(NO)(CO)_3$ derivatives (M' = Sn, Ge, Pb, Hg) which contain metal-metal bonds *(20,298-300)*.

The crystal structure of the dinitrosyl compound $[Fe(NO)_2I]_2$ reveals it to be dimeric with iodo bridges *(301,302)*. The chloro compound, namely $[Fe(NO)_2Cl]_2$, reacts with dppe to form $[Fe(NO)_2Cl]_2(\mu\text{-dppe})$ *(303,304)*, which is analogous to other $Fe(NO)_2I(L)$ (L = PPh_3, $AsPh_3$) compounds that have been found to undergo exchange reactions with ^{15}NO *(305)*. Nevertheless, both $[Fe(NO)_2I]_2$ and $Fe(NO)_2I(PPh_3)$ are completely ionized in THF to $[Fe(NO)_2]^+$ and $[Fe(NO)_2PPh_3]^+$, respectively. Both cationic products are electrochemically reducible to the catalytically active

"$Fe(NO)_2$" group (see also Chapter 7) *(306)*. The $[Ru(NO)_2Cl(PPh_3)_2]^+$ complex undergoes a one-electron reduction followed by intermolecular ligand scrambling to produce ultimately the stable $Ru(NO)_2(PPh_3)_2$ complex *(307)*.

The $[M(NO)_2XL_2]^+$ dinitrosyl cations represent a unique class of compounds in which the geometry of the $M(NO)_2$ group varies with coordination geometry of the complex as a whole. Thus, both linear and bent NO ligands exist in the complexes that display overall square-pyramidal geometries, particular examples being the Os-OH *(308)*, the Ru-OH *(92)*, and the Ru-Cl *(309,310)* compounds (see also Section 3.6.1). In contrast, the Os-Cl complex possesses a trigonal-bipyramidal structure with two linear, equatorial NO ligands *(311)*.

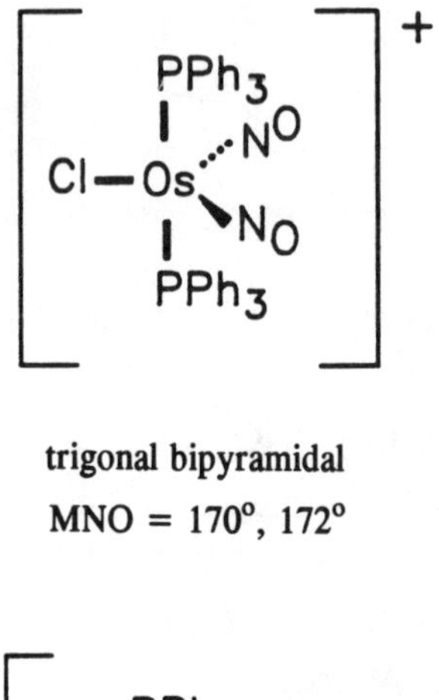

trigonal bipyramidal
MNO = 170°, 172°

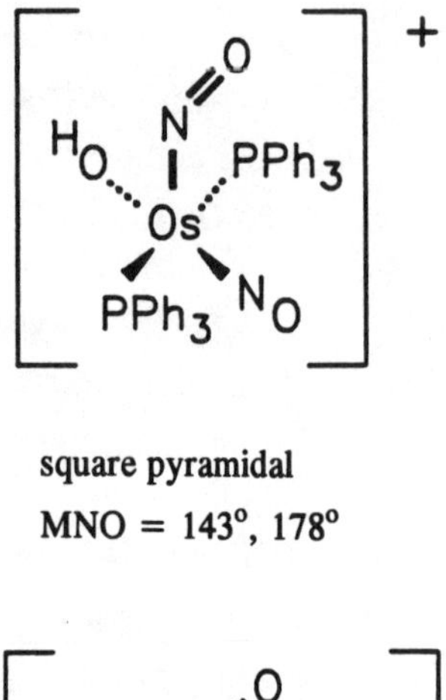

square pyramidal
MNO = 143°, 178°

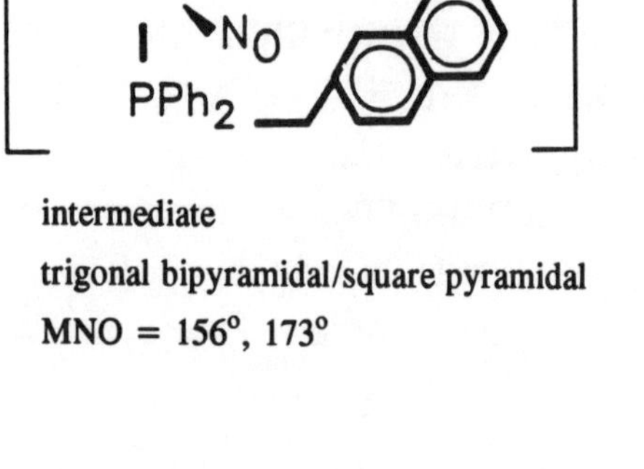

intermediate
trigonal bipyramidal/square pyramidal
MNO = 156°, 173°

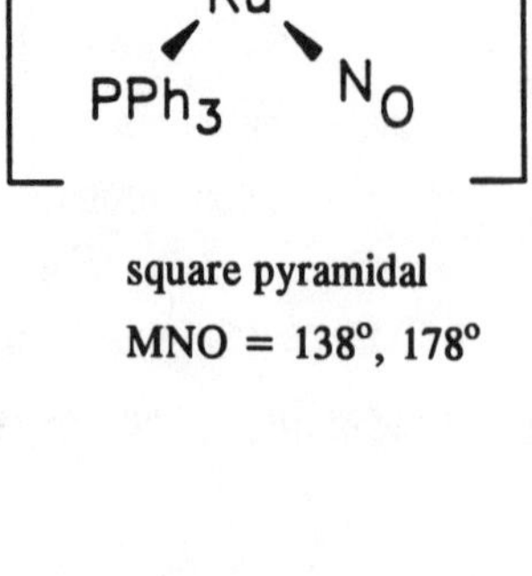

square pyramidal
MNO = 138°, 178°

Finally, the use of a chelating phosphine in the Ru-Cl analogue results in this complex possessing a structure that is intermediate between a square-pyramid and trigonal-bipyramid, with the RuNO geometries also being intermediate between linear and bent *(312)*.

The trinitrosyl $Fe(NO)_3Cl$ is prepared by the treatment of $Fe(CO)_5$ or $Fe(NO)_2(CO)_2$ with nitrosyl chloride *(15,313)*. It is unstable to vacuum, readily losing NO *(313)*. The bromide and iodide analogues have also been prepared *(314,315)*.

3.4.6: Cobalt, Rhodium and Iridium

The $[Rh(NO)X_2]_n$ compounds (X = Cl, Br) and $Ir(NO)Cl_2(COD)$ react with Lewis bases to form the corresponding $M(NO)X_2L_2$ derivatives, a general class of compounds that are known for all three metals in this group *(93,316-320)*. Indeed, several of these latter compounds have been crystallographically characterized *(321)*. Curiously, $Co(NO)Cl_2(PPh_2Me)_2$ exhibits a distorted trigonal-bipyramidal structure with *trans* phosphine ligands and a linear, equatorial NO group *(322)*, whereas $Ir(NO)Cl_2(PPh_3)_2$ exhibits approximately square-pyramidal geometry with an apical, bent NO ligand *(323)*. Spectroscopic evidence suggests that in solution, the trigonal bipyramidal structure of the Co complex is in equilibrium with a square-pyramidal structure that also possess an apical, bent NO group *(322)*.

Interestingly, the dibromide complex, $Rh(NO)Br_2\{P(OR)Ph_2\}_2$ (R = Me, Pr), adopts a trigonal-bipyramidal structure with a bent, apical NO ligand *(316)*, and the related $Rh(NO)Br[\{P(O)Ph_2\}_2H]_2$ dimeric compound possesses two bent NO groups as shown below *(317)*.

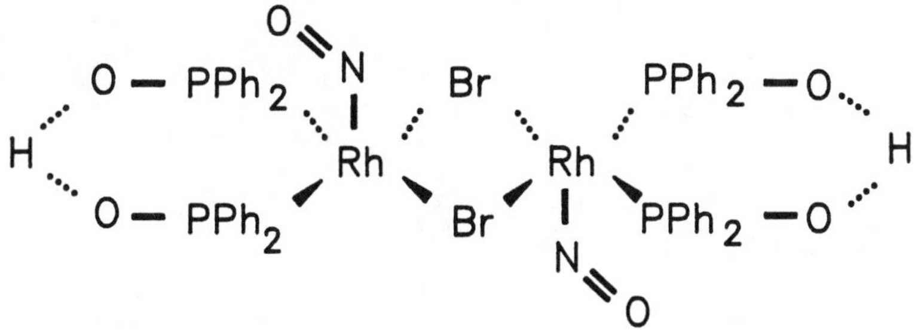

Cationic Co and Ir complexes of the form $[Co(NO)XL_3]^+$, $[Ir(NO)XL_2]^+$ and $[Ir(NO)X_3L_2]^+$ (X = halide, hydroxide, alkoxide) have all been prepared *(324,327-329)*, and the iridium complexes $[Ir(NO)(PPh_3)_2(CO)X]^+$ (X = Cl, I) possess distinctly bent NO ligands *(325,326)*.

The dinitrosyl compounds, $[Co(NO)_2X]_n$ (X = Cl, Br, I), are well-known as nitrosyl-transfer agents (see Chapter 2), and the polymeric nature of these complexes has been established by the X-ray structure determination of $[Co(NO)_2I]_n$, which reveals the existence of an infinite chain formed by "$Co(NO)_2I_2$" tetrahedra sharing the corners of both iodo atoms *(301,302,314,315,330)*. The Rh chloro congener, namely $[Rh(NO)_2Cl]_n$, is also polymeric *(331)*. Nevertheless, $[Co(NO)_2I]_n$ reacts with phosphines to produce $Co(NO)_2IL$ *(332)*. While not much chemistry has been done with these complexes, the related $Co(NO)_2Cl(L')$ compound (L = bicyclic phosphorane) has been shown to bind carbon dioxide reversibly *(333)*.

3.4.7: Nickel, Palladium and Platinum

The polymeric $[M(NO)X]_n$ compounds for Ni and Pd are known *(302,314,315,334-337)* as are the dimeric $[M(NO)XL]_2$ and monomeric $M(NO)XL_2$ derivatives for all three metals in this group *(334,338-341)*. The halide ligand in $Ni(NO)Br(PPh_3)_2$ undergoes substitution by carbonylmetallates such as $[Co(CO)_4]^-$ and $[Fe(NO)(CO)_3]^-$ to afford metal-metal bonded species *(342)*. Curiously, $[Pd(NO)X]_n$ reacts with halide anions to produce $[Pd_2X_4(NO)_2]^{2-}$ *(336)*. The crystal structure of the related $[Pt_2(NO)_2Cl_6]^{2-}$ reveals an unusual coordination of both a bridging NO group and a bent NO group to the same metal center *(343)*, i.e.,

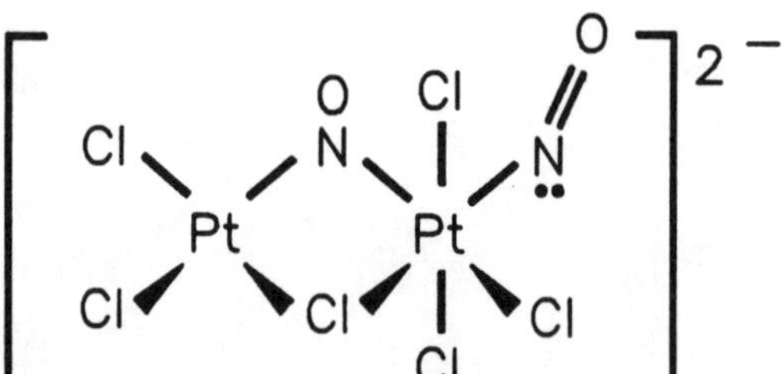

Heterobimetallic complexes of PdIr, PtIr, and PtMn *(344,345,346)*, and a homobimetallic complex of Pt_2 *(347)* have also been synthesized, and each of these bimetallic complexes contains bridging NO ligands.

The nitrosyl halides/pseudohalides of the Group 10 elements also include examples such as $Ni(NO)(OH)_x(OR)_{3-x}$, $[Ni(NO)Cl_2]_n$, $Pd_2(NO)(OAc)_3$ and the dinitrosyls $M(NO)_2Cl_2$ (M = Ni, Pd) *(172,331,341,348,349)*.

3.4.8: Copper

When cupric halides are reacted with NO, complexes of the form $Cu(NO)X_2$ result *(350)*. These latter complexes are known for all the halides, with the bromo and chloro complexes being the most stable *(351)*. Interestingly, the chloro compound $CuCl_2.2H_2O$ forms a 1:1 adduct with $[Ir(NO)(dppn)(PPh_3)_3]^{2+}$ to form a heterobimetallic crystalline CuIr complex containing a μ-NO ligand *(352)*. The homobimetallic complex

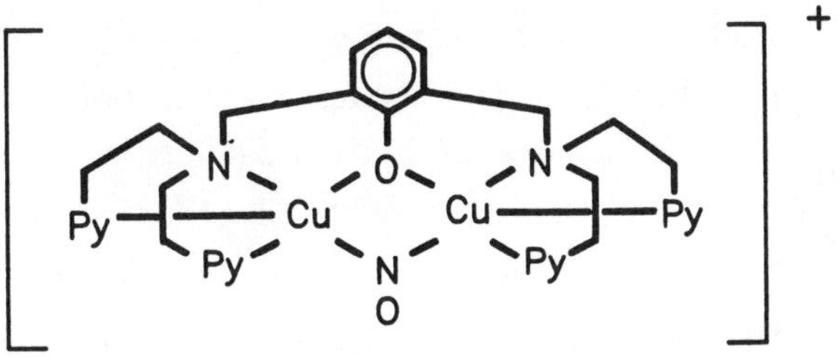

has been synthesized and characterized by X-ray crystallography *(353)*.

3.4.9: Tin and Lead

The $Sn(NO)Cl_2$ and $Pb(NO)F_2$ complexes have been detected in Ar matrices at 10 K and identified by IR spectroscopy *(354)*.

3.5: Nitrosyl Complexes Having Other Nitrogen-Containing Ligands

Nitrogen bases coordinate to metal centers due to presence of a lone pair of electrons on the nitrogen, and such coordination may be monodentate as with MeCN, bidentate to one metal center as found in $V(NO)(bipy)_3$ *(355)*, or bidentate-bridging to two metal centers as found in some IrCu, IrPd and IrPt bimetallic complexes *(344,345,352)*. Polydentate *N*-ligands are treated under a separate heading in the latter part of this section.

3.5.1: Amines and Related Complexes

Many complexes of the form *trans*-$[M(NO)(NH_3)_4L]^{n+}$ have been prepared and studied crystallographically *(264,295,356-375)*. In these complexes, the four NH_3 ligands all lie in the equatorial plane as shown below.

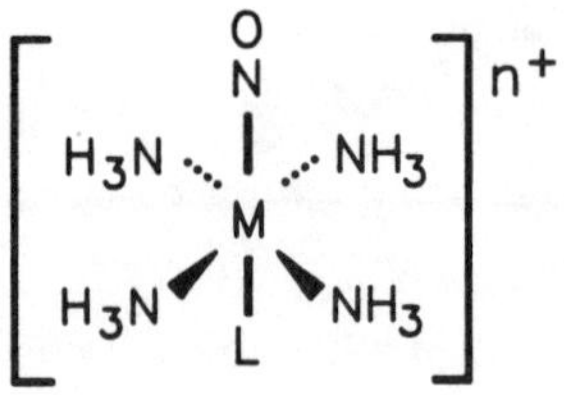

Varying the nature of L in the *trans*-$[Os(NO)(NH_3)_4L]^{2+}$ compounds does not seem to affect the magnitude of the force constants of the equatorial Os-NH_3 bonds, although it does affect the magnitude of the ν_{NO} of these complexes *(371)*. The amine ligand in $[LRe(NO)(CO)(NMe_3)]^{2+}$ (L = 1,4,7-triazacyclononane) is generated when the dicarbonyl precursor is reacted with Me_3NO *(376)*.

3.5.2: Dialkyl- and Diarylamides

Nitrosyl complexes containing NR_2 ligands are known, and such complexes include $Cr(NO)(NR_2)_{3-x}(OBu)_x$ (x = 0, 2; R = Pr, $SiMe_3$) *(377)*, $[Fe(NO)_2(\mu\text{-}NPh_2)]_2$ *(378)*, $Co(NO)_2(NPh_2)$ *(379)* and $Ni(NO)(NPh_2)(PPh_3)_2$ *(380)*.

3.5.3: Dinitrogen

The photolysis of $Mn(NO)(CO)_4$ in N_2 matrices at 20 K generates the spectroscopically observable $Mn(NO)(CO)_3(N_2)$ complex *(381)*. Similarly, the photolysis of $Mn(NO)_3(CO)$ in N_2 matrices generates the $Mn(NO)_3(N_2)$ species *(14)*.

3.5.4: Azides

Only a few metal nitrosyl azide complexes have been isolated and crystallographically characterized. The azide ligands are bound in a terminal η^1 fashion in complexes of Ni *(382)*, Fe *(383)*, Ru *(367)* and Mo *(384,385)*.

3.5.5: Aryldiazonium, Triazenide and Tetrazene

Although the aryldiazonium group (N_2Ar) is formally valence isoelectronic with the nitrosyl group, only a few mixed aryldiazonium-nitrosyl complexes are known *(386,387)*. For example, the $Fe(NO)(N_2Ar)(CO)(P^*)$ compound contains a terminal aryldiazonium group *(84,85)* and is thus analogous to other $Fe(NO)_2L_2$ compounds. The bimetallic $[\{Ir(NO)(PPh_3)\}_2(\mu\text{-}O)(\mu\text{-}N_2Ar)]^+$ cation forms when the oxo-bridged $[Ir(NO)(PPh_3)]_2(\mu\text{-}O)$ is reacted with N_2Ar^+ *(388)*. Aryldiazonium-nitrosyl complexes of Mo and W of the form $(L)M(NO)(N_2Ph)Cl$ (L = $HB(Pz)_3$ or Cp) have also been prepared *(389)*.

The triazenide group (RN_3R) bridges the two metals in $Ru_2(\mu\text{-}N_3Ph_2)_4(NO)_2$ *(390)*, whereas it chelates the single metal center in $[Rh(NO)(N_3R_2)(PPh_3)_2]^+$ *(391)*. In the tetrazene complexes $M(NO)(N_4R_2)(PPh_3)$ (M = Rh, Ir; R = $SO_2C_6H_4Me$) the $M(N_4R_2)$ moiety forms a pentaatomic ring with the R substituents placed on the α N atoms *(318)*.

3.5.6: Hydroxylamine and Related Ligands

Many metal nitrosyl complexes containing HNO *(392-394)*, H_2NO *(161,385,394-396)*, Me(H)NO *(396)* and H_2NOH *(394,396)* ligands are known, and their methods of preparation are discussed in Chapter 7.

3.5.7: Nitrites and Nitrates

As will be pointed out in Chapter 7, the nitrite-nitrosyl and nitrate-nitrosyl conversions are important processes in the nitrosyl-assisted oxidations of organic substrates. It is mainly for this reason that the study of metal complexes containing both NO and NO_x (x = 2, 3) groups has become important over the last decade. Such complexes include those of the Group 6 *(397)*, Group 8 *(398-405)* and Group 10 *(172,406-408)* elements. Curiously, the $Pt(NO)(NO_3)(PPh_3)_2$ compound undergoes an intramolecular isomerization in solution to $Pt(NO_2)_2(PPh_3)_2$ *(408)*.

3.5.8: N_2S_2

The planar μ-N_2S_2 group is found in

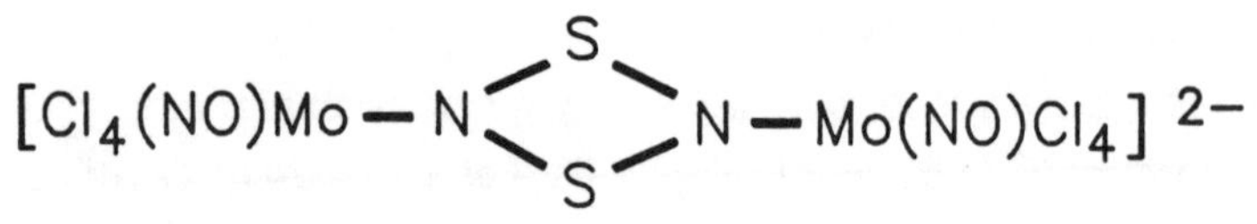

$$[Cl_4(NO)Mo-N(S)_2N-Mo(NO)Cl_4]^{2-}$$

In this compound, the NO groups are located *trans* to the N atoms of the N_2S_2 bridging group *(409)*.

3.5.9: Tridentate Nitrogen Ligands

The 1,4,7-triazacyclononane molecule (tacn) or its tribenzyl derivative (tacn')

R N N R
N
R

(R = H,tacn)

coordinate in a facial arrangement in cationic metal nitrosyl complexes such as $[(tacn)Re(NO)(CO)_2]^{2+}$, $[(tacn')M(N)(CO)_2]^{+}$ and $[(tacn')M(NO)X_2]^{+}$ (M = Mo, W; X = halide, alkoxide) *(376,410-412)*. The sulfur-substituted derivative (tscn = 1,thia-4,7-diazacyclononane) coordinates in a similar fashion in the $[(tscn)Mo(NO)_2Cl]^{+}$ and $[(tscn)Mo(NO)(CO)_2]^{+}$ complexes *(413)*.

The tri-2-pyridylmethane complexes $[\{HC(py)_3\}M(NO)_2(CO)]^{2+}$ of Mo and W are strongly electrophilic and form adducts with carbonyl-containing organic compounds *(414)*. The dipyridylalkane complexes $\{(R)HC(py)_2\}W(NO)(CO)_2Cl$ (R = alkyl) are also known *(415)*.

The tripodal, anionic hydrotris(3,5-dimethylpyrazolyl)borate ligand is found in nitrosyl complexes of the Group 6 elements *(389,416-418)*. As with their cyclopentadienyl analogues (see Chapter 4), the $HB(Me_2pz)_3M(NO)(CO)_2$ compounds for Mo and W are readily derivatized to their mononitrosyl dihalides, $HB(Me_2pz)_3M(NO)X_2$ (X = F, Cl, Br, I), or their dinitrosyl chlorides, $HB(Me_2pz)_3M(NO)_2Cl$. The halide ligands in the mononitrosyl dihalides may be exchanged with alkoxide, amine, amide, thiolate, acetate or azide groups *(416,419-432)*. Other groups that have been employed for such metathesis reactions include crown-ethers *(433)*, porphyrins *(434)* or ferrocenyl- or other metal-containing molecules *(435-437)*.

Many homo- and hetero-bimetallic complexes of Mo and W containing bifunctional bridging groups have been obtained *(430,438-440)*, and a number of these undergo two successive one-electron reductions indicative of substantial electronic interactions between the M(NO) and M'(NO) redox centers *(438,441,442)*. Interestingly, the Mo complex

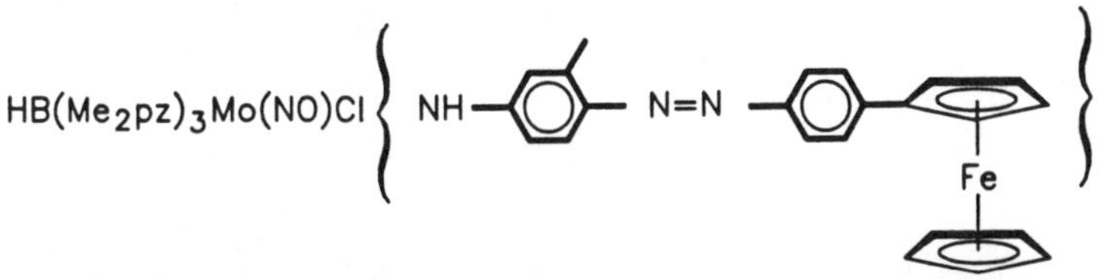

$HB(Me_2pz)_3Mo(NO)Cl$ { NH–(C$_6$H$_3$Me)–N=N–(C$_6$H$_4$)–(C$_5$H$_4$)Fe(C$_5$H$_5$) }

and some of its analogues have been identified as materials that display non-linear optical properties *(434,443)*.

The unsymmetrical Gapz group is flexible in its binding to metal centers in either a *fac* or *mer* conformation, and has been employed for the synthesis of some metal nitrosyl complexes such as L'Ni(NO), $L'Fe(NO)_2$, $L'Mn(NO)_2$ and $L'M(NO)(CO)_2$ (M = Mo, W) *(444-448)*. In the Fe dinitrosyl complex, this tridentate ligand (X = O; Y = NMe_2; R = Me) adopts a *mer* conformation in the formally 19-electron complex *(447)*.

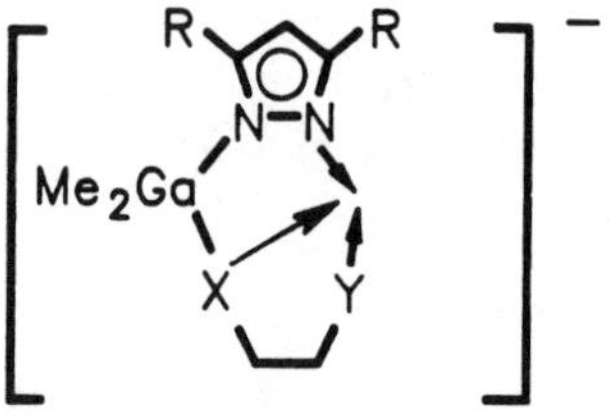

X = O,S
Y = NR_2,SR,py

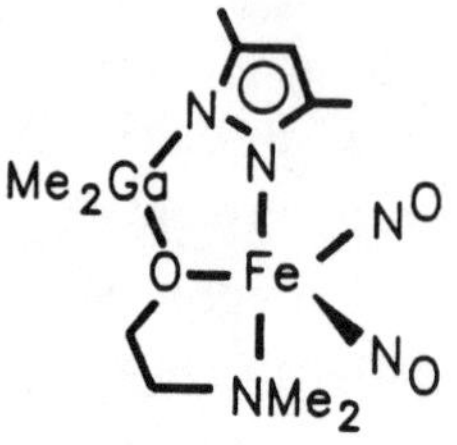

(Gapz)Fe$(NO)_2$

3.5.10: Tetradentate Nitrogen Ligands

A plethora of five-coordinate complexes of the form $M(NO)L_4$ exist, where L_4 is a tetradentate and essentially planar macrocycle *(449-453)*. Although many of these contain four donor N atoms, some contain heteroatomic donor atoms *(454-462)*, e.g.,

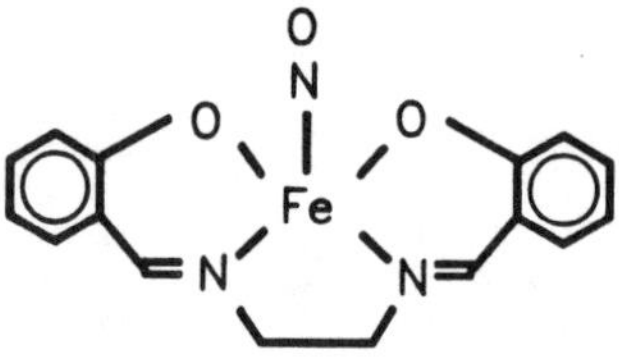

Fe(NO)(Salen)

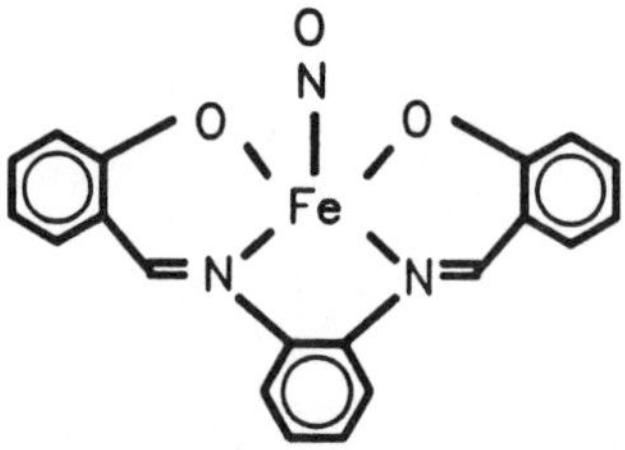

Fe(NO)(Salphen)

The salen complex, Fe(NO)(salen), exhibits a temperature-dependent S = 3/2 - 1/2 spin crossover at 175 K, with the higher spin state being favored at temperatures above 175 K. Some structural changes accompany this spin crossover: above 175 K, the average FeNO bond angle is 147°, whereas below this temperature it is 127° *(463)*. Although the 5-Cl-salen derivative of this complex shows no sharp spin crossover transition *(464)*, many other complexes of the general form $Fe(NO)L_4$ do *(453,465,466)*.

The vanadium complex $Na[V(NO)\{N(CH_2CH_2O)_3\}]$ possesses an approximately trigonal bipyramidal geometry with a *trans* ON-V-N group and the three oxygens of the tetradentate ligand in the equatorial plane *(467)*.

3.5.11: Miscellaneous Nitrogen Ligands

Diazadienes (DAD = RN=C(R)C(R)=NR) form strongly colored complexes of the type $(DAD)Fe(NO)_2$ and (DAD)Co(NO)(CO) *(468)*. The related diiminobenzene group (dib = $1,2\text{-}(NH)_2C_6H_4$) forms a bridge in $[Ru(CO)_2(PPh_3)]_2(\mu\text{-dib})(\mu\text{-NO})$ *(469)*. Finally, *L*-histidinato (L_3) and dipyridyltriazatridecadiene (L_5) groups have been employed for the syntheses of $L_3M(NO)(CO)_2$ complexes of Mo and W *(470)*, and $[L_5Mn(NO)]^+$ *(471)*, respectively.

3.6: Nitrosyl Complexes Containing *O*-donor Ligands

3.6.1: Water, Hydroxide and Alcohol-Containing Species

Many six-coordinate nitrosyl complexes containing coordinated water (or alcohol) are known and have been crystallographically characterized *(138,161,180,406,472-475)*. Monomeric hydroxonitrosyl and alkoxonitrosyl complexes are also known *(287,331,367,476-481)*, and in many respects resemble their halide analogues. Thus, the molecular structure of $Mo(NO)(dppe)_2(OH)$ reveals that the NO and OH ligands are trans to one another in this six-coordinate complex *(137)*. Interestingly, both $[M(NO)_2(PPh_3)(OH)]^+$ cations of Ru and Os contain both linear and bent NO groups (Section 3.4.5) *(92)*. The tetrameric $[Mo(NO)(CO)_2(OH)]_4$ compound contains a μ_3-hydroxo ligand placed in a Mo_4 tetrahedron, with each metal bonded to one NO and two CO ligands *(482)*.

The Group 6 nitrosyl alkoxides $[M(NO)(OR)_3]_2$ (M = Cr, Mo, W; R = alkyl) are cleaved by Lewis bases (e.g., py) to afford the monomeric $M(NO)(OR)_3L$ species *(482-484)*. The molecular structure of a representative example, namely $W(NO)(O\text{-}t\text{-}Bu)_3(py)$, reveals that this complex possesses a slightly distorted trigonal bipyramidal structure with apical NO and py ligands *(485)*.

3.6.2: Oxo-Nitrosyls

The syntheses of $[MoOCl_2(PPh_3)_2]_2(\mu\text{-}NO)$ and $[Mo(PPh_3)_2(\mu\text{-}O)(\mu\text{-}NO)]_2$ have been reported, but their formulations await confirmation by X-ray methods *(486)*. Mercuric halides react with the μ-oxo complex $[Ir(NO)(PPh_3)]_2(\mu\text{-}O)$ to produce $[Ir(NO)(PPh_3)X]_2(\mu\text{-}O)$ (X = Cl, Br, I), and the iodo product is also obtained by the reaction of the former complex with iodine *(487-489)*. Nevertheless, the bridging oxo ligand is removed by PPh_3 to form $Ir(NO)(PPh_3)_3$ and triphenylphosphine oxide *(488)*. The O_2 ligands in the ruthenium complexes $Ru(NO)(PPh_3)_2X(O_2)$ (X = Cl, NO_3) adopt side-on peroxo structures *(92,287)*.

3.6.3: Carbonates, Carboxylates, Diketonates and Related Ligands.

Insertion of CO_2 into the W-H bond of $HW(NO)(CO)_2(PPh_3)_2$ results in the formation of the formato η^1-OC(O)H ligand which equilibrates with its η^2-O_2CH isomer *(127)*. In a formal sense, the O_2CR group is related to the carbonate (O_2CO) group. However, although quite a few nitrosyl complexes containing O_2CR ligands are known *(490,491)*, there appears to be only one reported example of a nitrosyl-carbonate compound, namely $Os(NO)(\eta^2\text{-}CO_3)Cl(PPh_3)_2$ *(479)*.

The carboxylate O_2CR group differs from the well-known 1,3-diketonates (e.g., acetylacetonate) by the number of carbon atoms making up the metallacyclic ring in its complexes. As expected, this difference should result in different molecular structures for complexes containing these ligands, since different steric environments will be present in their compounds. For example, the $Ru_2(\mu\text{-}O_2CR)_4(NO)_2$ bimetallic compound possesses terminal NO groups *(390,492)*, whereas the $Ru_2(acac)_4(NO)_2$ complex possesses bridging NO groups *(493)*,

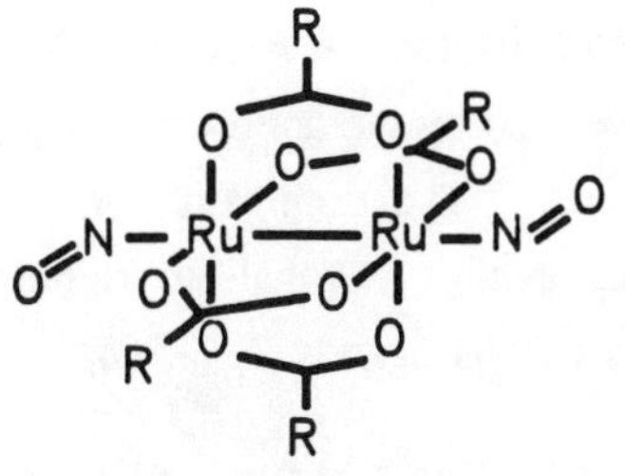

$Ru_2(\mu\text{-}O_2CR)_4(NO)_2$
$(R=Et, CF_3)$

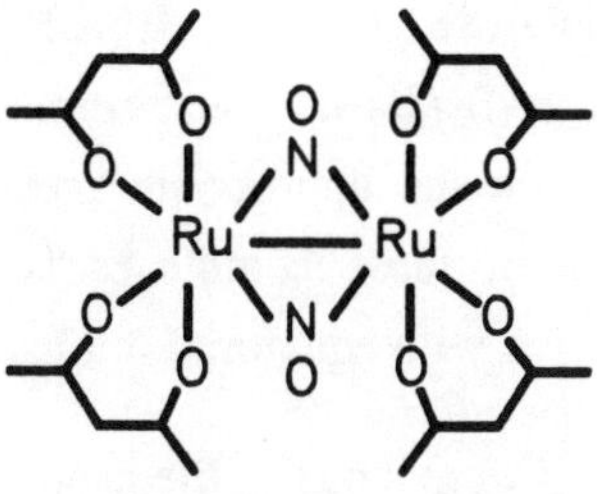

$Ru_2(acac)_4(\mu\text{-}NO)_2$

Other monomeric nitrosyl-acac complexes have also been synthesized *(48,262,494-497)*, and nitrosyl complexes containing $O_2C_2O_2$ *(498)*, O_2PX_2 (X = F, Cl) *(48, 499)* or catechol ligands *(500,501)* have also been reported in the literature.

3.7: Sulfur Ligands

3.7.1: Thioethers, Thiols and Thiolates

As expected, thioethers function as Lewis bases via the lone pair of electrons of the sulfur atoms. Surprisingly, however, only a few nitrosyl complexes containing SR_2 ligands are known *(282,502)*. In the case of $Ru(NO)Br_3(SEt_2)_2$, the coordinated thioether undergoes photochemical oxidation to produce $Ru(NO)Br_3(SEt_2)_{2-x}(OSEt_2)_x$ (x = 1, 2) *(282)*. The Os analogue of the latter complex (x = 1) has also been synthesized and characterized by X-ray methods *(503)*.

The thiol group (SH), like its hydroxide analogue, essentially acts as a pseudohalide in its complexes. Thus, complexes such as $[Fe(NO)_2(SH)_2]^-$ *(504)*, $Fe_2(NO)_4(SH)_2$ *(505)* and $Ir(NO)(SH)_2(PPh_3)_2$ *(481)* (for which their halide analogues are known) have been synthesized. The seleno complex $Os(NO)(SeH)(CS)(PPh_3)_2$ has also been prepared *(144)*.

Many nitrosyl complexes containing thiolate groups have been reported in the literature, although the majority of these belong to the Groups 6 to 8 metals. Most of the Group 6 thiolate complexes are in the form of the five-coordinate neutral $Mo(NO)(SR)_3(NH_3)$ or anionic $[Mo(NO)(SR)_4]^-$ and $[M(NO)(SR)_3Cl]^-$ complexes (M = Mo, W; R = alkyl, aryl) *(506)*. These compounds are generated by the action of thiolate anions on the appropriate precursor complexes. The dominant common feature of these complexes is that they all exhibit trigonal-bipyramidal structures with equatorial thiolate and axial linear NO ligands *(506-508)*.

Compounds of the general formula $Re(NO)(SAr)_4$ (Ar = aryl groups with *ortho* substituents) are formed from the reaction of $Re(NO)Cl_2(OMe)(PPh_3)_2$ with thiophenols, and they also possess trigonal bipyramidal geometries with axial linear NO groups *(507,509)*. The use of thiophenols without *ortho* substituents in these reactions produces the bimetallic $[Re_2(NO)_2(SAr)_7]^-$ compounds instead (Ar = Ph, *p*-tol) which are more appropriately described as $[\{Re(NO)(SAr)_2\}_2(\mu\text{-}SAr)_3]^-$ *(268,509)*. The technetium thiolate complex, $Tc(NO)Cl(SC_6Me_4H)_3$, also exhibits trigonal bipyramidal geometry with equatorial SAr ligands and an axial, linear NO group *(265)*. The nitrosyl thiolatc complexes of the Group 8 metals are considered under the section of Roussin salts.

On occasion, the SR group may also contain heteroatoms (e.g., N) in the R chain which can function as Lewis bases to metal centers to result in a polydentate bonding mode of the thiolate group *(510)*. Thiolate groups bonded in a monodentate fashion to one metal center may also act as Lewis bases via the lone pairs of electrons on the sulfur atoms to other metal centers, resulting in the formation of homo- or heterobimetallic complexes such as $Cp_2Ti(\mu\text{-}ER)_2Fe(NO)L$ (E =S, Se; R = Me, Ph) *(511)*, $[Co(NO)(PPh_3)(\mu\text{-}SPh)]_2$ and $Fe(NO)_2(\mu\text{-}SPh)_2Co(NO)(PPh_3)$ *(512)*.

3.7.2: Sulfides and Polysulfides

An amazing range of structural possibilities exists for nitrosyl complexes containing sulfide and polysulfide ligands. The monosulfide ligand (essentially an isolated sulfur atom) can adopt a terminal bonding mode as in the bimetallic $[Mo(NO)(S)Br_2]_2(\mu\text{-}Br)_2$ complex *(539)* or may also adopt a bridging bonding mode between two, three or even four metal atoms (vide infra). The disulfido ligand (S_2^{2-}) can coordinate to a single metal center in a side on η^2-S_2 fashion, or can bridge two metal centers in the μ-η^2,η^1 or μ-η^2,η^2 forms.

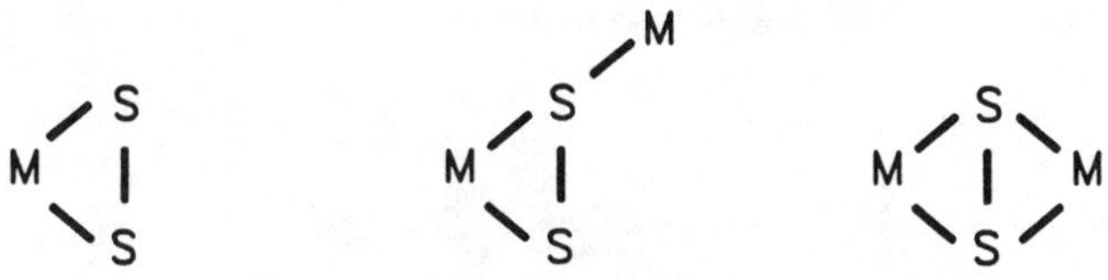

η^2 μ,η^2,η^1 μ,η^2,η^2

The simple side-on bonding mode to a single metal center can be found in the product of the reaction described by eq 3.7 *(535)*.

$$Os(NO)(PPh_3)_2Cl + 1/4\ S_8 \longrightarrow Os(NO)(PPh_3)_2Cl(\eta^2\text{-}S_2) \qquad (3.7)$$

The other bonding modes can be found in the molybdenum sulfur clusters described in the next section.

3.7.2.1: Molybdenum Sulfur Clusters

The $[Mo_4(NO)_4(S_{13})]^{4-}$ tetraanion displays both the η^2 and the μ-η^2,η^1 modes of disulfido bonding, in addition to containing one μ_4-S and two μ_3-S ligands *(513)*.

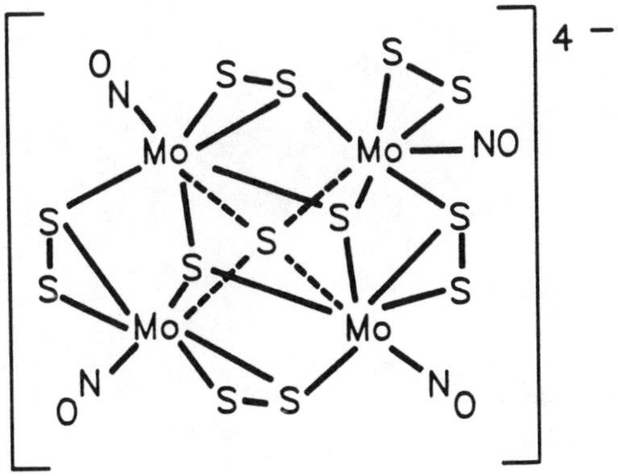

In this compound the μ_4-S atom lies outside the Mo_4 system.

The μ-η^2,η^2-S_2^{2-} ligand is found in the related $[Mo_4(NO)_4(S_2)_6O]^{2-}$ anion, where each Mo experiences a pentagonal bipyramidal environment, and the μ_4-O atom is inside the Mo_4 system *(514,515)*.

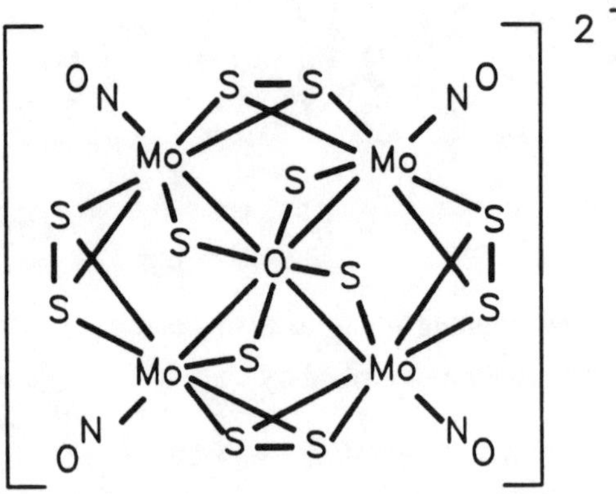

This complex reacts with KCN to produce $K_8[Mo_4(NO)_4S_4(CN)_8].4H_2O$, which contains linear NO ligands displaying a very low ν_{NO} of 1450 cm^{-1} *(516)*.

The μ-S_5^{2-} bidentate ligand is found in the $[Mo_2(NO)_2(S_2)_3(S_5)OH]^{3-}$ complex *(517)*.

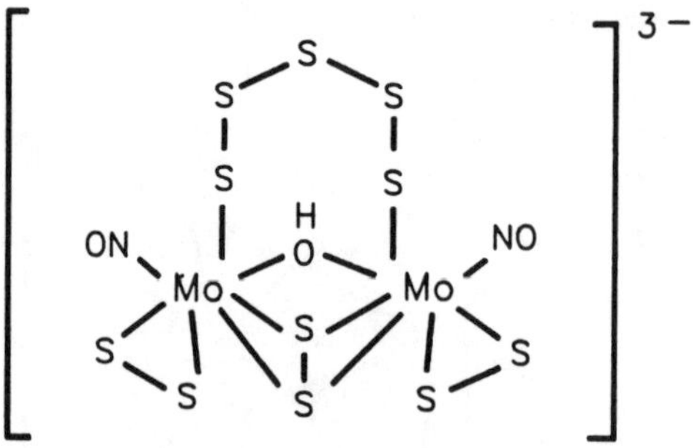

$[Mo_2(NO)_2(S_2)_3(S_5)OH]^{3-}$

The related $[Mo_2(NO)_2(O)(S_2)_3S_5]^{3-}$ trianion is also reported to possess an identical structure *(518,519)*.

The bonding in a wide range of Mo nitrosyl sulfur clusters has been analyzed by the use of extended Hückel calculations *(520)*. In general, the HOMOs of these Mo-S clusters are not concentrated of the metal centers, but rather in the Mo-NO bonding interaction. Consequently, oxidation of these complexes should result in the weakening of the MoNO interaction and lead to the expulsion of NO. Furthermore, the LUMOs of these complexes are concentrated in the (S_2) ligands and are generally antibonding.

3.7.2.2: Iron Sulfur Clusters

The importance of understanding the complex chemistry of the iron sulfur nitrosyl clusters cannot be underestimated since these clusters are fascinating not only because of their structural complexities, but also because they participate in biological processes in nature *(512)*. For a more detailed presentation of the chemistry of these clusters, the reader is advised to consult a recent review on the subject *(512)*.

The salts of the $[Fe_4(NO)_7S_3]^-$ and $[Fe_2(NO)_4S_2]^{2-}$ anions are referred to as Roussin's black and red salts, respectively, so-named after their discoverer and their colors.

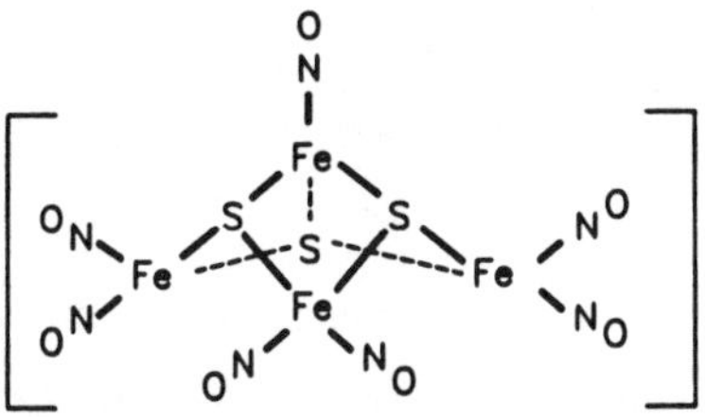

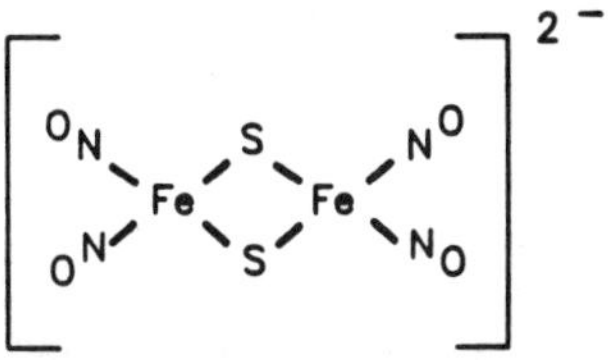

$[Fe_4(NO)_7S_3]^-$ $[Fe(NO)_2S]_2^{2-}$

The $[Fe_4(NO)_7S_3]^-$ anion (of which the Se and Te analogues are also known) possesses approximate C_{3v} symmetry with the $Fe(NO)_2$ groups arranged in an *attracto* conformation *(521)*. It is obtained from the reaction of iron(II) sulfate with sodium nitrite and ammonium sulfide, and it undergoes NO exchange at the basal $Fe(NO)_2$ groups but not at the apical FeNO group *(522)*. Interestingly, this tetrametallic anionic complex undergoes three sequential and reversible one-electron reductions to yield the di-, tri- and tetra-anionic derivatives *(523)*.

The bimetallic $[Fe(NO)_2S]_2^{2-}$ is obtained from $[Fe_4(NO)_7S_3]^-$ by reaction of the latter complex with aqueous sodium hydroxide:

$$[Fe_4(NO)_7S_3]^- \xrightarrow{OH^-} [Fe(NO)_2S]_2^{2-} \qquad (3.7)$$

This bimetallic $[Fe(NO)_2S]_2^{2-}$ dianion may also be conveniently prepared from appropriate carbonyl precursors *(524)*. It reacts with electrophiles such as R^+ (R = H, alkyl, aryl, $CpFe(CO)_2$) or organometallic halides to result in the generation of the neutral "Roussin esters" $[Fe(NO)_2(SR)]_2$. However, esters such as $[Fe(NO)_2(SMe)]_2$ are also obtainable from the direct reaction of $Fe(NO)_2(CO)_2$ with MeSSMe. These Roussin esters may exist in either of two forms, namely

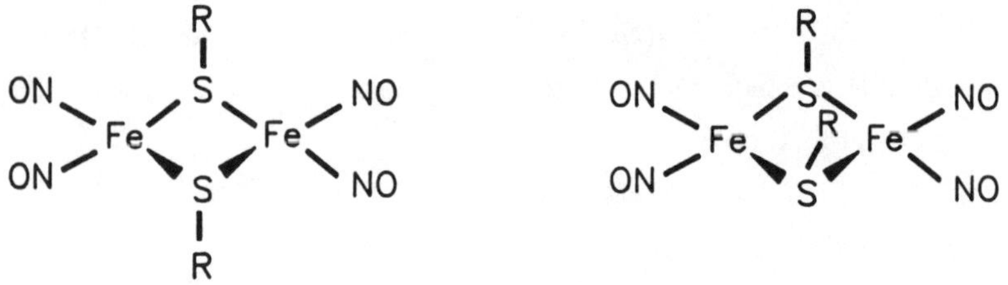

trans *cis*

In the solid state, the $[Fe(NO)_2(SR)]_2$ esters adopt a *trans*-conformation of SR ligands *(525)*, although in solution both the *trans* and *cis* forms may coexist *(526-528)*. In the presence of coordinating solvents, the $[Fe(NO)_2(SR)]_2$ complexes convert to the paramagnetic $[Fe(NO)_2(solv)_2]^+$ species with *g*-values at *ca.* 2.03, and these solvated dinitrosyl complexes have been appropriately named the "2.03 complexes" (Section 6.2.5.1). The Roussin methyl ester $[Fe(NO)_2(SMe)]_2$ reacts with bromide or iodide to produce the paramagnetic $[Fe(NO)_2X_2]^-$ derivatives *(398)*. When R = alkyl, reaction of $[Fe(NO)_2(SR)]_2$ with tetrathiomolybdate anion $[MoS_4]^-$ produces a mixture of $[Fe(NO)_2(SR)_2]^-$, $[Fe(NO)_2(MoS_4)]^-$ and $[Fe(NO)(MoS_4)_2]^{2-}$ *(529)*. Electrochemical studies on the $[Fe(NO)_2(SR)]_2$ complexes (R = alkyl) reveal that they undergo two successive reversible one-electron reductions to their $[Fe(NO)_2(SR)]_2^-$ monoanions and $[Fe(NO)_2(SR)]_2^{2-}$ dianions, respectively *(523)*.

The tetrametallic $[Fe(NO)S]_4$ is prepared via the reaction of $[Fe(NO)(CO)_3]^-$ with sulfur *(530)*

$$Hg[Fe(NO)(CO)_3]_2 \xrightarrow{S_8} [Fe(NO)S]_4$$

and consists of an almost perfect tetrahedron of Fe atoms with each face of this tetrahedron triply-bridged by a sulfur atom:

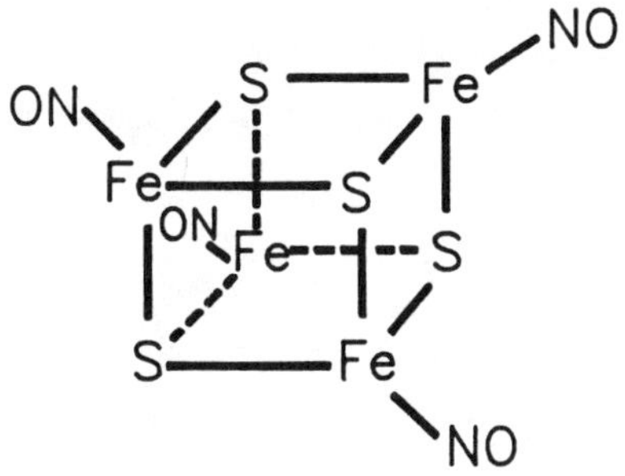

$Fe_4S_4(NO)_4$

Electrochemical reduction of this cubane cluster results in its conversion to its monoanion and then to its dianion *(531)*. The monoanion has been structurally characterized, and it has been observed that the addition of the extra electron to the neutral compound causes a significant reduction in symmetry of the cubane, an observation that is consistent with the inference that the added electron in the monoanion occupies a triply degenerate orbital of largely Fe_4 antibonding character *(531)*. Other cubane-type clusters are known in which the FeNO groups are formally replaced by other metal-containing moieties as in $[Fe(NO)]_2[Cp^*M]_2(\mu_3\text{-}S)_4$ (M = Cr, Mo) *(532,533)*. Interestingly, molecular cluster salts with low lattice energies of the form $[\{Cp'M(\mu_3\text{-}S)\}_4][\{Fe(NO)(\mu_3\text{-}S)\}_4]$ (M = Cr, Mo, Fe; Cp' = Cp, MeCp, PrCp) are formed when equimolar quantities of the individual components are mixed in toluene solution. The formation of these salts results from intermolecular electron-transfer processes, and such salts have been found to possess semiconductor properties *(534)*.

To conclude this section, it must be noted that the Se and Te analogues of the iron sulfur clusters are also known, as are other metal-containing compounds such as $[Co(NO)_2(SR)]_2$. A more detailed discussion of such complexes can be found elsewhere *(512)*.

3.7.2.3: Other Polysulfides as Ligands

The $Co(NO)_2(S_3N)$ compound is obtained from the reaction of $Co_2(CO)_8$ with S_4N_4 and NO gas. The Co atom in this compound is in a pseudotetrahedral environment and is part of a planar CoS_3N metallacycle *(536)*.

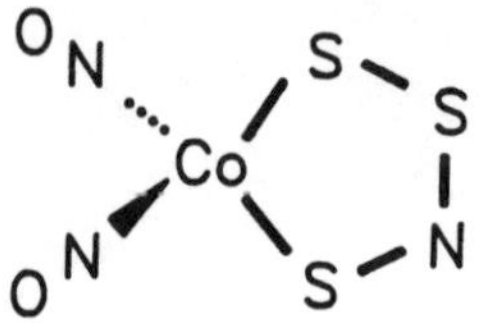

$Co(NO)_2(S_3N)$

Other S_3N complexes of Ru have been reported in the literature, but these await complete structural characterization *(537)*. The tetrasulfido ligand is also found in the recently characterized $[Ru(NO)(NH_3)(S_4)_2]^-$ monoanion *(538)*.

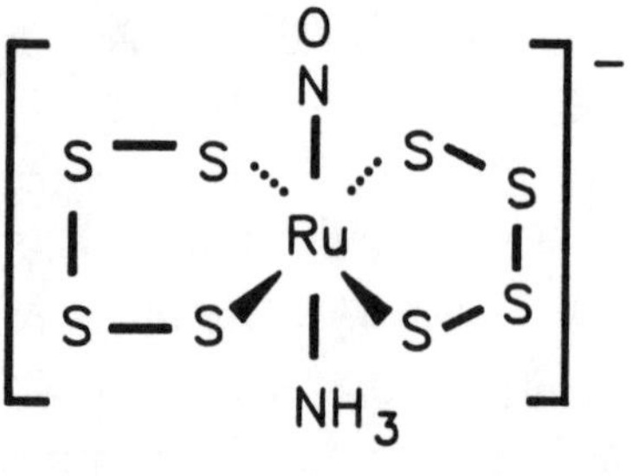

$[Ru(NO)(NH_3)(S_4)_2]^-$

3.7.3: Tetrathiometallates as Ligands

The tetrathiotungstate dianion WS_4^{2-} has been employed as a ligand in the diamagnetic complexes $[Fe(NO)(WS_4)_2]^{2-}$ and $[Fe(NO)_2(WS_4)]^-$ and in related cobalt complexes *(540)*.

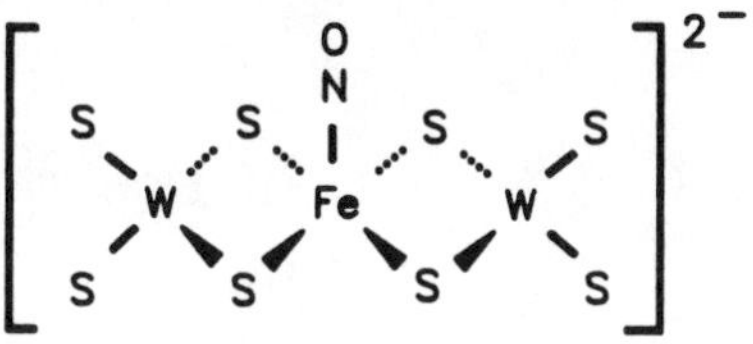

$[Fe(WS_4)_2NO]^{2-}$

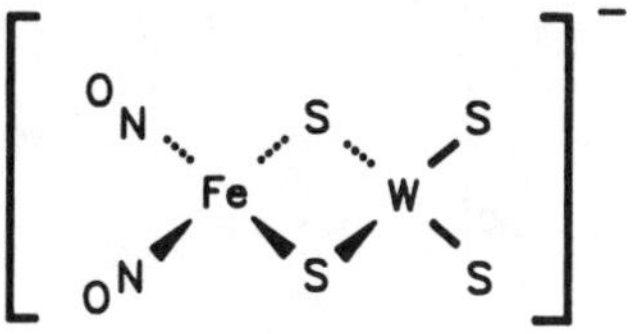

$[Fe(WS_4)(NO)_2]^-$

Indeed, the $[Fe(NO)_2(MS_4)]^{2-}$ compounds (M = Mo, W) are also known, and the Mo complex has been characterized by X-ray crystallography *(541)*. Interestingly, the related $S_2MoO_2^{2-}$ dianion has been employed in the synthesis of $(NEt_4)[Co(NO)_2(S_2MoO_2)]$ whose molecular structure indicates that the $S_2MoO_2^{2-}$ ligand binds via the two sulfur atoms and not the two oxygen atoms *(540)*.

The reaction of $(PPh_4)_2MoS_4$ with $Mo(NO)_2Br_2$ leads to the binuclear $(PPh_4)[S_2Mo(\mu\text{-}S_2)Mo(NO)_2Br_2]$ compound in which the two metals are linked by sulfido bridges *(539)*.

3.7.4: Carbon Disulfide, Dithiocarbamates and Related Ligands

The η^2-CS_2 ligand in $Os(NO)Cl(PPh_3)_2(CS_2)$ is readily converted by methylation to η^1 or η^2 dithioester species which yield the thiocarbonyl ligand upon reduction as shown at the top of the next page *(144)*. Also, CS_2 inserts in to the W-H bond of $HW(NO)(CO)_2(PPh_3)_2$ to produce the η^1-SC(S)H dithioformate ligand *(127)*.

Metal nitrosyl complexes containing dithiocarbamate (S_2CNR_2) ligands abound in the literature, and examples of these

MeI

$MeOSO_2CF_3$

(i) EtOH, PF_6^-
(ii) BH_4^-
(iii) HCl

TeH^-

dithiocarbamate-nitrosyl complexes include those containing Group 6 *(37,542-551)*, Group 7 *(545)*, Group 8 *(403,552-558)*, Group 9 *(552,559)* and Group 10 *(560,561)* metals. The mononitrosyl complexes are generally five-coordinate as in $M(NO)(S_2CNR_2)_2$ (M = Cr, Fe, Co), six-coordinate as in $[M(NO)(S_2CNR_2)_2X]^{+/0}$ (M = Mo, Mn, Fe, Ru, Ni) or seven-coordinate as in $Cr(NO)(S_2CNR_2)_3$. On the other hand, the dinitrosyl complexes are of the form *cis*-$M(NO)_2(S_2CNR_2)_2$ (M = Cr, Mo, W). Monothiocarbamates also chelate to metal centers in nitrosyl complexes as in $M(NO)_2(OSCNMe_2)_2$ for M = Cr and Mo *(543)*. The seleno analogue of the chromium complex, namely $Cr(NO)_2(OSeCNMe_2)_2$, has also been synthesized.

Other chelating dithioanions containing Group 15 elements, such as $[S_2AsMe_2]^-$ and $[S_2PMe_2]^-$ have also been employed for the syntheses of complexes of the form $M(NO)_2(chelate)_2$ (M = Cr, Mo, W) *(543)* and $Fe(NO)[S_2P(OR)_2]_2$ *(545,553,562)*. Related compounds such as $Cr(NO)_2(S_2CMe)_2$ *(543,545)* and $Mo_2(NO)(S_2CR)_4$ (R = Ph, Fc) have also been reported.

3.7.5: Dithiolenes and related compounds

Quite a number of nitrosyl complexes containing dithiolene ligands are known.

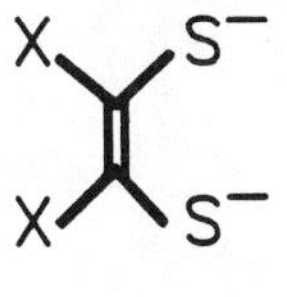

mnt; X = CN 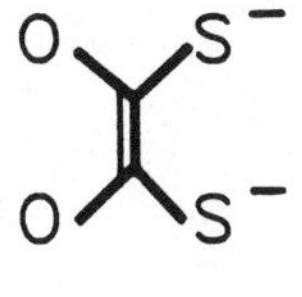   dtox 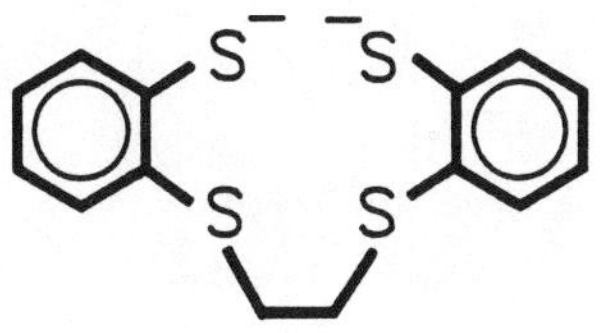   dttd

Thus, the neutral $M(NO)(mnt)L_2$ complexes of Fe and Co *(563,564)*, the anionic $[M(NO)(mnt)]^{n-}$ complexes of Co *(565)*, and the dianionic *cis*-$[M(NO)_2(mnt)_2]^{2-}$ complexes of Cr, Mo and W have all been prepared *(37,565)*, as has the ionic $[Fe(NO)(mnt)_2][Fe(NO)(bipy)_2]$ compound *(566)*. The related $[Fe(NO)(S_2C_2X_2)]^{n-}$ compounds (X = Ph, CF_3, Cl, CN, Me) are also known to contain the $S_2C_2X_2$ ligands chelating to the iron center via the two sulfur atoms *(565)*.

A bis-(dtox) nitrosyl complex of iron has been synthesized as its potassium salt, and its crystal structure reveals its unusual dimeric nature *(567)*.

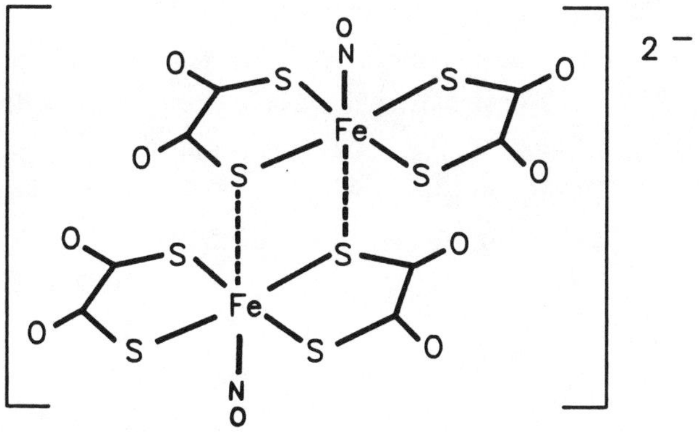

The dttd group is tetradentate through the four sulfur atoms, and complexes of the Group 6 and Group 8 metals containing this and related groups are

known *(568-580)*. Ethylene extrusion from the dttd backbone of Mo(NO)(dttd)Cl occurs upon electroreduction of the neutral complex to generate $[Mo(NO)(1,2\text{-}S_2C_6H_4)_2]^{3-}$ *(568)*.

3.7.6: Sulfur oxides as ligands

The disulfurmonoxide ligand, S_2O, may be generated by the reaction described by eq 3.9.

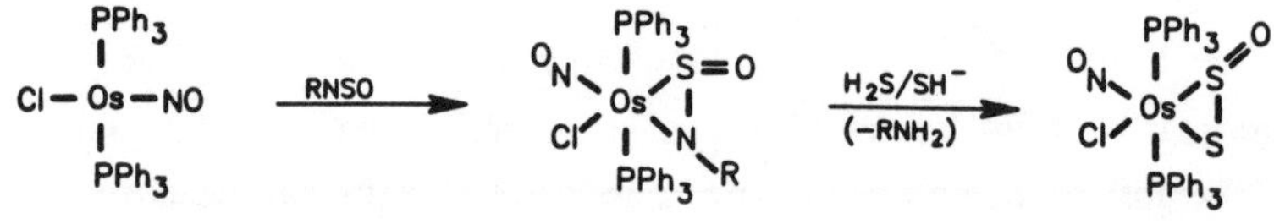

$Cl{-}Os(PPh_3)_2{-}NO$ —RNSO→ $Os(NO)Cl(PPh_3)_2(S(=O)NR)$ —H_2S/SH^- ($-RNH_2$)→ $Os(NO)Cl(PPh_3)_2(S_2O)$ (3.9)

$(R = p\text{-}SO_2C_6H_4Me)$

or by the oxidation of the bound η^2-S_2 ligand in $Os(NO)Cl(PPh_3)_2(\eta^2\text{-}S_2)$ *(535)*. The iminooxosulfurane complex, where R = C_6H_4Me in the compound in the middle of eq 3.9, has been crystallographically characterized *(581)*.

The reaction of $Ru(NO)_2(PPh_3)_2$ with sulfur dioxide yields the $Ru(NO)_2(PPh_3)_2(SO_2)$ adduct, which possesses both linear and bent NO ligands *(94)*. This adduct reacts with oxygen to produce the sulfato $Ru(NO)(PPh_3)(O_2SO_2)$ complex, and analogous complexes of Ru and Rh are also produced in a similar manner. Finally, the photolysis of $[Mo(NO)(OSO_3)(CO)_2(dmpe)_2]$ results in its conversion to $[Mo(NO)(O_2SO_2)(CO)(dmpe)_2]$ via CO loss *(582)*.

3.8: References

1. Johnson, B. F. G.; Haymore, B. L.; Dilworth, J. R. In *Comprehensive Coord. Chem.*; Wilkinson, G.; Gillard, R. D.; McCleverty, J. A., Eds.; Pergamon: Oxford, 1987; Vol. 2, pp 99-118.

2. Connelly, N. G. *Inorg. Chim. Acta* **1972**, *6*, 47.
3. Johnson, B. F. G.; McCleverty, J. A. *Prog. Inorg. Chem.* **1966**, *7*, 277.
4. Satija, S, K.; Swanson, B. I. *Inorg. Synth.* **1976**, *16*, 1.
5. Satija, S. K.; Swanson, B. I.; Crichton, O.; Rest, A. J. *Inorg. Chem.* **1978**, *17*, 1737.
6. Loubriel, G. *J. Vac. Sci. Technol.* **1980**, *17*, 169.
7. Plummer, E. W.; Loubriel, G.; Rajoria, D.; Albert, M. R.; Sneddon, L. G.; Salaneck, W. R. *J. Electron Spectrosc. Relat. Phenom.* **1980**, *19*, 35.
8. Hedberg, L.; Hedberg, K.; Satija, S. K.; Swanson, B. I. *Inorg. Chem.* **1985**, *24*, 2766.
9. Guest, M. F.; Hillier, I. H.; Vincent, M.; Rosi, M. *J. Chem. Soc., Chem. Commun.* **1986**, 438.
10. Bauschlicher, C. W., Jr.; Siegbahn, P. E. M. *J. Chem Phys.* **1986**, *85*, 2802.
11. Doeff, M. M.; Pearson, R. G.; Barrett, P. H. *Inorg. Chim. Acta* **1986**, *117*, 151.
12. Griffith, W. P.; Lewis, J.; Wilkinson, G. *J. Chem. Soc.* **1958**, 3993.
13. Sabherwal, I. H.; Burg, A. B. *J. Chem. Soc., Chem. Commun.* **1970**, 1001.
14. Crichton, O.; Rest, A. J. *J. Chem. Soc., Dalton Trans.* **1978**, 202.
15. Herberhold, M.; Klein, R. *Angew. Chem., Int. Ed. Engl.* **1978**, *17*, 454.
16. Chen, H. W.; Jolly, W. L. *Inorg. Chem.* **1979**, *18*, 2548.
17. Hedberg, K.; Hedberg, L.; Hagen, K.; Ryan, R. R.; Jones, L. H. *Inorg. Chem.* **1985**, *24*, 2771.
18. Fjare, K. L.; Ellis, J. E. *J. Am. Chem. Soc.* **1983**, *105*, 2303.
19. Mantell, D. R.; Gladfelter, W. L. *J. Organomet. Chem.* **1988**, *347*, 333.
20. King, R. B. *Organometallic Syntheses*; Academic: New York, 1965; Vol. 1.
21. Chen, Y. S.; Ellis, J. E. *J. Am. Chem. Soc.* **1983**, *105*, 1689.
22. Foffani, A.; Poletti, A.; Cataliotti, R. *Spectrochim. Acta* **1968**, *24A*, 1437.
23. Schiemann, J.; Weiss, E. *J. Organomet. Chem.* **1982**, *232*, 229.
24. Schiemann, J.; Weiss, E.; Naeumann, F.; Rehder, D. *J. Organomet. Chem.* **1982**, *232*, 219.
25. Rehder, D.; Ihmels, K.; Wenke, D.; Oltmanns, P. *Inorg. Chim. Acta* **1985**, *100*, L11.

26. Toefke, S.; Behrens, U. *Acta Crystallogr., Sect. C: Cryst. Struct. Commun.* **1986**, *C42*, 161.
27. Shi, Q.; Richmond, T. G.; Trogler, W. C.; Basolo, F. *Inorg. Chem.* **1984**, *23*, 957.
28. Naeumann, F.; Rehder, D. *Z. Naturforsch., B: Anorg. Chem., Org. Chem.* **1984**, *39B*, 1654.
29. Herberhold, M.; Trampisch, H. *Inorg. Chim. Acta* **1983**, *70*, 143.
30. Dawes, H. M.; Hursthouse, M. B.; Del Paggio, A. A.; Muetterties, E. L.; Parkins, A. W. *Polyhedron* **1985**, *4*, 379.
31. Honeychuck, R. V.; Hersh, W. H. *Inorg. Chem.* **1987**, *26*, 1826.
32. Kundel, P.; Berke, H. *Z. Naturforsch., B: Chem. Sci.* **1987**, *42*, 993.
33. Bond, A. M.; Colton, R.; Kevekordes, J. E.; Panagiotidou, P. *Inorg. Chem.* **1987**, *26*, 1430.
34. Robinson, W. R.; Wigley, D. E.; Walton, R. A. *Inorg. Chem.* **1985**, *24*, 918.
35. Lloyd, M. K.; McCleverty, J. A. *J. Organomet. Chem.* **1973**, *61*, 261.
36. Ashford, P. K.; Baker, P. K.; Connelly, N. G.; Kelly, R. L.; Woodley, V. A. *J. Chem. Soc., Dalton Trans.* **1982**, 477.
37. Clamp, S.; Connelly, N. G.; Taylor, G. E.; Louttit, T. S. *J. Chem. Soc., Dalton Trans.* **1980**, 2162.
38. Herberhold, M.; Haumaier, L. *Chem. Ber.* **1982**, *115*, 1399.
39. Connelly, N. G.; Gardner, C. *J. Chem. Soc., Dalton Trans.* **1979**, 609.
40. Honeychuck, R. V.; Hersh, W. H. *Inorg. Chem.* **1989**, *28*, 2869.
41. Hersh, W. H. *J. Am. Chem. Soc.* **1985**, *107*, 4599.
42. Honeychuck, R. V.; Hersh, W. H. *J. Am. Chem. Soc.* **1989**, *111*, 6056.
43. Hersh, W. H. *Inorg. Chem.* **1990**, *29*, 713.
44. Honeychuck, R. V.; Bonnesen, P. V.; Farahi, J.; Hersh, W. H. *J. Org. Chem.* **1987**, *52*, 5293.
45. Bonnesen, P. V.; Puckett, C. L.; Honeychuck, R. V.; Hersh, W. H. *J. Am. Chem. Soc.* **1989**, *111*, 6070.
46. Ballivet-Tkatchenko, D.; Boughriet, A.; Bremard, C. *Inorg. Chem.* **1986**, *25*, 826.
47. Ballivet-Tkatchenko, D.; Boughriet, A.; Bremard, C. *J. Electroanal. Chem. Interfacial Electrochem.* **1985**, *196*, 315.
48. Legzdins, P.; Oxley, J. C. *Inorg. Chem.* **1984**, *23*, 1053.
49. Thomas, R. R.; Sen, A. *Inorg. Synth.* **1990**, *28*, 63.

50. Mantell, D. R.; Gladfelter, W. L. In *Organometallic Syntheses*; King, R. B.; Eisch, J. J., Eds.; Elsevier: New York, 1988; Vol. 4, p 56.
51. Frenz, B. A.; Enemark, J. H.; Ibers, J. A. *Inorg. Chem.* **1969**, *8*, 1288.
52. Dobbie, R. C.; Mason, P. R. *J. Chem. Soc., Dalton Trans.* **1976**, 189.
53. Hieber, V. W.; Tengler, H. *Z. Anorg. Allg. Chem.* **1962**, *318*, 136.
54. Moll, M.; Behrens, H.; Trummer, K. H.; Merbach, P. *Z. Naturforsch., B: Anorg. Chem., Org. Chem.* **1983**, *38B*, 411.
55. Lionel, T.; Morton, J. R.; Preston, K. F. *J. Phys. Chem.* **1982**, *86*, 367.
56. Laing, M; Reimann, R. H.; Singleton, E. *Inorg. Chem.* **1979**, *18*, 2666.
57. Linder, K. E.; Davison, A.; Dewan, J. C.; Costello, C. E.; Maleknia, S. *Inorg. Chem.* **1986**, *25*, 2085.
58. Schumann, H.; Meissner, M. *Z. Naturforsch., B: Anorg. Chem., Org. Chem.* **1980**, *35B*, 863.
59. Wilson, R. D.; Bau, R. *J. Organomet. Chem.* **1980**, *191*, 123.
60. Pannell, K. H.; Chen, Y. S.; Belknap, K.; Wu, C. C.; Bernal, I.; Creswick, M. W.; Huang, H. N. *Inorg. Chem.* **1983**, *22*, 418.
61. Dehmlow, E. V. *Z. Naturforsch., B: Anorg. Chem., Org. Chem.* **1982**, *37B*, 1216.
62. Belousov, Y. A.; Kolosova, T. A. *Polyhedron* **1987**, *6*, 1959.
63. Roustan, J. L. A.; Forgues, A.; Merour, J. Y.; Venayak, N. D.; Morrow, B. A. *Can. J. Chem.* **1983**, *61*, 1339.
64. Pilloni, G.; Zotti, G.; Zecchin, S. *J. Electroanal. Chem. Interfacial Electrochem.* **1981**, *125*, 129.
65. Roustan, J. L. A.; Forgues, A. *J. Organomet. Chem.* **1980**, *184*, C13.
66. Ahmed, F. R.; Roustan, J. L. A.; Al-Janabi, M. Y. *Inorg. Chem.* **1985**, *24*, 2526.
67. Roustan, J. L.; Ansari, N.; Charland, J. P.; Le Page, Y. *Can. J. Chem.* **1989**, *67*, 2016.
68. Albertin, G.; Bordignon, E. *Inorg. Chem.* **1984**, *23*, 3822.
69. Sacconi, L. *Proc. Conf. Coord. Chem.* **1978**, *(7)*, 209.
70. Couture, C.; Morton, J. R.; Preston, K. F.; Strach, S. J. *J. Magn. Reson.* **1980**, *41*, 88.
71. Roustan, J. L.; Ansari, N.; Lee, F.; Charland, J. P. *Inorg. Chim. Acta* **1989**, *155*, 11.
72. Mordenti, L.; Roustan, J. L.; Riess, J. G. *Inorg. Chem.* **1984**, *23*, 4503.
73. Ellerman, J.; Wend, W. *J. Organomet. Chem.* **1983**, *258*, 21.

74. Schumann, H.; Koehlricht, K. H.; Meissner, M. *Z. Naturforsch., B: Anorg. Chem., Org. Chem.* **1983**, *38B*, 705.
75. Mordenti, L.; Roustan, J. L.; Riess, J. G. *Organometallics* **1983**, *2*, 843.
76. Plankey, B. J.; Rund, J. V. *Inorg. Chem.* **1979**, *18*, 957.
77. Schumann, H.; Koehricht, K. H. *J. Organomet. Chem.* **1989**, *373*, 307.
78. Johnson, B. F. G.; Bhaduri, S.; Connelly, N. G. *J. Organomet. Chem.* **1972**, *40*, C36.
79. Le Borgne, G.; Mordenti, L.; Riess, J. G.; Roustan, J. L. *Nouv. J. Chim.* **1986**, *10*, 97.
80. Trenkle, A.; Vahrenkamp, H. *Chem. Ber.* **1981**, *114*, 1366.
81. Trenkle, A.; Vahrenkamp, H. *Chem. Ber.* **1981**, *114*, 1343.
82. Ellerman, J.; Dorn, K. *Chem. Ber.* **1968**, *101*, 643.
83. Ellerman, J.; Dorn, K. *Z. Naturforsch., B: Anorg. Chem., Org. Chem.* **1967**, *23B*, 420.
84. Brunner, H.; Miehling, W. *Angew. Chem., Int. Ed. Engl.* **1983**, 22, 164.
85. Carroll, W. E.; Deeney, F. A.; Lalor, F. J. *J. Organomet. Chem.* **1980**, *198*, 189.
86. Chau, C. N.; Yu, Y. F.; Wojcicki, A.; Calligaris, M.; Nardin, G.; Balducci, G. *Organometallics* **1987**, *6*, 308.
87. Yu, Y. F.; Chau, C. N.; Wojcicki, A. *Inorg. Chem.* **1986**, *25*, 4098.
88. Wojcicki, A. *Inorg. Chim. Acta* **1985**, *100*, 125.
89. Yousif-Ross, S. A.; Wojcicki, A. *Inorg. Chim. Acta* **1990**, *171*, 115.
90. McCleverty, J. A.; Nennes, C. W.; Wolochowicz, I *J. Chem. Soc., Dalton Trans.* **1986**, 743.
91. Haymore, B.; Huffman, J. C.; Dobson, A.; Robinson, S. D. *Inorg. Chim. Acta* **1982**, *65*, L231.
92. Grundy, K. R.; Laing, K. R.; Roper, W. R. *J. Chem. Soc., Chem. Commun.* **1970**, 1500.
93. Ahmad, N.; Levison, J. J.; Robinson, S. D.; Uttley, M. F. *Inorg. Synth.* **1974**, *15*, 45.
94. Bhaduri, S.; Johnson, B. F. G.; Khair, A.; Ghatak, I.; Mingos, D. M. P. *J. Chem. Soc., Dalton Trans.* **1980**, 1572.
95. Piazza, G.; Innorta, G. *Inorg. Chim. Acta* **1983**, *70*, 111.
96. Atherton, N. M.; Morton, J. R.; Preston, K. F.; Vuolle, M. J. *Chem. Phys. Lett.* **1980**, *70*, 4.

97. Ellermann, J.; Kock, E.; Zimmerman, H.; Gomm, M. *J. Organomet. Chem.* **1988**, *345*, 167.
98. Clegg, W.; Morton, S. *J. Chem. Soc., Dalton Trans.* **1978**, 1452.
99. Ellermann, J.; Will, N.; Knoch, F. *J. Organomet. Chem.* **1989**, *366*, 197.
100. Bianco, T.; Rossi, M. Uva, L. *Inorg. Chim. Acta* **1969**, *3*, 443.
101. Collman, J. P.; Hoffman, N. W.; Morris, D. E. *J. Am. Chem. Soc.* **1969**, *91*, 5659.
102. Albano, V. G.; Bellon, P.; Sansoni, M. *J. Chem. Soc. (A) 1971*, 2420.
103. Dolcetti, G.; Gandolfi, O.; Ghedini, M.; Hoffman, N. W. *Inorg. Synth.* **1976**, *16*, 32.
104. Rayner, D. M.; Nazran, A. S.; Drouin, M.; Hackett, P. A. *J. Phys. Chem.* **1986**, *90*, 2882.
105. Moody, D. C.; Ryan, R. R.; Larson, A. C. *Inorg. Chem.* **1979**, *18*, 227.
106. Mann, K. R.; DiPierro, M. J. *Cryst. Struct. Commun.* **1982**, *11*, 1049.
107. Carroll, W. E.; Green, M.; Galas, A. M. R.; Murray, M.; Turney, T. W.; Welch, A. J.; Woodward, P. *J. Chem. Soc., Dalton Trans.* **1980**, 80.
108. Zotto, A. D.; Rigo, P. *Inorg. Chim. Acta* **1988**, *147*, 55.
109. Del Zotto, A.; Mezzetti, A.; Rigo, P. *Inorg. Chim. Acta* **1990**, *171*, 61.
110. Albertin, G.; Bordignon, E.; Mazzocchin, G. A.; Orio, A. A.; Seeber, R. *J. Chem. Soc., Dalton Trans.* **1981**, 2127.
111. Gall, R. S.; Connelly, N. G.; Dahl, L. F. *J. Am. Chem. Soc.* **1974**, *96*, 4017.
112. Dobson, A.; Moore, D. S.; Robinson, S. D.; Galas, A. M. R.; Hursthouse, M. B. *J. Chem. Soc., Dalton Trans.* **1985**, 611.
113. Boyar, E. B.; Higgins, W. G.; Robinson, S. D. *Inorg. Chim. Acta* **1983**, *76*, L293.
114. Mazanec, T. J.; Tau, K. D.; Meek, D. W. *Inorg. Chem.* **1980**, *19*, 85.
115. Ghedini, M.; Longeri, M.; Neve, F. *J. Chem. Soc., Dalton Trans.* **1986**, 2669.
116. Morton, J. R.; Preston, K. F.; Strach, S. J. *J. Phys. Chem.* **1980**, *84*, 2478.
117. Khair, A. *Dhaka Univ. Stud.* **1985**, *33B*, 261. CA107(12):103604g.
118. Roustan, J. L.; Ansari, N. Ahmed, F. R. *Inorg. Chim. Acta* **1987**, *129*, L11.
119. Alnaji, O.; Peres, Y.; Dahan, F.; Dartiguenave, M.; Dartiguenave, Y. *Inorg. Chem.* **1986**, *25*, 1383.

120. Seeber, R.; Albertin, G.; Mazzochin, G. A. *J. Chem. Soc., Dalton Trans.* **1982**, 2561.
121. Albertin, G.; Bordignon, E.; Canovese, L.; Orio, A. A. *Inorg. Chim. Acta* **1980**, *38*, 77.
122. Bhaduri, S.; Johnson, B. F. G.; Savory, C. J.; Segal, J. A.; Walter, R. H. *J. Chem. Soc., Chem. Commun.* **1974**, 809.
123. Kaduk, J. A.; Ibers, J. A. *Inorg. Chem.* **1975**, *14*, 3070.
124. Rahman, A. F. M. M.; Salem, G.; Stephens, F. S.; Wild, S. B. *Inorg. Chem.* **1990**, *29*, 5225.
125. Elbaze, G.; Dahan, F.; Dartiguenave, M.; Dartiguenave, Y. *Inorg. Chim. Acta* **1984**, *87*, 91.
126. Ratliff, K. S.; DeLaet, D. L.; Gao, J.; Fanwick, P. E.; Kubiak, C. P. *Inorg. Chem.* **1990**, *29*, 4022.
127. Hillhouse, G. L.; Haymore, B. L. *Inorg. Chem.* **1987**, *26*, 1876.
128. La Monica, G.; Freni, M.; Cenini, S. *J. Organomet. Chem.* **1974**, *71*, 57.
129. Giusto, D.; Ciani, G.; Manassero, M. *J. Organomet. Chem.* **1976**, *105*, 91.
130. Kruck, T.; Lang, W. *Chem. Ber.* **1966**, *99*, 3794.
131. Cygler, M.; Ahmed, F. R.; Forgues, A.; Roustan, J. L. A. *Inorg. Chem.* **1983**, *22*, 1026.
132. Sanchez-Delgado, R. A.; Wilkinson, G. *J. Chem. Soc., Dalton Trans.* **1977**, 804.
133. Bradley, J. S.; Wilkinson, G. *Inorg. Synth.* **1977**, *17*, 73.
134. (a) Wilson, S. T.; Osborn, J. A. *J. Am. Chem. Soc.* **1971**, *93*, 3068. (b) Werner, H.; Michenfelder, A.; Schulz, M. *Angew. Chem., Int. Ed. Engl.* **1991**, *30*, 596.
135. Johnson, B. F. G.; Lewis, J.; Nelson, W. J. H.; Puga, J.; McPartlin, M.; Sironi, A. *J. Organomet. Chem.* **1983**, *253*, C5.
136. Berke, H.; Kundel, P. *Z. Naturforsch., B: Anorg. Chem. Org. Chem.* **1986**, *41B*, 527.
137. Kan, C. T.; Hitchcock, P. B.; Richards, R. L. *J. Chem. Soc., Dalton Trans.* **1982**, 79.
138. Chen, J. Y.; Grundy, K. R.; Robertson, K. N. *Can. J. Chem.* **1989**, *67*, 1187.
139. Grundy, K. R.; Robertson, K. N. *Inorg. Chem.* **1985**, *24*, 3898.

140. Cameron, T. S.; Grundy, K. R.; Robertson, K. N. *Inorg. Chem.* **1982**, *21*, 4149.

141. Svetlov, A. A.; Sinitsyn, N. M. *Russ. J. Inorg. Chem.* **1986**, *31*, 1667.

142. Boyar, E. B.; Dobson, A.; Robinson, S. D.; Haymore, B. L.; Huffman, J. C. *J. Chem. Soc., Dalton Trans.* **1985**, 621.

143. Pierpont, C. G.; Eisenberg, R. *Inorg. Chem.* **1972**, *11*, 1094.

144. Herberhold, M.; Hill, A. F.; McAuley, N.; Roper, W. R. *J. Organomet. Chem.* **1986**, 310, 95.

145. Mingos, D. M. P.; Ibers, J. A. *Inorg. Chem.* **1971**, *10*, 1479.

146. Clark, G. R.; Waters, J. M.; Whittle, K. R. *Inorg. Chem.* **1974**, *13*, 1628.

147. Reed, C. A.; Roper, W. R. *J. Chem. Soc. A* **1970**, 3054.

148. Lin, J. T.; Wang, S. Y.; Huang, P. S.; Hsiao, Y. M.; Wen, Y. S.; Yeh, S. K. *J. Organomet. Chem.* **1990**, *388*, 151.

148. Albertin, G.; Amendola, P.; Antoniutti, S.; Bordignon, E. *J. Chem. Soc., Dalton Trans.* **1990**, 2979.

149. Lee, M.-K; Huang, P. S.; Wen, Y. S.; Lin, J. T. *Organometallics* **1990**, *9*, 2181.

150. Love, R. A.; Chin, H. B.; Koetzle, T. F.; Kirtley, S. W.; Whittlesty, B. R.; Bau, R. *J. Am. Chem. Soc.* **1976**, *98*, 4491.

151. Dziegielewski, J. O.; Filipek, K.; Jezowska-Trzebiatowska, B. *Polyhedron* **1991**, *10*, 429.

152. Stevens, R. E.; Fjare, D. E.; Gladfelter, W. L. *J. Organomet. Chem.* **1988**, *347*, 373.

153. Henrick, K.; Johnson, B. F. G.; Lewis, J.; Mace, J.; McPartlin, M.; Morris, J. *J. Chem. Soc., Chem. Commun.* **1985**, 1617.

154. Smieja, J. A.; Stevens, R. E.; Fjare, D. E.; Gladfelter, W. L. *Inorg. Chem.* **1985**, *24*, 3206.

155. Johnson, B. F. G.; Lewis, J.; Mace, J. M. *J. Chem. Soc., Chem. Commun.* **1984**, 186.

156. Johnson, B. F. G.; Raithby, P. R.; Zuccaro, C. *J. Chem. Soc., Dalton Trans.* **1980**, 99.

157. Smieja, J. A.; Gladfelter, W. L. *J. Organomet. Chem.* **1985**, *297*, 349.

158. Braga, D.; Johnson, B. F. G.; Lewis, J.; Mace, J. M.; McPartlin, M.; Puga, J.; Nelson, W. J. H.; Raithby, P. R.; Whitmire, K. H. *J. Chem. Soc. Chem. Commun.* **1982**, 1081.

159. Gadd, G. E.; Upmacis, R. K.; Poliakoff, M.; Turner, J. J. *J. Am. Chem. Soc.* **1986**, *108*, 2547.
160. Volatron, F.; Jean, Y.; Lledos, A. *New J. Chem.* **1987**, *11*, 651.
161. Mueller, A.; Eltzner, W.; Sarkar, S.; Boegge, H.; Aymonino, P. J.; Mohan, N.; Seyer, U.; Subramanian, P. *Z. Anorg. Allg. Chem.* **1983**, *503*, 22.
162. Ciani, G.; Giusto, D.; Manassero, M.; Sansoni, M. *Inorg. Chim. Acta* **1975**, *14*, L25.
163. Casey, J. A.; Murmann, R. K. *J. Am. Chem. Soc.* **1970**, *92*, 78.
164. Salomov, A. S.; Mikhailov, Y. N.; Kanishceva, A. S.; Svetlov, A. A.; Sinitsyin, N. M.; Porai-Koshits, M. A.; Parpiev, M. A. *Russ. J. Inorg. Chem.* **1989**, *34*, 216.
165. Rogalevich, N. L.; Bobkova, E. Y.; Novitskii, G. G.; Skutov, I. K.; Svetlov, A. A.; Sinitsyn, N. M. *Russ. J. Inorg. Chem.* **1986**, *31*, 694.
166. Mikhailov, Y. N.; Kanishcheva, A. S.; Svetlov, A. A. *Zh. Neorg. Khim.* **1989**, *34*, 2803.
167. Buslaeva, T. M.; Sinitsyn, N. M.; Romm, I. P.; Koteneva, N. A.; Malynov, I. V.; Shcherbakova, E. S. *Zh. Neorg. Khim.* **1989**, *34*, 1796.
168. Bobkova, E. Y.; Svetlov, A. A.; Rogalevich, N. L.; Novitskii, G. G.; Borkovskii, N. B. *Russ. J. Inorg. Chem.* **1990**, *35*, 549.
169. Svetlov, A. A.; Sinitsyn, M. N.; Fal'kengof, A. T.; Kokunova, Y. V. *Russ. J. Inorg. Chem.* **1990**, *35*, 1007.
170. Bottomley, F. *J. Chem. Soc., Dalton Trans.* **1975**, 2538.
171. Griffith, W. P. In *Comprehensive Inorganic Chemistry*; Pergamon: Oxford, 1973; Vol. 4, p 105.
172. Griffith, W. P.; Lewis, J.; Wilkinson, G. *J. Chem. Soc.* **1961**, 775.
173. Salomov, A. S.; Mikhailov, Y. N.; Kanishcheva, A. S.; Svetlov, A. A.; Sinitsyn, N. M.; Porai-Koshits, M. A.; Parpiev, N. A. *Russ. J. Inorg. Chem.* **1988**, *33*, 1496.
174. Salomov, A. S.; Sharipov, K. T.; Parpiev, N. A.; Porai-Koshits, M. A.; Mikhailov, Y. N.; Kanishcheva, A. S.; Sinitsyn, N. M.; Svetlov, A. A. *Koord. Khim.* **1984**, *10*, 1285.
175. Mikhailov, Y. N.; Kanishcheva, A. S.; Svetlov, A. A. *Russ. J. Inorg. Chem.* **1989**, *34*, 1603.
176. Dehnicke, K.; Liebelt, A.; Weller, F. *Z. Anorg. Allg. Chem.* **1981**, *474*, 83.

177. Sinitsyn, N. M.; Septsova, N. M.; Svetlov, A. F. *Russ. J. Inorg. Chem.* **1983**, *28*, 1464.

178. Bhattacharyya, R.; Saha, A. M.; Ghosh, P. N.; Mukherjee, M.; Mukherjee, A. K. *J. Chem. Soc., Dalton Trans.* **1991**, 501.

179. Veal, J. T.; Hodgson, D. J. *Inorg. Chem.* **1972**, *11*, 1420.

180. Nevskii, N. N.; Sinitsyn, N. M.; Svetlov, A. A. *Russ. J. Inorg. Chem.* **1990**, *35*, 653.

181. Kravchenko, E. A.; Burtsev, M. Y.; Morgunov, V. G.; Svetlov, A. A.; Sinitsyn, M. N.; Kokunov, Y. V.; Buslaev, Y. A. *Koord. Khim.* **1988**, *14*, 49. CA108(14):123130y.

182. Tarasov, V. P.; Kirakosyan, G. A.; Svetlov, A. A.; Sinitsyn, N. M.; Buslaev, Y. A. *Koord. Khim.* **1982**, *8*, 817.

183. Sinitsyn, M. N.; Svetlov, A. A.; Kokunov, Y. V.; Fal'kengof, A. T.; Larin, G. M.; Minin, V. V.; Buslaev, Y. A. *Dokl. Akad. Nauk SSSR* **1987**, *293*, 1144. (Engl. Trans. p 199.)

184. Coombe, V. T.; Heath, G. A.; Stephenson, T. A.; Tocher, D. A. *J. Chem. Soc., Chem. Commun.* **1983**, 303.

185. Weber, R.; Dehnicke, K. *Z. Naturforsch., B: Anorg. Chem., Org. Chem.* **1984**, *39B*, 262.

186. Czeska, B.; Weller, F.; Dehnicke, K. *Z. Anorg. Allg. Chem.* **1983**, *498*, 121.

187. Sharpe, A. G. *The Chemistry of Cyano Complexes of the Transition Metals*; Academic: London, U.K., 1976.

188. Wasielewska, E. *Proc. Conf. Coord. Chem.* **1985**, *(10)*, 493. CA104(26):236210k.

189. Bottomley, F.; Grein, F. *J. Chem. Soc., Dalton Trans.* **1980**, 1359.

190. Vannerberg, N.; Jagner, S. *Chem. Scripta* **1974**, *6*, 19.

191. Bhattacharyya, R.; Bhattacharjee, G. P.; Roy, P. S.; Ghosh, N. *Inorg. Synth.* **1985**, *23*, 182.

192. Sarkar, S.; Müller, A. *Z. Naturforsch., B: Anorg. Chem., Org. Chem.* **1978**, *33B*, 1053.

193. Griffith, W. P.; Kiernan, P. M.; Bregeault, J. M. *J. Chem. Soc., Dalton Trans.* **1978**, 1411.

194. Fenske, R. F.; DeKock, R. L. *Inorg. Chem.* **1972**, *11*, 437.

195. Cotton, F. A.; Monchamp, R. R.; Henry, R. J. M.; Young, R. C. *J. Inorg. Nucl. Chem.* **1959**, *10*, 28.

196. Bottomley, F.; White, P. S. *Acta Crystallogr., Sect. B* **1979**, *B35*, 2193.
197. Olabe, J. A.; Gentil, L. A.; Rigotti, G.; Navaza, A. *Inorg. Chem.* **1984**, *23*, 4297.
198. Bottomley, F.; Brooks, W. V. F.; Clarkson, S. G.; Tong, S.-B. *Chem. Commun.* **1973**, 919.
199. Griffith, W. P.; Lewis, J.; Wilkinson, G. *J. Chem. Soc.* **1959**, 1632.
200. Ribas, J.; Monfort, M.; Casabo, J. *Transition Met. Chem. (Weinheim)* **1984**, *9(11)*, 407.
201. Skorsepa, J.; Gyoryova, K. *Proc. Conf. Coord. Chem.* **1989**, *(12)*, 355. CA112(22):209019p.
202. Rueda, F. J. M. V.; Perez, R. P.; Diez, L. M. P. *An. Quim., Ser. B* **1989**, *85*, 77.
203. Soria, D. B.; Gentil, L. A.; Aymonino, P. J. *J. Crystallogr. Spectrosc. Res.* **1988**, *18 (2)*, 133.
204. De Villena Rueda, F. J. M.; Diez, L. M. P.; Perez, R. P. *Analyst (London)* **1988**, *113*, 573.
205. Picard-Bersellini, A.; Cheikh, M.; Broquier, M. *Chem. Phys.* **1989**, *133*, 461.
206. Gyoryova, K.; Mohai, B. *Thermochim. Acta* **1985**, *92*, 771.
207. Swinehart, J. H. *Coord. Chem. Rev.* **1967**, *2*, 385.
208. Kluefers, P.; Haussuehl, S. *Z. Kristallogr.* **1985**, *170*, 289.
209. Mullica, D. F.; Sappenfield, E. L.; Tippin, D. B.; Leschnitzer, D. H. *Inorg. Chim. Acta* **1989**, *164*, 99.
210. Rigotti, G.; Punte, G.; Rivero, B. E.; Castellano, E. E. *Acta Crystallogr., Sect. B* **1980**, *B36*, 1475.
211. Punte, G.; Rigotti, G.; Rivero, B. E.; Podjarny, A. D.; Castellano, E. E. *Acta Crystallogr., Sect. B* **1980**, *B36*, 1472.
212. Vedova, C. O. D.; Lesk, J. H.; Varetti, E. L.; Aymonino, P. J.; Piro, O. E.; Rivero, B. E.; Castellano, E. E. *J. Mol. Struct.* **1981**, *70*, 241.
213. Mullica, D. F.; Tippin, D. B.; Sappenfield, E. L. *Inorg. Chim. Acta* **1990**, *174*, 129.
214. Castellano, E. E.; Rivero, B. E.; Piro, O. E.; Amalvy, J. I. *Acta Crystallogr., Sect. C: Cryst. Struct. Commun.* **1989**, *C45*, 1207.
215. Antipin, M. Y.; Tsirel'son, V. G.; Flyugge, M. P.; Struchkov, Y. T.; Ozerov, R. P. *Koord. Khim.* **1987**, *13*, 121.

216. Navaza, A.; Chevrier, G.; Alzari, P. M.; Aymonino, P. J. *Acta Crystallogr., Sect. C.: Cryst. Struct. Commun.* **1989**, *C45*, 839.
217. Vergara, M. M.; Varetti, E. L. *An. Asoc. Quim. Argent.* **1988**, *76(5)*, 329.
218. Baran, E. J.; Etcheverry, S. B.; Mercader, R. C. *Z Anorg. Allg. Chem.* **1985**, *531*, 199.
219. Gonzalez, S. R.; Piro, O. E.; Aymonino, P. J.; Castellano, E. E. *Phys. Rev. B: Condens. Matter* **1986**, *33*, 5818.
220. Guida, J. A.; Piro, O. E.; Aymonino, P. J. *Solid State Commun.* **1986**, *57*, 175.
221. Butler, A. R.; Glidewell, C.; Hyde, A. R.; McGinnis, J. *Inorg. Chem.* **1985**, *24*, 2931.
222. Gonzalez, S. R.; Aymonino, P. J.; Piro, O. E. *J. Chem Phys.* **1984**, *81*, 625.
223. Murgich, J.; Ambrosetti, R. *J. Magn. Reson.* **1987**, *74*, 344.
224. Cyvin, B. N.; Cyvin, S. J.; Zabokrzycka, A.; Kedzia, B. B. *Spectrosc. Lett.* **1983**, *16*, 249.
225. Golebiewski, A.; Wasielewska, E. *J. Mol. Struct.* **1980**, *67*, 183.
226. Guida, J. A.; Piro, O. E.; Castellano, E. E.; Aymonino, P. J. *J. Chem. Phys.* **1989**, *91*, 4265.
227. Corbella, M.; Monfort, M.; Ribas, J. *Z. Anorg. Allg. Chem.* **1986**, *543*, 233.
228. Ribas, J.; Julia, J. M.; Solans, X.; Font-Altaba, M.; Isalgue, A.; Tejeda, X. *Transition Met. Chem. (Weinheim)* **1984**, *9(2)*, 57.
229. Woike, T.; Krasser, W.; Bechtold, P. S.; Haussuehl, S. *Phys. Rev. Lett.* **1984**, *53*, 1767.
230. Zoellner, H.; Woike, T.; Krasser, W.; Haussuehl, S. *Z. Kristallogr.* **1989**, *188*, 139. CA112(6):42891c.
231. Yang, Y. Y.; Zink, J. I. *J. Am. Chem. Soc.* **1985**, *107*, 4799.
232. Perng, J. H.; Zink, J. I. *Inorg. Chem.* **1988**, *27*, 1403.
233. Fiedler, J.; Masek, J. *Inorg. Chim. Acta* **1986**, *111*, 39.
234. Glidewell, C.; Johnson, I. L. *Inorg. Chim. Acta* **1987**, *132*, 145.
235. Symons, M. C. R.; Wilkinson, J. G.; West, D. X. *J. Chem. Soc., Dalton Trans.* **1982**, 2041.
236. Bhattacharyya, R.; Roy, P. S.; Dasmahaptra, A. K. *J. Organomet. Chem.* **1984**, *267*, 293.

237. Moran, M. Gayoso, M. *Z. Naturforsch., B: Anorg. Chem., Org. Chem.* **1981**, *36B*, 434.
238. Davis, R.; Johnson, B. F. G.; Al-Obaidi, K. H. *J. Chem. Soc., Dalton Trans.* **1972**, 508.
239. Carmona, E.; Gutierrez-Puebla, E.; Monge, A.; Perez, P. J.; Sanchez, L. J. *Inorg. Chem.* **1989**, *28*, 2120.
240. Bond, A. M.; Colton, R.; Panagiotidou, P. *Organometallics* **1988**, *7*, 1774.
241. Crease, A. E.; Egglestone, H.; Taylor, N. *J. Organomet. Chem.* **1982**, *238*, C5.
242. Seyferth, K.; Taube, R. *J. Organomet. Chem.* **1982**, *229*, 275.
243. Kolobova, N. E.; Zdanovich, V. I.; Lobanova, I. A. *Koord. Khim.* **1979**, *5*, 86. CA90(16):132072t.
244. Sulfab, Y.; Basolo, F.; Rheingold, A. L. *Organometallics* **1989**, *8*, 2139.
245. Lin, J. T.; Shan, C. H.; Fang, D.; Liu, L. K. Hsiou, Y. *J. Chem. Soc., Dalton Trans.* **1988**, 1397.
246. Cerveau, G.; Colomer, E.; Corriu, R. J. P. *J. Organomet. Chem.* **1982**, *236*, 33.
247. Isaacs, E. E.; Graham, W. A. G. *J. Organomet. Chem.* **1975**, *99*, 119.
248. Liu, L. K.; Lin, J. T.; Fang, D. *Inorg. Chim. Acta* **1989**, *161*, 239.
249. Hunter, A. D.; Legzdins, P. *Inorg. Chem.* **1984**, *23*, 4198.
250. Banks, R. E.; Dickinson, N.; Morrissey, A. P.; Richards, A. *J. Fluorine Chem.* **1984**, *26*, 87.
251. Schumacher, C.; Schmidt, R. E.; Dehnicke, K. *Z. Anorg. Allg. Chem.* **1985**, *520*, 25.
252. Schumacher, C.; Weller, F.; Dehnicke, F. *Z. Anorg. Allg. Chem.* **1984**, *508*, 79.
253. Ballivet-Tkatchenko, D.; Bremard, C. *J. Chem. Soc., Dalton Trans.* **1983**, 1143.
254. Ballivet-Tkatchenko, D.; Bremard, C.; Abraham, F.; Nowogrocki, G. *J. Chem. Soc., Dalton Trans.* **1983**, 1137.
255. Bencze, L.; Kohan, J.; Mohai, B. *Acta Chim. Hung.* **1983**, *113*, 183.
256. Frankenau, A.; Willing, W.; Mueller, U.; Dehnicke, K. *Z. Anorg. Allg. Chem.* **1987**, *550*, 149.
257. Nimry, T.; Urbancic, M. A.; Walton, R. A. *Inorg. Chem.* **1979**, *18*, 691.

258. Berg, A.; Dehnicke, K. *Z. Naturforsch., B: Anorg. Chem., Org. Chem.* **1985**, *40B*, 842.
259. Barratt, D. S.; McAuliffe, C. A. *J. Chem. Soc., Chem. Commun.* **1984**, 594.
260. Pearlstein, R. M.; Davis, W. M.; Jones, A. G.; Davison, A. *Inorg. Chem.* **1989**, *28*, 3332.
261. Kirmse, R.; Stach, J.; Abram, U. *Polyhedron* **1985**, *4*, 1275.
262. Brown, D. S.; Newman, J. L.; Thornback, J. R.; Pearlstein, R. M.; Davison, A.; Lawson, A. *Inorg. Chim. Acta* **1988**, *150*, 193.
263. Di Castro, V.; Giusto, D.; Mattogno, G. *J. Microsc. Spectrosc. Electron.* **1979**, *4(3)*, 251.
264. Yang, G. C.; Heitzmann, M. W.; Ford, L. A.; Benson, W. R. *Inorg. Chem.* **1982**, *21*, 3242.
265. de Vries, N.; Cook, J.; Davison, A.; Nicholson, T.; Jones, A. G. *Inorg. Chem.* **1990**, *29*, 1062.
266. Kirmse, R.; Abram, U. *Z. Anorg. Allg. Chem.* **1989**, *573*, 63.
267. Laing, M.; Reiman, R. H.; Singleton, E. *Inorg. Chem.* **1979**, *18*, 1648.
268. Blower, P. J.; Dilworth, J. R.; Hutchinson, J. P.; Zubieta, J. *Transition Met. Chem. (Weinheim)* **1982**, *7*, 354.
269. Adams, R. W.; Chatt, J.; Hooper, N. E.; Leigh, G. J. *J. Chem. Soc., Dalton Trans.* **1974**, 1075.
270. Rakshit, S.; Sen, B. K.; Bandyopadhyay, P. *Z. Anorg. Allg. Chem.* **1978**, *445*, 245.
271. Jana, T. K.; Rakshit, S.; Bandyopadhyay, P. Sen, B. K. *Z. Anorg. Allg. Chem.* **1981**, *477*, 229.
272. Mronga, N.; Mueller, U.; Dehnicke, K. *Z. Anorg. Allg. Chem.* **1981**, *482*, 95.
273. Mronga, V. N.; Müller, U.; Dehnicke, K. *Z. Anorg. Allg. Chem.* **1981**, *482*, 95.
274. Fenske, D.; Mronga, N.; Dehnicke, K. *Z. Anorg. Allg. Chem.* **1983**, *498*, 131.
275. Vogler, S.; Dehnicke, K.; Fenske, D. *Z. Naturforsch., B.: Chem. Sci.* **1989**, *44*, 1393.
276. Mronga, N.; Dehnicke, K.; Fenske, D. *Z. Anorg. Allg. Chem.* **1982**, *491*, 237.
277. Zingales, F.; Trovati, A.; Uguagliati, P. *Inorg. Chem.* **1971**, *10*, 510.

278. Dolcetti, G.; Norton, J. R. *Inorg. Synth.* **1976**, *16*, 35.

279. Steimann, M.; Nagel, U.; Grenz, R.; Beck, W. *J. Organomet. Chem.* **1983**, *247*, 171.

280. Fenske, D.; Demant, U.; Dehnicke, K. *Z. Naturforsch., B: Anorg. Chem., Org. Chem.* **1985**, *40B*, 1672.

281. Dehnicke, K.; Loessberg, R. *Chem. -Ztg.* **1981**, *105 (10)*, 305.

282. Coll, R. K.; Fergusson, J. E.; McKee, V.; Page, C. T.; Robinson, W. T.; Teou, S. K. *Inorg. Chem.* **1987**, *26*, 106.

283. Torroni, S.; Innorta, G.; Foffani, A.; Modelli, A.; Scagnolari, F. *J. Organomet. Chem.* **1981**, *221*, 309.

284. Page, C. T.; Fergusson, J. E. *Aust. J. Chem.* **1983**, *36*, 855.

285. Southern, T. G.; Dixneuf, P. H.; Le Marouville, J. Y.; Grandjean, D. *Inorg. Chim. Acta* **1978**, *31*, L415.

286. Innorta, G.; Modelli, A. *Inorg. Chim. Acta* **1978**, *31*, L367.

287. Laing, K. R.; Roper, W. R. *J. Chem. Soc., Chem. Commun.* **1968**, 1556.

288. Aràneo, A.; Valenti, V.; Cariati, F. *J. Inorg. Nucl. Chem.* **1970**, *32*, 1877.

289. Haymore, B. L.; Ibers, J. A. *Inorg. Chem.* **1975**, *14*, 3060.

290. Robinson, S. D.; Uttley, M. F. *J. Chem. Soc., Dalton Trans.* **1972**, 1.

291. Irving, R. J.; Laye, P. G. *J. Chem. Soc. A* **1966**, 161.

292. Gandolfi, O.; Giovannitti, B.; Ghedini, M.; Dolcetti, G. *J. Organomet. Chem.* **1976**, *104*, C41.

293. Giovannitti, B.; Gandolfi, O.; Ghedini, M.; Dolcetti, G. *J. Organomet. Chem.* **1977**, *129*, 207.

294. Czeska, B.; Dehnicke, K.; Fenske, D. *Z. Naturforsch., B: Anorg. Chem., Org. Chem.* **1983**, *38B*, 1031.

295. Kimura, T.; Sakurai, T.; Shima, M.; Togano, T.; Mukaida, M. Nomura, T. *Inorg. Chim. Acta* **1983**, *69*, 135.

296. Brant, P.; Feltham, R. D. *Inorg. Chem.* **1980**, *19*, 2673.

297. Holderegger, R.; Venanzi, L. M.; Bachechi, F.; Mura, P.; Zambonelli, L. *Helv. Chim. Acta* **1979**, *62*, 2159.

298. Cleland, A. J.; Fieldhouse, S. A.; Freeland, B. H.; Mann, C. D. M.; O'Brien, R. J. *J. Chem. Soc. (A)* **1971**, 736.

299. Casey, M.; Manning, A. R. *J. Chem. Soc. (A)* **1971**, 256.

300. Behrens, H.; Moll, M.; Merbach, P.; Trummer, K.-H. *Z. Naturforsch., B: Anorg. Chem., Org. Chem.* **1986**, *41B*, 845.

301. Dahl, L. F.; de Gil, E. R.; Feltham, R. D. *J. Am. Chem. Soc.* **1969**, *91*, 1653.
302. Haymore, B.; Feltham, R. D. *Inorg. Synth.* **1973**, *14*, 81.
303. Wah, H. L. K.; Postel, M.; Pierrot, M. *Inorg. Chim. Acta* **1989**, *165*, 215.
304. Guillaume, P.; Wah, H. L. K.; Postel, M. *Inorg. Chem.* **1991**, *30*, 1828.
305. Innorta, G.; Torroni, S.; Foffani, A. *J. Organomet. Chem.* **1974**, *66*, 459.
306. Piazza, G.; Innorta, G. *J. Organomet. Chem.* **1982**, *240*, 257.
307. Innorta, G.; Piazza, G. *J. Electroanal. Chem. Interfacial Electrochem.* **1980**, *106*, 137.
308. Clark, G. R.; Waters, J. M.; Whittle, K. R. *J. Chem. Soc., Dalton Trans.* **1975**, 463.
309. Pierpont, C. G.; Eisenberg, R. *Inorg. Chem.* **1972**, *11*, 1088.
310. Reed, J.; Pierpont, C. G.; Eisenberg, R. *Inorg. Synth.* **1976**, *16*, 21.
311. Mingos, D. M. P.; Sherman, D. J.; Bott, S. G. *Transition Met. Chem. (London)* **1987**, *12(5)*, 471.
312. Mingos, D. M. P.; Sherman, D. J.; Williams, I. D. *Transition Met. Chem. (London)* **1987**, *12(6)*, 493.
313. Legzdins, P.; Malito, J. T. *Inorg. Chem.* **1975**, *14*, 1875.
314. Hieber, W.; Beck, W. *Z. Naturforsch., B: Anorg. Chem., Org. Chem.* **1958**, *13B*, 194.
315. Hieber, W.; Jahn, A. *Z. Anorg. Allg. Chem.* **1959**, *301*, 301.
316. English, R. B.; Steyn, M. M. de V.; Haines, R. J. *Polyhedron* **1987**, *6*, 1503.
317. English, R. B.; Steyn, M. M. V. *S. Afr. J. Chem.* **1984**, *37*, 177.
318. Monica, G. L.; Sandrini, P.; Zingales, F.; Cenini, S. *J. Organomet. Chem.* **1973**, *50*, 287.
319. Crooks, G. R.; Johnson, B. F. G. *J. Chem. Soc. A* **1970**, 1662.
320. Dolcetti, G.; Ghedini, M.; Reed, C. A. *Inorg. Synth.* **1976**, *16*, 29.
321. Alnaji, O.; Peres, Y.; Dartiguenave, M.; Dahan, F.; Dartiguenave, Y. *Inorg. Chim. Acta* **1986**, *114*, 151.
322. Brock, C. P.; Collman, J. P.; Dolcetti, G.; Farnham, P. H.; Ibers, J. A.; Lester, J. E.; Reed, C. A. *Inorg. Chem.* **1973**, *12*, 1304.
323. Mingos, D. M. P.; Ibers, J. A. *Inorg. Chem.* **1971**, *10*, 1035.
324. Seeber, R.; Mazzocchin, G. A.; Albertin, G.; Bordignon, E. *J. Chem. Soc., Dalton Trans.* **1980**, 979.

325. Hodgson, D. J.; Payne, N. C.; McGinnety, J. A.; Pearson, R. G.; Ibers, J. A. *J. Am. Chem. Soc.* **1968**, *90*, 4486.
326. Hodgson, D. J.; Ibers, J. A. *Inorg. Chem.* **1969**, *8*, 1282.
327. Reed, C. A.; Roper, W. R. *J. Chem. Soc., Chem. Commun.* **1969**, 1459.
328. Angoletta, M.; Beringhelli, T.; Morazzoni, F. *Spectrochim. Acta* **1982**, *38A*, 1177.
329. Fitzgerald, R. J.; Lin, H.-M. W. *Inorg. Synth.* **1976**, *16*, 41.
330. Gwost, D.; Caulton, K. G. *Inorg. Synth.* **1976**, *16*, 16.
331. Griffith, W. P.; Lewis, J.; Wilkinson, G. *J. Chem. Soc.* **1959**, 1775.
332. Haymore, B. L.; Huffman, J. C.; Butler, N. E. *Inorg. Chem.* **1983**, *22*, 168.
333. Aresta, M.; Ballivet-Tkatchenko, D.; Bonnet, M. C.; Faure, R.; Loiseleur, H. *J. Am. Chem. Soc.* **1985**, *107*, 2994.
334. Kravtsova, E. A.; Mazalov, L. N.; Kurasov, S. S.; Mikhailov, V. A. *Zh. Strukt. Khim.* **1987**, *28 (5)*, 163. CA108(14):121416x.
335. Zavorokhina, Z. M.; Levchenko, L. V. *Izv. Akad. Nauk Kaz. SSR, Ser. Khim* **1983**, *(5)*, 69. CA99(26):224155m.
336. Kurasov, S. S.; Troyan, N. N.; Eremenko, N. K.; Mikhailov, V. A. *Izv. Sib. Otd. Akad. Nauk SSSR, Ser. Khim. Nauk* **1980**, *(6)*, 31. CA94(16):131425f.
337. Smidt, J.; Jira, R. *Chem. Ber.* **1960**, *93*, 162.
338. Ambach, E.; Beck, W. *Z. Naturforsch., B: Anorg. Chem., Org. Chem.* **1985**, *40B*, 288.
339. Hidai, M.; Kokura, M.; Uchida, Y. *Bull. Chem. Soc. Jpn.* **1973**, *46*, 686.
340. Brunner, H. *Z. Naturforsch., B: Anorg. Chem., Org. Chem.* **1969**, *24B*, 275.
341. Addison, C. C.; Johnson, B. F. G. *Proc. Chem. Soc.* **1962**, 305.
342. Braunstein, P.; Dehand, J.; Munchenbach, B. *J. Organomet. Chem.* **1977**, *124*, 71.
343. Epstein, J. M.; White, A. H.; Wild, S. B.; Willis, A. C. *J. Chem. Soc., Dalton Trans.* **1974**, 436.
344. Tiripicchio, A.; Camellini, M. T.; Neve, F.; Ghedini, M. *J. Chem. Soc., Dalton Trans.* **1990**, 1651.
345. Neve, F.; Ghedini, M. *Inorg. Chim. Acta* **1990**, *175*, 111.
346. Carr, S. W.; Shaw, B. L. *J. Chem. Soc., Dalton Trans.* **1986**, 1815.

347. Ghedini, M.; Neve, F.; Mealli, C.; Tiripicchio, A.; Ugozzoli, F. *Inorg. Chim. Acta* **1990**, *178*, 5.

348. Podberezskaya, N. V.; Bakakin, V. V.; Kuznetsova, N. I.; Danilyuk, A. F.; Likholobov, V. A. *Dokl. Akad, Nauk SSSR* **1981**, *256*, 870. CA94(18):149428u.

349. Iqbal, Z.; Waddington, T. C. *J. Chem. Soc. A* **1969**, 1092.

350. Fraser, R. T. M.; Dasent, W. E. *J. Am. Chem. Soc.* **1960**, *82*, 348.

351. Mercer, M.; Fraser, R. T. M. *J. Inorg. Nucl. Chem.* **1963**, *25*, 525.

352. Tiripicchio, A.; Lanfredi, A. M. M.; Ghedini, M.; Neve, F. *J. Chem. Soc., Chem. Commun.* **1983**, 97.

353. (a) Partha, P. P.; Tyeklar, Z.; Farooq, A.; Karlin, K. D.; Liu, S.; Zubieta, J. *J. Am. Chem. Soc.* **1990**, *112*, 2430. (b) Partha, P. P.; Karlin, K. D. *J. Am. Chem. Soc.* **1991**, *113*, 6331.

354. Tevault, D.; Nakamoto, K. *Inorg. Chem.* **1976**, *15*, 1282.

355. Quirk, J.; Wilkinson, G. *Polyhedron* **1982**, *1*, 209.

356. Bobkova, E. Y.; Rogalevich, N. L.; Borkovskii, N. B.; Novitskii, G. G.; Kokunova, V. N.; Sinitsyn, N. M. *Russ. J. Inorg. Chem.* **1989**, *34*, 524.

357. Svetlov, A. A.; Sinitsyin, M. M.; Kravchenko, V. V. *Russ. J. Inorg. Chem.* **1989**, *34*, 535.

358. Sinitsyn, N. M.; Kokunova, V. N.; Novittskii, G. G.; Bobkova, E. Y. *Russ. J. Inorg. Chem.* **1988**, *33*, 1172.

359. Sinitsyn, N. M.; Kokunova, V. N.; Svetlov, A. A. *Russ. J. Inorg. Chem.* **1988**, *33*, 1336.

360. Kravtsova, E. A.; Mazalov, L. N. *Zh. Strukt. Khim.* **1987**, *28(5)*, 68. CA108(10):84659f.

361. Salomov, A. S.; Parpiev, N. A.; Sharipov, K. T.; Kokunova, V. N.; Sinitsyn, N. M.; Porai-Koshits, M. A. *Russ. J. Inorg. Chem.* **1984**, *29*, 1633.

362. Radonovich, L. J.; Hoard, J. L. *J. Phys. Chem.* **1984**, *88*, 6711.

363. Sinitsyn, N. M.; Kokunova, V. N.; Svetlov, A. A. *Russ. J. Inorg. Chem.* **1982**, *27*, 1317.

364. Noell, J. O.; Morokuma, K. *Inorg. Chem.* **1979**, *18*, 2774.

365. Wishart, J. F.; Taube, H.; Breslauer, K. J.; Isied, S. S. *Inorg. Chem.* **1986**, *25*, 1479.

366. Salomov, A. S.; Parpiev, N. A.; Sharipov, K. T.; Sinitsyn, N. M.; Porai-Koshits, M. A.; Svetlov, A. A. *Russ. J. Inorg. Chem.* **1984**, *29*, 1492.

367. Nishimura, H.; Matsuzawa, H.; Togano, T.; Mukaida, M.; Kakihana, H.; Bottomley, F. *J. Chem. Soc., Dalton Trans.* **1990**, 137.

368. Sinitsyn, M. N.; Svetlov, A. A.; Kanishcheva, A. S.; Mikhailov, Y. N.; Sadikiv, G. G.; Kokunov, Y. V.; Buslaev, Y. A. *Zh. Neorg. Khim.* **1989**, *34*, 2795. CA112(20):190639s.

369. Nagao, H.; Nishimura, H.; Funato, H.; Ichikawa, Y.; Howell, F. S.; Mukaida, M.; Kakihana, H. *Inorg. Chem.* **1989**, *28*, 3955.

370. Sinitsyn, N. M.; Svetlov, A. A.; Kanishcheva, A. S.; Mikhailov, Y. N.; Sadikov, G. G.; Kokunov, Y. V.; Buslaev, Y. A. *Russ. J. Inorg. Chem.* **1989**, *34*, 1599.

371. Bobkova, E. Y.; Svetlov, A. A.; Rogalevich, N. L.; Novitskii, G. G.; Borkovskii, N. B.; Sinitsyn, M. N. *Russ. J. Inorg. Chem.* **1990**, *35*, 546.

372. Kanishcheva, A. S.; Mikhailov, Y. N.; Svetlov, A. A. *Russ. J. Inorg. Chem.* **1990**, *35*, 1003.

373. Bottomley, F.; Tong, S. B. *Inorg. Synth.* **1976**, *16*, 9.

374. Schreiner, A. F.; Lin, S. W. *Inorg. Synth.* **1976**, *16*, 13.

375. Schreiner, A. F.; Lin, S. W.; Hauser, P. J.; Hopcus, E. A.; Hamm, D. J.; Gunter, J. D. *Inorg. Chem.* **1972**, *11*, 880.

376. Pomp, C.; Weighardt, K.; Nuber, B.; Weiss, J. *Inorg. Chem.* **1988**, *27*, 3789.

377. Bradley, D. C.; Newing, C. W. *J. Chem. Soc., Chem. Commun.* **1970**, 219.

378. Froehlich, H. O.; Roemhild, W. *Z. Chem.* **1980**, *20(4)*, 154.

379. Froehlich, H. O.; Roemhild, W. *Z. Chem.* **1979**, *19(11)*, 414.

380. Seidel, W.; Geinitz, D. *Z. Chem.* **1979**, *19(11)*, 413.

381. Crichton, O.; Rest, A. J. *J. Chem. Soc., Dalton Trans.* **1978**, 208.

382. Enemark, J. H. *Inorg. Chem.* **1971**, *10*, 1952.

383. Pohl, K.; Weighardt, K.; Nuber, B.; Weiss, J. *J. Chem. Soc., Dalton Trans.* **1987**, 187.

384. Beck, J.; Straehle, J. *Z. Naturforsch., B: Anorg. Chem., Org. Chem.* **1985**, *40B*, 891.

385. Weighardt, K.; Backes-Dahmann, G.; Swiridoff, W.; Weiss, J. *Inorg. Chem.* **1983**, *22*, 1221.

386. Sutton, D. *Chem. Soc. Rev.* **1975**, *4*, 443.

387. Ferguson, G.; Ruhl, B. L.; Parvez, M.; Lalor, F. J.; Deane, M. E. *J. Organomet. Chem.* **1990**, *381*, 357.

388. Norman, J. G., Jr.; Osborne, J. H. *Inorg. Chem.* **1982**, *21*, 3241.
389. Deane, M.; Lalor, F. J. *J. Organomet. Chem.* **1973**, *57*, C61.
390. Lindsay, A. J.; Wilkinson, G.; Motevalli, M.; Hursthouse, M. B. *J. Chem. Soc., Dalton Trans.* **1987**, 2723.
391. Carriedo, C.; Connelly, N. G.; Hettrich, R.; Orpen, A. G.; White, J. M. *J. Chem. Soc., Dalton Trans.* **1989**, 745.
392. Weighardt, K.; Holzbach, W.; Weiss, J. *Z. Naturforsch., B: Anorg. Chem., Org. Chem.* **1982**, *37B*, 680.
393. Wieghardt, K.; Holzbach, W. *Angew. Chem., Int. Ed. Engl.* **1979**, *18*, 549.
394. Grundy, K. R.; Reed, C. A.; Roper, W. R. *J. Chem. Soc., Chem. Commun.* **1970**, 1501.
395. Wieghardt, K.; Holzbach, W.; Weiss, J.; Nuber, B.; Prikner, B. *Angew. Chem., Int. Ed. Engl.* **1979**, *18*, 548.
396. Wieghardt, K.; Quilitzsch, U. *Z. Anorg. Allg. Chem.* **1979**, *457*, 75.
397. Lukehart, C. M.; Troup, J. M. *Inorg. Chim. Acta* **1977**, 22, 81.
398. Glidewell, C.; Johnson, I. L. *Polyhedron* **1988**, *7*, 1371.
399. Sato, T. *J. Radioanal. Nucl. Chem.* **1986**, *104*, 151.
400. Sinitsyn, N. M.; Kokunova, V. N.; Kravchenko, V. V. *Russ. J. Inorg. Chem.* **1985**, *30*, 396.
401. Blasius, E.; Luxenburger, H. J.; Neumann, W. *Fresenius' Z. Anal. Chem.* **1984**, *319*, 38. CA102(8):71575b.
402. Sinitsyn, N. M.; Svetlov, A. A. *Russ. J. Inorg. Chem.* **1982**, *27*, 555.
403. Dubrawski, J. V.; Feltham, R. D. *Inorg. Chem.* **1980**, *19*, 355.
404. Ochkin, A. V.; Obruchnikov, A. V.; Smelov, V. S.; Chubukov, V. V. *Radiokhimiya* **1989**, *31(6)*, 143. CA112(12):106129v.
405. Sato, T. *Radiochim. Acta* **1989**, *46(4)*, 213.
406. Peterson, E. S.; Larsen, R. D.; Abbott, E. H. *Inorg. Chem.* **1988**, *27*, 3514.
407. Kriege-Simondsen, J.; Elbaze, G.; Dartiguenave, M.; Feltham, R. D.; Dartiguenave, Y. *Inorg. Chem.* **1982**, *21*, 230.
408. Bhaduri, S. A.; Bratt, I.; Johnson, B. F. G.; Khair, A.; Segal, J. A.; Walters, R.; Zuccaro, C. *J. Chem. Soc., Dalton Trans.* **1981**, 234.
409. Frankenau, A.; Dehnicke, K.; Fenske, D. *Z. Anorg. Allg. Chem.* **1987**, *554*, 101.
410. Pomp, C.; Weighardt, K. *Inorg. Chem.* **1988**, *27*, 3796.

411. Backes-Dahmann, G.; Weighardt, K. *Inorg. Chem.* **1985**, *24*, 4044.

412. Beissel, T.; Vedova, B. S. P. C. D.; Wieghardt, K.; Boese, R. *Inorg. Chem.* **1990**, *29*, 1736.

413. Hoffmann, P.; Mattes, R. *Inorg. Chem.* **1989**, *28*, 2092.

414. Faller, J. W.; Ma, Y. *J. Am. Chem. Soc.* **1991**, *113*, 1579.

415. Frauendorfer, E.; Castrillo, T. *Acta Cient. Venez.* **1982**, *33*, 385. CA100(10):78873q.

416. McCleverty, J. A. *Chem. Soc. Rev.* **1983**, *12(3)*, 331.

417. Trofimenko, S. *Inorg. Chem.* **1969**, *8*, 2675.

418. Trofimenko, S. *Inorg. Chem.* **1971**, *10*, 504.

419. Al Obaidi, N.; Jones, C. J.; McCleverty, J. A. *Polyhedron* **1989**, *8*, 1033.

420. Coe, B. J.; Jones, C. J.; McCleverty, J. A.; Bruce, D. W. *Polyhedron* **1990**, *9*, 687.

421. Cano, M.; Heras, J. V.; Trofimenko, S.; Monge, A.; Gutierrez, E.; Jones, C. J.; McCleverty, J. A. *J. Chem. Soc., Dalton Trans.* **1990**, 3577.

422. Al Obaidi, N. J.; Jones, C. J.; McCleverty, J. A. *J. Chem. Soc., Dalton Trans.* **1990**, 3329.

423. Roberts, S. A.; Enemark, J. H. *Acta Crystallogr., Sect. C: Cryst. Struc. Commun.* **1989**, *C45*, 1292.

424. Al Obaidi, N.; Edwards, A. J.; Jones, C. J.; McCleverty, J. A.; Neaves, B. D.; Mabbs, F. E.; Collison, D. *J. Chem. Soc., Dalton Trans.* **1989**, 127.

425. Al Obaidi, N.; Hamor, T. A.; Jones, C. J.; McCleverty, J. A.; Paxton, K. *Polyhedron* **1988**, *7*, 1931.

426. Al Obaidi, N.; Charsley, S. M.; Hussain, W.; Jones, C. J.; McCleverty, J. A.; Neaves, B. D.; Reynolds, S. J. *Transition Met. Chem. (London)* **1987**, *12(2)*, 143.

427. Al Obaidi, N.; Jones, C. J.; McCleverty, J. A.; Howes, A. J.; Hursthouse, M. B. *Polyhedron* **1988**, *7*, 235.

428. McCleverty, J. A.; Wlodarczyk, A. *Polyhedron* **1988**, *7*, 449.

429. Wlodarczyk, A.; Edwards, A. J.; McCleverty, J. A. *Polyhedron* **1988**, *7*, 103.

430. Charsley, S. M.; Jones, C. J.; McCleverty, J. A.; Neaves, B. D.; Reynolds, S. J.; Denti, G. *J. Chem. Soc., Dalton Trans.* **1988**, 293.

431. Young, C. G.; Minelli, M.; Enemark, J. H.; Hussain, W.; Jones, C. J.; McCleverty, J. A. *J. Chem. Soc., Dalton Trans.* **1987**, 619.

432. Reynolds, S. J.; Smith, C. F.; Jones, C. J.; McCleverty, J. A. *Inorg. Synth.* **1985**, *23*, 4.
433. Beer, P. D.; Jones, C. J.; McCleverty, J. A.; Salam, S. S. *J. Inclusion Phenom.* **1987**, *5 (4)*, 521.
434. McCleverty, J. A. *Polyhedron* **1989**, *8*, 1669.
435. Beer, P. D.; Jones, C. J.; McCleverty, J. A.; Sidebotham, R. P. *J. Organomet. Chem.* **1987**, *325*, C19.
436. Jones, C. J.; McCleverty, J. A.; Reynolds, S. J. *Transition Met. Chem. (Weinheim)* **1986**, *11*, 138.
437. Sidebotham, R. P.; Beer, P. D.; Hamor, T. A.; Jones, C. J.; McCleverty, J. A. *J. Organomet. Chem.* **1989**, *371*, C31.
438. Al Obaidi, N.; Hamor, T. A.; Jones, C. J.; McCleverty, J. A.; Paxton, K. *J. Chem. Soc., Dalton Trans.* **1987**, 2653.
439. Al Obaidi, N. J.; Jones, C. J.; McCleverty, J. A. *Polyhedron* **1990**, *9*, 693.
440. Chiappetta, S.; Denti, G.; McCleverty, J. A. *Transition Met. Chem. (London)* **1989**, *14(6)*, 449.
441. Charsley, S. M.; Jones, C. J.; McCleverty, J. A.; Neaves, B. D.; Reynolds, S. J. *J. Chem. Soc., Dalton Trans.* **1988**, 301.
442. McWhinnie, S. L. W.; Jones, C. J.; McCleverty, J. A.; Collison, D.; Mabbs, F. E. *J. Chem. Soc., Chem. Commun.* **1990**, 940.
443. Coe, B. J.; Jones, C. J.; McCleverty, J. A.; Bloor, D.; Kolinski, P. V.; Jones, R. J. *J. Chem. Soc., Chem. Commun.* **1989**, 1485.
444. Rettig, S. J.; Storr, A.; Trotter, J.; Uhrich, K. *Can. J. Chem.* **1984**, *62*, 2783.
445. Rettig, S. J.; Storr, A.; Trotter, J. *Can. J. Chem.* **1981**, *59*, 2391.
446. Chong, K. S.; Rettig, S. J.; Storr, A.; Trotter, J. *Can. J. Chem.* **1979**, *57*, 3107.
447. Chong, K. S.; Rettig, S. J.; Storr, A.; Trotter, J. *Can. J. Chem.* **1979**, *57*, 3113.
448. Breakell, K. R.; Rettig, S. J.; Storr, A.; Trotter, J. *Can. J. Chem.* **1979**, *57*, 139.
449. Littlejohn, D.; Chang, S. G. *Ind. Eng. Chem. Res.* **1987**, *26*, 1232.
450. (a) Corazza, F.; Solari, E.; Floriani, C.; Chiesi-Villa, A. *J. Chem. Soc., Chem. Commun.* **1986**, 1562. (b) Larkworthy, L. F.; Sengupta, S. K.

Inorg. Chim. Acta **1991**, *179*, 157. (c) Earnshaw, A.; King, E. A.; Larkworthy, L. F. *J. Chem. Soc. A* **1969**, 2459.

451. Griffiths, E. A.; Chang, S. G. *Ind. Eng. Chem. Fundam.* **1986**, *25*, 356.
452. Riley, D. P.; Busch, D. H. *Inorg. Chem.* **1984**, *23*, 3235.
453. Numata, Y.; Kubokura, K.; Nonaka, Y.; Okawa, H.; Koda, S. *Inorg. Chim. Acta* **1980**, *43*, 193.
454. Englert, U.; Straehle, J. *Gazz. Chim. Ital.* **1988**, *118*, 845.
455. Gallagher, M.; Ladd, M. F. C.; Larkworthy, L. F.; Povey, D. C. Salib, K. A. R. *J. Crystallogr. Spectros. Res.* **1986**, *16*, 967.
456. Leeuwenkamp, O. R.; Plug, C. M. Bult, A. *Polyhedron* **1987**, *6*, 295.
457. Berno, P.; Floriani, C.; Chiesi-Villa, A.; Guastini, C. *J. Chem. Soc., Dalton Trans.* **1988**, 1409.
458. Weighardt, K.; Kleine-Boymann, M.; Swiridoff, W.; Nuber, B.; Weiss, J. *J. Chem. Soc., Dalton Trans.* **1985**, 2493.
459. Larkworthy, L. F.; Povey, D. C. J. *Crystallogr. Spectrosc. Res.* **1983**, *13*, 413.
460. Girandon, J.-M.; Mandon, D.; Sala-Pala, J.; Guerchais, J. E.; Kerbaol, J.-M.; Mest, Y. L.; L'Haridon, P. *Inorg. Chem.* **1990**, *29*, 707.
461. Larkworthy, L. F.; Povey, D. C.; Smith, G. W.; Tucker, B. J. *J. Crystallogr. Spectrosc. Res.* **1989**, *19*, 439.
462. Calligaris, M.; Randaccio, L. In *Comprehensive Coordination Chemistry*; Wilkinson, G.; Gillard, R. D.; McCleverty, J. A., Eds.; Pergamon: Oxford, 1987; Chapter 20.
463. Haller, K. J.; Johnson, P. L.; Feltham, R. D.; Enemark, J. H.; Ferraro, J. R.; Basile, L. J. *Inorg. Chim. Acta* **1979**, *33*, 119.
464. Wells, F. V.; McCann, S. W.; Wickman, H. H.; Kessel, S. L.; Hendrickson, D. N.; Feltham, R. D. *Inorg. Chem.* **1982**, *21*, 2306.
465. Koenig, E.; Ritter, G.; Waigel, J.; Larkworthy, L. F.; Thompson, R. M. *Inorg. Chem.* **1987**, *26*, 1563.
466. Hodges, K. D.; Wollmann, R. G.; Kessel, S. L.; Hendrickson, D. N.; Van Derveer, D. G.; Barefield, E. K. *J. Am. Chem. Soc.* **1979**, *101*, 906.
467. Kitagawa, S.; Munakata, M.; Ueda, M. *Inorg. Chim. Acta* **1989**, *164*, 49.
468. Tom Dieck, H.; Bruder, H.; Kuehl, E.; Junghans, D.; Hellfeldt, K. *New J. Chem.* **1989**, *13*, 259.
469. Anillo, A.; Cabeza, J. A.; Obeso-Rosete, R.; Riera, V. *J. Organomet. Chem.* **1990**, *393*, 423.

470. Beck, W.; Petri, W.; Meder, J. *J. Organomet. Chem.* **1980**, *191*, 73.
471. Cooper, D. J.; Ravenscroft, M. D.; Stotter, D. A.; Trotter, J. *J. Chem. Res., Synop.* **1979**, *(9)*, 287.
472. Weighardt, K.; Quilitzsch, U.; Weiss, J. *Inorg. Chim. Acta* **1984**, *89*, L43.
473. Jhanji, A. K.; Gould, E. S. *Inorg. Chem.* **1990**, *29*, 3890.
474. Armstrong, R. A.; Taube, H. *Inorg. Chem.* **1976**, *15*, 1904.
475. Griffith, W. P. *J. Chem. Soc.* **1963**, 3286.
476. Nishimura, H.; Nagao, H.; Howell, F. S.; Mukaida, M.; Kakihana, H. *Chem. Lett.* **1988**, *(3)*, 491.
477. Kamata, Y.; Miki, E.; Mizumachi, K.; Ishimori, T. *Bull. Chem. Soc. Jpn.* **1986**, *59*, 1597.
478. Walsh, J. L.; Durham, B. *Inorg. Chem.* **1982**, *21*, 329.
479. Laing, K. R.; Roper, W. R. *J. Chem. Soc., Chem. Commun.* **1968**, 1568.
480. Ishiyama, T.; Matsumura, T. *Bull. Chem. Soc. Jpn.* **1979**, *52*, 619.
481. Reed, C. A.; Roper, W. R. *J. Chem. Soc., Dalton Trans.* **1973**, 1014.
482. Bradley, D. C.; Newing, C. W.; Chisholm, M. H.; Kelly, R. L.; Haitko, D. A.; Little, D.; Cotton, F. A.; Fanwick, P. E. *Inorg. Chem.* **1980**, *19*, 3010.
482. Albano, V.; Bellon, P.; Ciani, G.; Manassero, M. *J. Chem. Soc., Chem. Commun.* **1969**, 1242.
483. Chisholm, M. H.; Huffman, J. C.; Kelly, R. L. *Inorg. Chem.* **1980**, *19*, 2762.
484. Chisholm, M. H.; Cotton, F. A.; Extine, M. W.; Kelly, R. L. *J. Am. Chem. Soc.* **1978**, *100*, 3354.
485. Chisholm, M. H.; Cotton, F. A.; Extine, M. W.; Kelly, R. L. *Inorg. Chem.* **1979**, *18*, 116.
486. Dziegielewski, J. O.; Filipek, K.; Jezowska-Trzebiatowska, B. *Inorg. Chim. Acta* **1990**, *171*, 89.
487. Carty, P.; Walker, A.; Mathew, M.; Palenik, G. J. *J. Chem. Soc., Chem. Commun.* **1969**, 1374.
488. Brownlee, G. S.; Carty, P.; Cash, D. N.; Walker, A. *Inorg. Chem.* **1975**, *14*, 323.
489. Cheng, P. T.; Nyburg, S. C. *Inorg. Chem.* **1975**, *14*, 327.
490. Tatsumi, T.; Sekizawa, K.; Tominaga, H. *Bull. Chem. Soc. Jpn.* **1980**, *53*, 2297.

491. Carvill, A.; Higgins, P.; McCann, G. M.; Ryan, H.; Shiels, A. *J. Chem. Soc., Dalton Trans.* **1989**, 2435.

492. Quelch, G. E.; Hillier, I. H.; Guest, M. F. *J. Chem. Soc., Dalton Trans.* **1990**, 3075.

493. Bottomley, F.; White, P. S.; Mukaida, M. *Acta Crystallogr.* **1982**, *B38*, 2674.

494. Brown, D. S.; Newman, J. L.; Thornback, J. R. *Acta Crystallogr., Sect. C: Cryst. Struct Commun.* **1988**, *C44*, 973.

495. Englert, U.; Straehle, J. *Z. Naturforsch., B: Chem. Sci.* **1987**, *42*, 959.

496. Keller, A.; Jezowska-Trzebiatowska, B. *Bull. Acad. Pol. Sci., Ser. Sci. Chim.* **1980**, *28*, 73.

497. Sarkar, S.; Subramanian, P. *J. Inorg. Nucl. Chem.* **1981**, *43*, 202.

498. Mueller, A.; Sarkar, S.; Mohan, N.; Bhattacharyya, R. G. *Inorg. Chim. Acta* **1980**, *45*, L245.

499. Liebelt, A.; Weller, F.; Dehnicke, K. *Z. Anorg. Allg. Chem.* **1981**, *480*, 13.

500. Shorthill, W. B.; Buchanan, R. M.; Pierpont, C. G.; Ghedini, M.; Dolcetti, G. *Inorg. Chem.* **1980**, *19*, 1803.

501. Ghedini, M.; Dolcetti, G.; Giovannitti, B.; Denti, G. *Inorg. Chem.* **1977**, *16*, 1725.

502. Rigo, P. *Inorg. Chim. Acta* **1980**, *44*, L223.

503. Fergusson, J. E.; Robinson, W. T.; Coll, R. K. *Inorg. Chim. Acta* **1991**, *181*, 37.

504. Baty, J. D.; Willis, R. G.; Burdon, M. G.; Butler, A. R.; Glidewell, C.; Johnson, I. L.; Massey, R. *Inorg. Chim. Acta* **1987**, *138*, 15.

505. Beck, W.; Grenz, R.; Guetzfried, F.; Vilsmaier, E. *Chem. Ber.* **1981**, *114*, 3184.

506. Bishop, P. T.; Dilworth, M. J. R.; Hutchinson, J.; Zubieta, J. *J. Chem. Soc., Dalton Trans.* **1986**, 967.

507. Blower, P. J.; Bishop, P. T.; Dilworth, J. R.; Hsieh, T. C.; Hutchinson, J.; Nicholson, T.; Zubieta, J. *Inorg. Chim. Acta* **1985**, *101*, 63.

508. Bishop, P. T.; Dilworth, J. R.; Hutchinson, J.; Zubieta, J. A. *Inorg. Chim. Acta* **1984**, *84*, L15.

509. Blower, P. J.; Dilworth, J. R.; Hutchinson, J. P.; Zubieta, J. A.; *J. Chem. Soc., Dalton Trans.* **1985**, 1533.

510. Baltusis, L. M.; Karlin, K. D.; Rabinowitz, H. N.; Dewan, J. C.; Lippard, S. J. *Inorg. Chem.* **1980**, *19*, 2627.

511. Sato, M.; Yoshida, T. *J. Organomet. Chem.* **1975**, *94*, 403.

512. Butler, A. R.; Glidewell, C.; Li, M. *Adv. Inorg. Chem.* **1988**, *32*, 335.

513. Müller, A.; Eltzner, W.; Mohan, N. *Angew. Chem., Int. Ed. Engl.* **1979**, *18*, 168.

514. Müller, A.; Eltzner, W.; Bogge, H.; Sarkar, S. *Angew. Chem., Int. Ed. Engl.* **1982**, *21*, 535.

515. Lu, S.; Wang, B.; Wu, X.; Huang, J. *Huaxue Xuebao* **1984**, *42*, 214. CA100(22):184792x.

516. Müller, A.; Eltzner, W.; Clegg, W.; Sheldrick, G. M. *Angew. Chem., Int. Ed. Engl.* **1982**, *21*, 536.

517. Müller, A.; Eltzner, W.; Boegge, H.; Krickemeyer, E. *Angew. Chem., Int. Ed. Engl.* **1983**, *22*, 884.

518. Lu, S.; Ni, B.; Wu, X.; He, M.; Huang, J. *Jiegou Huaxue* **1985**, *4(4)*, 307. CA106(8):60248d.

519. Lu, S.; Wu, X.; Huang, J.; Lu, J. *Jiegou Huaxue* **1984**, *3*, 217. CA103(24):204755e.

520. Butler, A. R.; Glidewell, C. *Inorg. Chim. Acta* **1986**, *120*, 85.

521. Glidewell, C.; Lambert, R. J.; Harman, M. E.; Hursthouse, M. B. *J. Chem. Soc., Dalton Trans.* **1990**, 2685.

522. Butler, A. R.; Glidewell, C.; Johnson, I. L. *Polyhedron* **1987**, *6*, 2091.

523. Crayston, J. A.; Glidewell, C.; Lambert, R. J. *Polyhedron* **1990**, *9*, 1741.

524. Glidewell, C.; Hyde, A. R.; McKechnie, J. S.; Pogorzelec, P. J. *J. Chem. Ed.* **1985**, *62*, 534.

525. Glidewell, C.; Harman, M. E.; Hursthouse, M. B.; Johnson, I. L.; Motevalli, M. *J. Chem. Res., Synop.* **1988**, *(7)*, 212.

526. Glidewell, C.; Johnson, I. L. *Chem. Scr.* **1987**, *27*, 441.

527. Butler, A. R.; Glidewell, C.; Johnson, I. L. *Polyhedron* **1987**, *6*, 1147.

528. Glidewell, C.; Hyde, A. R. *Polyhedron* **1985**, *4*, 1155.

529. Butler, A. J.; Glidewell, C.; Johnson, I. L.; Walton, J. C. *Polyhedron* **1987**, *6*, 2085.

530. Butler, A. R.; Glidewell, C.; Hyde, A. R.; McGinnis, J.; Seymour, J. E. *Polyhedron* **1983**, *2*, 1045.

531. Chu, C. T.-W.; Lo, F. Y.-K.; Dahl, L. F. *J. Am. Chem. Soc.* **1982**, *104*, 3409.

532. Brunner, H.; Grassl, R.; Wachter, J.; Nuber, B.; Ziegler, M. L. *J. Organomet. Chem.* **1990**, *393*, 119.
533. Brunner, H.; Kauermann, H.; Wachter, J. *Angew. Chem., Int. Ed. Engl.* **1983**, *22*, 549.
534. Green, M. L. H.; Hamnett, A.; Qin, J.; Baird, P.; Bandy, J. A.; Prout, K.; Marseglia, E.; Obertelli, S. D. *J. Chem. Soc., Chem. Commun.* **1987**, 1811.
535. Herberhold, M.; Hill, A. F. *J. Chem. Soc., Dalton Trans.* **1988**, 2027.
536. Herberhold, M.; Haumaier, L.; Schubert, U. *Inorg. Chim. Acta* **1981**, *49*, 21.
537. Pander, K. K.; Nehete, D. T.; Nassoudipur, M. *Inorg. Chim. Acta* **1987**, *129*, 253.
538. Müller, A.; Khan, M. I.; Krickemeyer, E.; Bögge, H. *Inorg. Chem.* **1991**, *30*, 2040.
539. Herdtweck, E.; Schumacher, C.; Dehnicke, K. *Z. Anorg. Allg. Chem.* **1985**, *526*, 93.
540. Mueller, A.; Stolz, P.; Boegge, H.; Sarkar, S.; Schmitz, K.; Fangmeier, A.; Bueker, H.; Twistel, W. *Z. Anorg. Allg. Chem.* **1988**, *559*, 57.
541. Coucouvanis, D.; Simhon, E. D.; Stremple, P. *Inorg. Chim. Acta* **1981**, *53*, L135.
542. Nakamoto, M.; Tanaka, K.; Tanaka, T. *Inorg. Chim. Acta* **1987**, *132*, 193.
543. Herberhold, M.; Haumaier, L. *Chem. Ber.* **1983**, *116*, 2896.
544. Labuda, J.; Mocak, J.; Bustin, D. I. *Chem. Zvesti* **1983**, *37*, 337.
545. Jezierski, A.; Jezowska-Trzebiatowska, B. *Nouv. J. Chim.* **1980**, *4(10)*, 599.
546. Ferguson, G.; Somogyvari, A. *J. Crystallogr. Spectrosc. Res.* **1983**, *13*, 49.
547. Alyea, E. C.; Ferguson, G.; Somogyvari, A. *Inorg. Chem.* **1982**, *21*, 1372.
548. Broomhead, J. A.; Budge, J. R. *Aust. J. Chem.* **1979**, *32*, 1187.
549. Connor, J. A.; James, E. J.; Overton, C.; Walshe, J, M. A.; Head, R. A. *J. Chem. Soc., Dalton Trans.* **1986**, 511.
550. Budge, J. R.; Broomhead, J. A.; Boyd, P. D. W. *Inorg. Chem.* **1982**, *21*, 1031.

551. Perpinan, M. F.; Ballester, L.; Santos, A.; Monge, A.; Ruiz-Valero, C.; Puebla, E. G. *Polyhedron* **1987**, *6*, 1523.

552. Ileperuma, O. A.; Feltham, R. D. *Inorg. Synth.* **1976**, *16*, 5.

553. Iliev, V. I.; Shopov, D. *Polyhedron* **1987**, *6*, 1497.

554. Feltham, R. D.; Crain, H. *Inorg. Chim. Acta* **1980**, *40*, 37.

555. Fitzsimmons, B. W.; Hume, A. R. *J. Chem. Soc., Dalton Trans.* **1979**, 1548.

556. Büttner, H.; Feltham, R. D. *Inorg. Chem.* **1972**, *11*, 971.

557. Ileperuma, O. A.; Feltham, R. D. *Inorg. Chem.* **1977**, *16*, 1876.

558. Butcher, R. J.; Sinn, E. *Inorg. Chem.* **1980**, *19*, 3622.

559. Enemark, J. H.; Feltham, R. D. *J. Chem. Soc., Dalton Trans.* **1972**, 718.

560. Fanariotis, I. A.; Christidis, P. C.; Rentzeperis, P. J. *Z. Kristallogr.* **1983**, *164*, 109. CA100(12):94870s.

561. Tsipis, C. A.; Kessissoglou, D. P.; Manoussakis, G. E. *Chem. Chron.* **1982**, *11*, 235.

562. Iordanov, N.; Shopov, D. *Inorg. Chim. Acta* **1978**, *31*, 31.

563. Stach, J.; Kirmse, R.; Dietzsch, W.; Thomas, P. *Z. Anorg. Allg. Chem.* **1981**, *480*, 60.

564. Thomas, P.; Rehorek, D.; Nefedov, V. I.; Zhumadilov, E. K. *Z. Anorg. Allg. Chem.* **1979**, *448*, 167.

565. McCleverty, J. A.; Atherton, N. M.; Locke, J.; Wharton, E. J.; Winscom, C. J. *J. Am. Chem. Soc.* **1967**, *89*, 6082.

566. Thomas, P.; Lippman, M.; Rehorek, D.; Hennig, H. *Z. Chem.* **1980**, *20(4)*, 155.

567. Qian, L.; Singh, P.; Ro, H.; Hatfield, W. E. *Inorg. Chem.* **1990**, *29*, 761.

568. Ibrahim, S. K.; Pickett, C. J. *J. Chem. Soc., Chem. Commun.* **1991**, 246.

569. Sellmann, D.; Poehlmann, G.; Knoch, F.; Moll, M. *Z. Naturforsch., B: Chem. Sci.* **1989**, *44*, 312.

570. Sellmann, D.; Zapf, L.; Keller, J.; Moll, M. *J. Organomet. Chem.* **1985**, *289*, 71.

571. Sellmann, D.; Ludwig, W.; Huttner, G.; Zsolnai, L. *J. Organomet. Chem.* **1985**, *294*, 199.

572. Sellmann, D.; Binker, G.; Schwarz, J.; Knoch, F.; Boese, R.; Huttner, G.; Zsolnai, L. *J. Organomet. Chem.* **1987**, *323*, 323.

573. Sellmann, D.; Waeber, M. *Z. Naturforsch., B: Anorg. Chem., Org. Chem.* **1986**, *41B*, 877.

574. Sellmann, D.; Kaeppler, O.; Knoch, F. *J. Organomet. Chem.* **1989**, *367*, 161.
575. Sellmann, D.; Kunstmann, H.; Moll, M.; Knoch, F. *Inorg. Chim. Acta* **1988**, *154*, 157.
576. Sellmann, D.; Binker, G.; Moll, M.; Herdtweck, E. *J. Organomet. Chem.* **1987**, *327*, 403.
577. Sellmann, D.; Reineke, U. *J. Organomet. Chem.* **1986**, *314*, 91.
578. Sellmann, D.; Barth, I.; Knoch, F.; Moll, M. *Inorg. Chem.* **1990**, *29*, 1822.
579. Sellmann, D. *J. Organomet. Chem.* **1989**, *372*, 99.
580. Sellmann, D.; Weiss, R.; Knoch, F. *Inorg. Chim. Acta* **1990**, *175*, 65.
581. Hill, A. F.; Clark, G. R.; Rickard, C. E. F.; Roper, W. R.; Herberhold, M. *J. Organomet. Chem.* **1991**, *401*, 357.
582. Connor, J. A.; Riley, P. I. *J. Chem. Soc., Dalton Trans.* **1979**, 1231.

4

Organometallic Compounds

The nitrosyl ligand often imparts unique chemistry to complexes containing it. This is especially so for organometallic nitrosyl complexes, since the strong π-accepting ability of NO affects both the nature and strength of the metal-carbon bonds in these compounds *(1)*. In this chapter, organometallic nitrosyl complexes will be surveyed according to the type of hydrocarbyl ligand present. The η^5-cyclopentadienyl group is present in most of these complexes, and is unreactive for the most part. Consequently, it will not be assigned a class of its own. Unless otherwise stated, the Cp' notation will be used to represent both the Cp (η^5-C_5H_5) and the Cp* (η^5-C_5Me_5) groups.

4.1: Nitrosyl Complexes Containing σ-Alkyls and -Aryls

Given the large number of metal nitrosyl halides currently available, it is somewhat surprising that there are only a limited number of organometallic nitrosyl complexes that contain only simple σ-bound alkyl or aryl ligands. Representative examples of this small class of nitrosyl compounds are listed in Table 4.1. Whereas alkyl derivatives of chromium *(2)*, manganese *(3)*, osmium *(4)*, iridium *(5,8)* and platinum *(7,9)* are known, only aryl derivatives of nickel *(6)* and platinum *(9)* have been reported. The $[(CF_3)Mn(NO)(CO)_3]^-$ anion has been obtained from the reaction of $[(CF_3)Mn(CO)_4]^-$ with NO *(10)*. Bridging methylidene complexes of the Group 8 metals are also known *(11)*, but most of these contain the Cp ligand as well (Section 4.3.5).

Table 4.1 Nitrosyl Complexes Containing σ-Alkyl or σ-Aryl Ligands

Compound	ν_{NO} (cm^{-1})	Ref
$MeCr(NO)(CO)_4$	1735	*2*
$PhCH_2Mn(NO)$	1775	*3*
$(CF_3)Os(NO)(PPh_3)_2Cl_2$	1856	*4a*
$(PhC_2)Os(NO)(H)(PPr_3)_2Cl$	1795	*4b*
$MeIr(NO)(PPh_3)_2I$	1525	*5*
$PhNi(NO)(PPh_3)_2$	1697	*6*
$(CF_3)Pt(NO)(PPh_3)_2$	1660	*7*

The η^1-Cp complexes (η^1-Cp)W(NO)$(CO)_2(PMe_3)_2$ and (η^1-Cp)Re(NO)(CO)$(PMe_3)_2$Me are produced when their precursor η^5-Cp complexes are reacted with trimethylphosphine *(12,13)*. A related cyclohexadienyl compound, namely (η^1-$C_6H_5Me_2$)Mo(NO)$(CO)_2(PMe_3)_2$, has been obtained as an intermediate in the reaction of (η^5-$C_6H_5Me_2$)Mo(NO)$(CO)_2$ with trimethylphosphine *(14)*. Formyl complexes of tungsten *(15)*, carbamoyl complexes of osmium *(16)* and acyl complexes of iron *(17)* have also been prepared by other routes.

4.2: Cyclopentadienyl Carbonyl Nitrosyls and Their Derivatives

4.2.1: Vanadium

The CO ligand in CpV$(NO)_2$CO is readily displaced by Lewis bases (L) to yield CpV$(NO)_2$L *(18,19)*. In THF, the CpV$(NO)_2$(THF) complex is produced initially, which subsequently converts to THF-solvated aggregates consisting of "CpV$(NO)_2$" units *(19,20)*. The CO ligand in the parent

$CpV(NO)_2CO$ is also substituted by NO^+ to produce $[CpV(NO)_3]^+$ *(21)*. The fulvene complex $(\eta^5\text{-}C_5H_4(CHNMe_2))V(NO)(CO)_2$ has been obtained by carbonyl substitution in $V(NO)(CO)_5$ by 8-dimethylaminofulvalene *(22)*.

Vanadocene and nitric oxide react to produce $Cp_2V(NO)$ as the initial product which then undergoes disproportionation to a mixture of compounds *(23)*.

4.2.2: Chromium, Molybdenum and Tungsten

The $CpM(NO)(CO)_2$ complexes of the Group 6 metals are well-known compounds which possess pseudooctahedral, three-legged piano-stool molecular geometries *(24)*. Many analogues containing substituted Cp derivatives such as Cp* *(25-27)*, indenyl *(28,29)*, fluorenyl *(29,30)* or even a polystyrene-supported Cp *(31)* are also known to possess similar geometries. Furthermore, the complexes of the form $(\eta^5\text{-}C_5H_4X)M(NO)(CO)_2$ where X is an alkyl, formyl, amide or ferrocenyl group have also been prepared *(32-39)*. The ability of the $CpCr(NO)(CO)_2$ group to stabilize carbonium ions in alkyl substituents of the Cp ring is partly responsible for the multitude of substituted derivatives available *(37,40)*. The cyclohexadienyl analogues for chromium and molybdenum *(14,41,42)* and the ylidic thiabenzene 1-oxide derivatives for chromium *(43-45)* are also known.

The electrochemical and chemical reductions of the $CpM(NO)(CO)_2$ complexes of chromium and molybdenum result in the generation of the corresponding radical monoanions *(46,47)* which possess bent MNO geometries, an observation that is consistent with these octahedral complexes converting from $\{MNO\}^6$ to $\{MNO\}^7$ species. Photolysis of the $CpM(NO)(CO)_2$ complexes of chromium and tungsten results in the expulsion of CO as the primary photoreaction, followed by intramolecular rearrangements to form isocyanate ligands *(19,48)* (Section 7.2.4). The CO ligand in the $CpM(NO)(CO)_2$ complexes may also be displaced thermally or photolytically by Lewis bases such as phosphines to yield the $CpM(NO)(CO)L$ and $CpM(NO)L_2$ derivatives *(49-51)*. For instance, it has been found that the $CpM(NO)(CO)_2$ compounds of Mo and W react with trimethylphosphine via an associative mechanism to ultimately yield CpM(NO)(CO)(PMe_3). In these reactions, the addition of trimethylphosphine does not result in a linear-to-bent NO geometry change, but rather results in a Cp ring-slippage to produce the six-coordinate (η^1-Cp)M(NO)$(CO)_2(PMe_3)_2$ *(13)*. The CpM(NO)(CO)L complexes (L = CS, CNR) have also been prepared *(52-54)*. The isonitrile complexes are prepared by treatment of the $[CpM(NO)(CO)(CN)]^-$ anions with electrophilic alkylating agents *(53)*. The disubstituted CpM(NO)L_2 compounds are, however, best synthesized by the reduction of their halide precursors in the presence of the appropriate Lewis base *(55)*. The cationic complexes $[CpCr(NO)(L_2)]^+$ (L_2 = bipy, phen) are also preparable from their dinitrosyl precursors via NO displacement by the incoming Lewis base *(56)*.

The coordinatively unsaturated $[Cp'M(NO)_2]^+$ dinitrosyl cations (Cp' = Cp, Cp*) are obtainable via halide-abstraction from their neutral chloride precursors *(57)*. A variety of Lewis bases may coordinate to these formally 16-electron cations of chromium and tungsten to produce the corresponding 18-electron $[CpM(NO)_2L]^+$ species. These latter species for all three Group 6 metals are also preparable by reaction of the CpM(NO)(CO)L precursors (L = CO *(59)*, CNMe *(53)*) with the nitrosonium cation. The chromium thiocarbonyl analogue (i.e., L = CS) is also preparable via a similar route *(52)*.

Interestingly, the reaction of $[CpW(NO)_2L]^+$ (L = phosphine or phosphite) with reducing agents results in the generation of the formally 19-electron, neutral $CpW(NO)_2L$ complexes which display a localization of the added electron density in the $W(NO)_2$ fragment *(60)*. The $[Cp'M(NO)_2]_2$ dinitrosyl dimers for chromium (Cp' = Cp, Cp*) and molybdenum (Cp' = Cp*) are obtained by the reduction of their $Cp'M(NO)_2Cl$ precursors in aromatic solvents *(61,62)*. The CpCr complex possesses a *trans* configuration in the solid state *(63)*, and undergoes a reversible one-electron reduction in solution to form the bimetallic radical monoanion $[CpCr(NO)_2]_2^{\cdot -}$, which has been isolated as its $[CpFe(C_6Me_6)]^+$ salt *(64)*. The electronic structures of $[CpCr(NO)_2]_2$ and its isoelectronic carbonyl analogue $[CpFe(CO)_2]_2$ have been compared and the results used to rationalize the differences in their electrochemical behavior *(65)*.

$$2\,[CpCr(NO)_2]^+ \xleftarrow{-2e^-} [CpCr(NO)_2]_2 \xrightleftharpoons{+e^-} [CpCr(NO)_2]_2^{\cdot -} \quad (4.1)$$

$$[CpFe(CO)_2]_2^{\cdot +} \xrightleftharpoons{-e^-} [CpFe(CO)_2]_2 \xrightarrow{+2e^-} 2\,[CpFe(CO)_2]^- \quad (4.2)$$

The HOMO of the Cr complex consists mainly of a Cr-Cr bonding interaction. Thus, removal of electrons from this orbital should result in the weakening of the metal-metal interaction. On the other hand, the LUMO of the Cr complex is largely of μ-NO 2π character, which means that the addition of an electron to the neutral dimer should result in the localization of electron density in the $Cr(\mu\text{-}NO)_2Cr$ bridge, a result that is observed experimentally as well *(64)*. In contrast, the HOMO of the Fe complex is calculated to be metal-based and Fe-Fe π^* in character. In other words, removal of electrons from the complex should strengthen the Fe-Fe interaction. However, the LUMO is strongly Fe-Fe σ^* in character: reduction would be expected to lead to a weaker Fe-Fe interaction.

The $[CpCr(NO)_2]_2$ dimer exhibits a wide range of reactivity patterns including the chemical reduction of its μ-NO ligands and its use for selective dehalogenation of *vic*-dihaloalkanes to produce the corresponding alkenes *(1)*. The bimetallic dinitrosyl complex $Cp^*W(NO)(\mu\text{-}NO)(\mu\text{-}CO)W(CO)_2Cp^*$ represents a rare example of a mixed carbonyl-nitrosyl dimer, and is prepared by the reaction of $[Cp^*W(CO)_2(\mu\text{-}H)]_2$ with NO *(26)*.

4.2.3: Manganese and Rhenium

The cationic $[Cp'M(NO)(CO)_2]^+$ complexes of Mn and Re are preparable by the reaction of their neutral tricarbonyl precursors with NO^+ *(66,67)*. As expected, the carbonyl ligands in these cationic complexes are readily substituted by Lewis bases *(66,68)* and are also attacked by nucleophiles such as amines *(69)*. Indeed, the reactions of the $[CpMn(NO)(CO)L]^+$

compounds with transition-metal anions produce heterobimetallic complexes as shown in eq 4.3 *(70,71)*.

$$[CpMn(CO)(NO)L]^+ \quad \xrightarrow[CS_2, L=CO]{Fp^-} \quad Fp{-}C(S)_2Mn(Cp)(NO) \qquad \xleftarrow[L=PMe_3]{CpCo(NO)^-} \quad CpCo(\mu\text{-}NO)_2Mn(Cp)L \tag{4.3}$$

That the carbonyl carbon in the $[Cp'M(NO)(CO)_2]^+$ compounds is sufficiently electrophilic is seen in the reactions of these cationic compounds with complexes containing electron-rich oxo ligands. Thus, net [2+2] cycloadditions occur in the reactions described by eq 4.4 to form μ,η^3-CO_2 complexes *(72,73)*.

$$[(C_5R_5)M(NO)(CO)_2]^+ \xrightarrow{L_nM'=O} [(C_5R_5)M(NO)(CO)(\mu\text{-}CO_2)M'L_n]^+ \tag{4.4}$$

(R_5 = H_4Me, H_5, Me_5; LnM' = Cp_2Mo, Cp_2W, tmtaa)

Other μ-carboxylate complexes of Re (where M' is a main-group element such as Ge, Sn or Pb) are known *(74)*. Furthermore, the complex in which L_nM' is Cp_2ZrCl has been prepared, but by a different route *(75)*.

The dichloromethane complexes $[Cp'Re(NO)(PPh_3)(ClCH_2Cl)]^+$ contain weak Re-Cl interactions, and are functional equivalents of the coordinatively unsaturated $[Cp'Re(NO)(PPh_3)]^+$ cations *(76,77)*. Thus, the coordinated dichloromethane ligand is readily replaced by Lewis bases (L) such as alkyl halides *(78,79)*, aldehydes and ketones *(80-82)*, amines *(83)*, phosphines *(84,85)*, alcohols *(86)*, ethers and thioethers *(87)*.

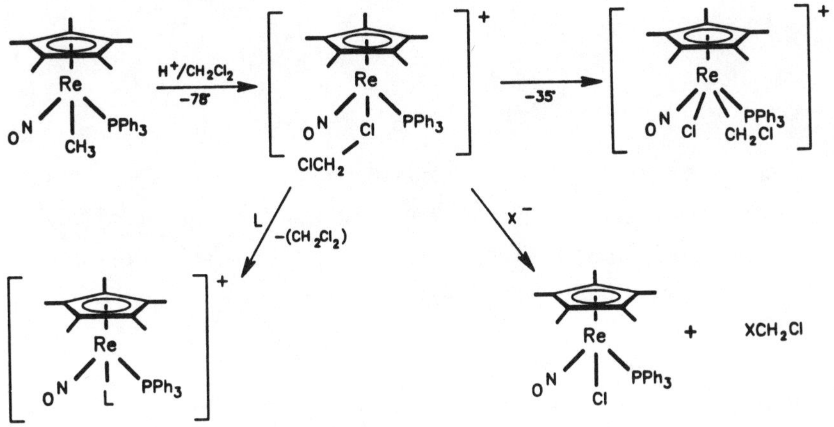

Attack of hydride on $[CpRe(NO)(CO)_2]^+$ produces the neutral formyl complex CpRe(NO)(CO)(CHO), which has a rich and varied chemistry of its own *(88)*. It has been proposed that attack of H^- on the phosphine analogue $[CpRe(NO)(CO)(PPh_3)]^+$ to produce the corresponding formyl proceeds initially via the attack of H^- on the nitrosyl nitrogen *(89)*, i.e.,

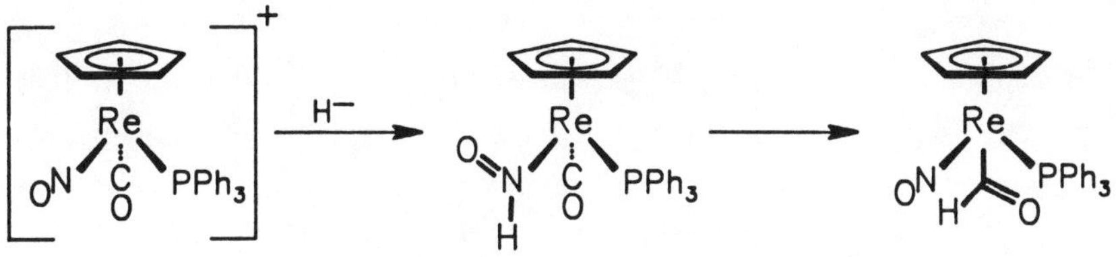

The $[Cp'Mn(NO)L]_2$ dimers (Cp' = Cp, MeCp; L = CO, CS) are prepared by a variety of routes *(90,91)*, and the structure of the carbonyl compound is best described as *(92)*:

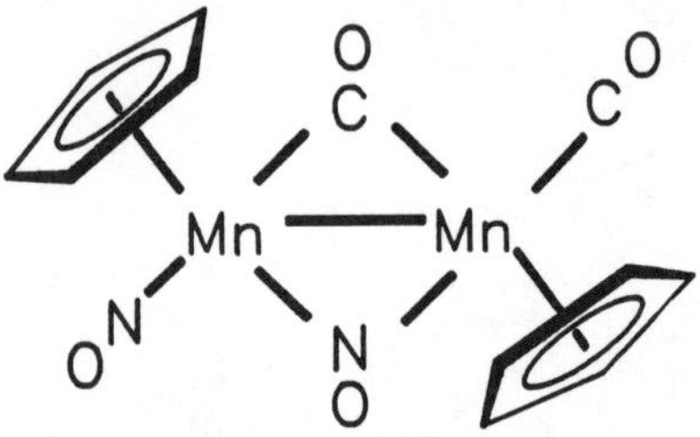

O
C
C O
Mn Mn
O N
N
O

Although the electrochemistry of this complex has not been reported, it has been predicted (based on molecular orbital calculations) that the reduction of this dimer should result in the production of the thermodynamically unstable and still unknown $[CpMn(NO)(CO)]^-$ anion *(65)*.

Cationic cyclohexadienyl complexes of Mn and Re of the form $[(\eta^5\text{-}C_6R_7)M(NO)L_2]^+$ (R = H, alkyl; L = CO, phosphine) have been prepared *(93-95)*, and whereas the Mn dicarbonyl complexes are reduced *irreversibly* at a platinum electrode, the Re analogues undergo *reversible* reductions at similar potentials and under similar conditions *(96)*.

The treatment of $[CpMn(NO)(CO)_2]^+$ with nitrite yields a crystalline product formulated as polymeric $[CpMn(NO)_2]_n$ *(97)*. The bimetallic $Cp_3Mn_2(NO)_3$ complex contains a $Mn(\mu\text{-}NO)_2Mn$ bridge with one of the Cp ligands displaying an η^1 bonding mode *(98)*.

4.2.4: Iron and Ruthenium

The electronic structures of the $[Cp'M(\mu\text{-}NO)]_2$ dimers of Fe and Ru have been the subject of much study *(99-104)*. Although these complexes are generally represented by

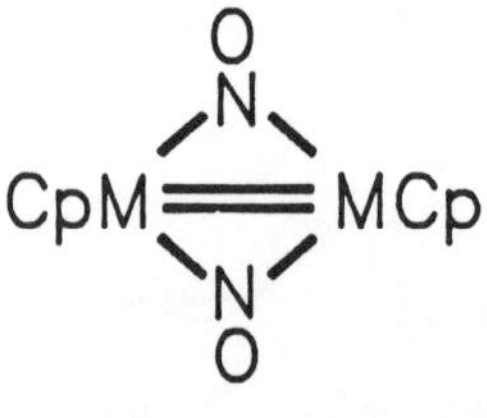

O
N
CpM = MCp
N
O

with a metal-metal double bond (mostly to satisfy the 18-electron rule), the occupied metal orbitals in these complexes are actually

$$(\sigma)^2(\delta_1)^2(\delta_2{}^*)^2(\pi_1)^2(\delta_1{}^*)^2(\delta_2)^2(\pi_2{}^*)^2(\sigma^*)^2$$

In other words, as many antibonding orbitals as bonding orbitals are occupied when symmetry interactions are taken into consideration, in effect producing a net zero formal metal-metal bond *(100)*. However, the bond energy is not zero: the "double-bonds" are not formed directly between the two metals, but are coupled through the nitrosyl bridges. This view is consistent with an overall delocalized $M(\mu\text{-}NO)_2M$ system in this and related complexes.

The $[Cp'Fe(NO)]_2$ dimers (Cp' = Cp, MeCp, Cp*) may be reduced electrochemically to their respective monoanions *(101)*. Upon such reduction, the Fe-Fe separation in the Cp complex increases by 0.052 Å (from 2.326 Å in the neutral complex to 2.378 Å in the monoanion) which suggests that the added electron occupies a HOMO composed of out-of-plane Fe-Fe antibonding character. Nevertheless, the "metal-metal bonds" in these complexes of Fe and Ru are attacked by diazoalkanes to yield μ-alkylidene complexes *(105-107)*.

Cationic complexes of the form $[Cp'M(NO)L_2]^{2+}$ (M = Fe, Ru; L = phosphine) and $[(C_5Ph_5)Ru(NO)(CO)L]^{2+}$ are known *(108-113)*, and the $[CpRu(NO)dppe]^{2+}$ complex undergoes two successive reversible one-electron reductions in the absence of oxygen to ultimately produce the neutral CpRu(NO)dppe compound. In the presence of oxygen, this compound decomposes to $CpRu(NO_2)dppe$ *(110)*. The neutral $(\eta^5\text{-}C_5Ph_5)Ru(NO)L$ complex has been obtained by the chemical reduction of $(\eta^5\text{-}C_5Ph_5)Ru(NO)L(Br)]^+$ with cobaltocene *(111)*. The related $Cp^*Ru(NO)L_2$ complexes (L = PMe_3, PPh_3, 1/2 dmpe) have also been prepared, and probably contain bent NO ligands *(114)*.

4.2.5: Cobalt, Rhodium and Iridium

The $[CpCo(NO)]_2$ dimer possesses a diamagnetic ground state *(115)*. As a solid, it contains a planar $Co(\mu\text{-}NO)_2Co$ core, and in the vapor phase, it possesses a bent $Co(\mu\text{-}NO)_2Co$ core with a direct Co-Co interaction *(116)*.

The family of $[Cp'M(NO)]_2$ compounds (Cp' = Cp, MeCp, Cp*, C_5Ph_5; M = Co, Rh) are reversibly oxidized to their mono- and dications *(101,117-118)*. The crystal structures of the $[CpCo(NO)]_2^{0/+}$ compounds reveal that a slight shortening of the Co-Co distance (by 0.02 Å) accompanies one-electron oxidation of the neutral complexes *(101,119)*.

Electrochemical reduction of $[CpCo(NO)]_2$ produces initially the dimeric radical anion, which then splits up into the neutral "CpCo(NO)" and anionic $[CpCo(NO)]^-$ fragments. Addition of another electron to the neutral component of this mixture then accounts for the overall two-electron reduction observed for the bimetallic complex *(116)*. The crystallographically characterized Na[(MeCp)Co(NO)] complex has been obtained by the sodium-amalgam reduction of the parent dimer *(70)*. In contrast, $[CpRh(NO)]_2$ is reduced to its bimetallic radical monoanion *(120)*.

Partial nitrosylation of $CpCo(CO)_2$ yields $[CpCo]_2(\mu\text{-}CO)(\mu\text{-}NO)$ *(121)*. This product may be oxidized or reduced to produce initially the monocation or monoanion respectively, which then undergo follow-up chemical reactions *(116,122)*. In general, the $Cp'M(CO)_2$ complexes of cobalt and rhodium react with nitrosonium salts to form the $[(Cp'M)_2(\mu\text{-}CO)(\mu\text{-}NO)]^+$ cationic species *(117,120)*. A variety of cationic complexes of the form $[Cp'M(NO)L]^+$ (M = Co, Rh; L = olefin, phosphine) have been synthesized from a number of different routes *(117,120,123-125)*. Whereas the neutral $(\eta^5\text{-}C_5Ph_5)Co(NO)L$ complexes (L = $P(OMe)_3$, $P(OPh)_3$) are unstable, the Rh congeners are stable and isolable *(118)*. Diazoalkanes attack the Rh-Rh bond in $[(Cp^*Rh)_2(CO)(NO)]^+$ to form bridging alkylidene complexes *(126)*.

Finally, the ethylene complex $[Cp^*Ir(NO)(C_2H_4)]^+$ is obtained by the treatment of $Cp^*Ir(C_2H_4)_2$ with NO^+ *(127)*.

4.2.6: Nickel, Palladium and Platinum

The one-legged piano-stool compounds CpM(NO) are well-known for nickel *(128-131)*, palladium *(132)* and platinum *(133)*. There is a considerable interest in the electronic structures of these species since they represent a unique class of neutral one-legged piano-stool compounds *(134)*. It has been proposed, for instance, that whereas the linear NiNO bond in CpNi(NO) is largely covalent, the CpNi bond is predominantly ionic *(135)*. Nevertheless, loss of NO from CpNi(NO) appears to dominate its chemistry *(136-139)*. A carborane analogue of CpNi(NO), namely (1,7-B_9H_9CHPMe)Ni(NO) has also been briefly mentioned in the literature *(141)*. The palladium complexes (η^5-C_5R_5)Pd(NO) (R_5 = Ph_5, *p*-tol_5, Ph_3Et_2) have also been prepared, and these compounds also possess one-legged piano-stool geometries *(140)*.

4.3: Cyclopentadienyl Nitrosyl Halides, Alkyls and Their Derivatives

4.3.1: Titanium

The paramagnetic Cp_2TiR compounds (R = Ph, CH_2Ph) both react with NO to afford a complex of the formula $Cp_3Ti_3O_4$(NO), whose ν_{NO} of 1550 cm^{-1} has been assigned to a bent TiNO group *(142)*.

4.3.2: Vanadium

The vanadocene nitrosyl halides Cp'_2V(NO)X (X = Cl, Br, I) are formally 19-electron species and are paramagnetic *(20,23,143)*. In the case of Cp_2V(NO)I, this molecule is fluxional in THF solution, displaying two NO streching frequencies at 1670 and 1590 cm^{-1}. However, it decomposes in this solvent at room temperature to $[Cp_2VI]_2[CpV(NO)]_2(\mu\text{-}O)_4$, which consists of an eight-membered ring of alternating V and O atoms, each set of "CpVI" or "CpV(NO)" groups located *trans* to each other *(23)*. The reaction of $Cp^*_2VBr_2$ with NO produces the salt $[Cp^*_2V(NO)Br]^+[Cp^*V(NO)Br_2]^-$. The cation in this complex is diamagnetic (formally 18-electrons) whereas the anion is paramagnetic and contains two unpaired electrons localized on the vanadium atom even though it is formally a 16-electron d^2 complex *(143)*.

4.3.3: Chromium, Molybdenum and Tungsten

The mononitrosyl Cp'M(NO)X_2 compounds (X = Cl, Br, I) of molybdenum and tungsten are obtainable by the reaction of their dicarbonyl precursors with the appropriate halogenating agent *(27,144-146)*. These compounds are monomeric in solution *(144)*, but in the solid state may exist as monomers or dimers depending on the nature of Cp', M or X. For example, $Cp^*W(NO)I_2$ is monomeric in the solid state *(27)*, whereas $Cp^*Mo(NO)Br_2$ is dimeric *(145)*, i.e.

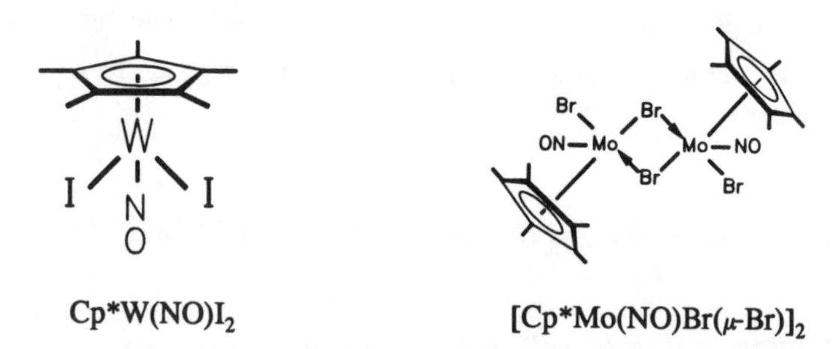

$Cp^*W(NO)I_2$ $[Cp^*Mo(NO)Br(\mu\text{-}Br)]_2$

In any event, the $Cp'M(NO)X_2$ compounds display a rich chemistry and are precursors to a host of other compounds of the same type. All the Cp'Mo compounds undergo reversible one-electron reductions to their 17-electron $[Cp'Mo(NO)X_2]^{\cdot-}$ radical anions which are isolable as their cobalticinium salts *(144)*. Interestingly, the unpaired electrons in these radical anions exhibit no hyperfine coupling to the nitrosyl nitrogen. That their MoNO links may still be essentially linear is consistent with the prediction that no distortion in this link is expected in going from the pseudooctahedral neutral $\{MoNO\}^4$ to the anionic $\{MoNO\}^5$ species.

Not surprisingly, the corresponding 18-electron $Cp'M(NO)X_2L$ derivatives (L = CO *(27,145,147)*, phosphine *(147-149)*, CNR *(150-152)* or alkyl- or arylhydrazines *(153)*) are also isolable. Addition of excess L may result in the displacement of halogen from the metals' coordination sphere *(148,149,152)*.

$$Cp'M(NO)X_2L + L \longrightarrow [Cp'M(NO)XL_2]X \qquad (4.5)$$

Indeed, $Cp^*Mo(NO)Br_2$ reacts with dppm or dppe (L_2) to afford the complex salts $[Cp^*Mo(NO)BrL_2]^+[Cp^*Mo(NO)Br_3]^-$ *(145)*. The reaction of $CpMo(NO)I_2$ with iodide affords the corresponding $[CpMo(NO)I_3]^-$ anion *(154)*. When the Lewis base in reaction 4.5 is PMe_3 or CNR, addition of an excess of it to the diiodo compounds results in the eventual displacement of the Cp ligand to form $M(NO)L_4I$ *(148,150)*.

Other derivatives of the form CpMo(NO)X(Y), where Y = cyanide *(149)*, acyl *(155)*, Cp' *(156-162)*, carboxylate *(162)*, hydroxide or thiol *(162)*, alkoxide *(162)*, triazenido *(163)* or acac *(163)* have all been prepared. More interestingly, the halide ligands in $Cp'M(NO)X_2$ may be metathesized by alkyl groups to afford the corresponding alkyl derivatives *(164-167)*.

$$Cp'M(NO)X_2 \xrightarrow{RMgX} Cp'M(NO)R_2 \qquad (4.6)$$

(M = Mo, W)

$$Cp'Mo(NO)X_2 \xrightarrow{AlMe_3} [Cp'Mo(NO)Me]_2(\mu\text{-}X)_2 \qquad (4.7)$$

The dialkyl complexes of eq 4.6 are monomeric, 16-electron species and react with a number of small molecules. Molecular orbital calculations on the $CpM(NO)Me_2$ compounds reveal that their LUMOs are (i) nonbonding, (ii) localized on the metal center, and (iii) contain no NO 2π character *(168)*. Thus, the 18-electron anionic $[CpM(NO)R_2X]^-$ and $[Cp'M(NO)R_2]^{2-}$ and the neutral $Cp'M(NO)R_2L$ analogues should be stable, since the added electrons are expected to reside in a nonbonding orbital *(169)*. Consistent with this prediction is the fact that the $Cp'W(NO)(CH_2SiMe_3)_2$ compounds are reduced reversibly to their 17-electron radical anions *(144)*.

$$Cp'W(NO)(CH_2SiMe_3)_2 \underset{}{\overset{+e^-}{\rightleftharpoons}} [Cp'W(NO)(CH_2SiMe_3)_2]^{\cdot -} \qquad (4.8)$$

and the $[CpMo(NO)Br_2Me]^-$, $[Cp^*Mo(NO)Br_3]^-$ and $[CpMo(NO)I_3]^-$ anions have also been isolated *(145,154,167)* (vide supra).

That the 18-electron analogues of the $Cp'M(NO)R_2$ compounds would be expected to be stable is reflected in the fact that in the bis(benzyl) complexes, the benzyl ligands adopt a unique η^1,η^2 geometrical conformation

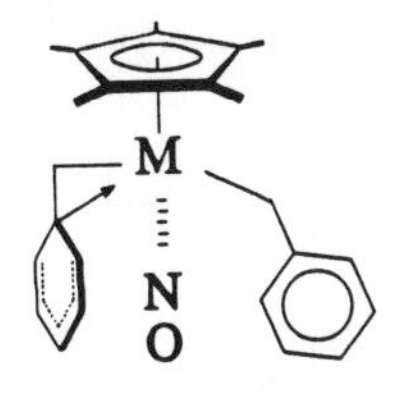

to result in a formally 18-electron configuration *(165,170)*. Furthermore, the complex shown below

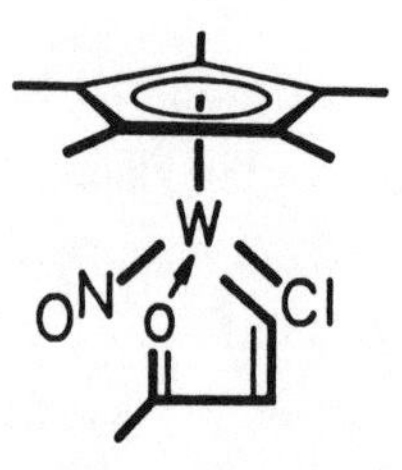

is also formally an 18-electron entity *(171)*.

The product of equation 4.7 reacts with Lewis bases to afford the monomeric CpMo(NO)Me(X)L species, which may be alkylated further to afford the CpMo(NO)(Me)(R)L derivatives (L = PPh_3; R = Me, C_2Ph, C_6F_5) *(167)*. Related alkyl hydride complexes of tungsten, namely Cp'W(NO)H(R)L are also stable, and activate C-H bonds intra- and intermolecularly *(172)*(Section 7.4.2.7).

Returning to the monomeric 16-electron species of eq 4.6, they react with small molecules via the initial coordination of these molecules to the metal centers followed by subsequent transformations of one or more of the M-R bonds in the complexes *(164)*. For example, reactions of Cp'M(NO)R_2 or [Cp*Mo(NO)Me]$_2$(μ-Br)$_2$ with air (or oxygen) afford the corresponding Cp'M(O)$_2$R and [Cp*Mo(O)]$_2$(μ-O) complexes, respectively *(173,174)*. Furthermore, *t*-BuNC and NO insert into one of the W-R bonds of CpW(NO)R_2 (R = CH_2SiMe_3) as shown in eqs 4.9 and 4.10 *(1645)*.

$$\mathrm{CpW(NO)R_2} + t\text{-}\mathrm{BuNC} \longrightarrow \mathrm{CpW(NO)(R)(\eta^2\text{-}C(R){=}N\text{-}}t\text{-}\mathrm{Bu)} \quad (4.9)$$

$$\mathrm{CpW(NO)R_2} + 2\ \mathrm{NO} \longrightarrow \mathrm{CpW(NO)(R)\ (\eta^2\text{-}ONN(R)O)} \quad (4.10)$$

Reaction of this CpW(NO)R_2 compound with elemental sulfur (or selenium) results in the sequential insertions described below *(164,175)*.

$$\underset{\mathrm{M-R}}{\overset{\mathrm{R}}{|}} \xrightarrow{\ \mathrm{S}\ } \underset{\mathrm{M-SR}}{\overset{\mathrm{R}}{|}} \xrightarrow{\ \mathrm{S}\ } \underset{\mathrm{M\overset{S}{\frown}SR}}{\overset{\mathrm{R}}{|}} \longrightarrow \underset{\mathrm{M-SR}}{\overset{\mathrm{RS}}{|}}$$

The related $CpMo(NO)(SPh)_2$ dithiolate compound is obtained from the disproportionation of $CpMo(NO)_2SPh$, and its molecular structure reveals an unusually short Mo-S distance of 2.34 Å (av) *(176)*. The results of Fenske-Hall molecular orbital calculations on this dithiolate complex are indicative of a $p\pi$-$d\pi$ bonding interaction between the lone pair of electrons on the thiolate ligands and the empty d_{xy} orbital of Mo in this formally 16-electron monomeric complex *(176)*. Not surprisingly, the monothiolate $[CpMo(NO)(SR)]_2$ complexes are dimeric in the solid state *(154,177,178)*, as are the other members of this general class of $[Cp'M(NO)X]_2$ compounds (M = Cr, Mo, W; X = halide, OR, SR, SeR, TeR) *(144,179-182)*.

The Cr compounds deserve mention, since the analogous dihalide $Cp'Cr(NO)X_2$ species remain unknown. Indeed, reaction of $Cp'Cr(NO)(CO)_2$ with X_2 results in the formation of dimeric $[Cp'Cr(NO)X]_2$ *(181a)*. Addition of Lewis bases to these monohalo dimers followed by alkylation results in the generation of the 17-electron Cp'Cr(NO)(R)L compounds *(181b)*. The molecular structure of the related $CpCr(NO)I(NPh_2)$ reveals a normal three-legged piano-stool formulation for this complex *(183)*. Since the hybridization of the N atom of the $Cr(NPh_2)$ group approximates to sp^2 (the N is in an almost planar arrangement) it is likely that some multiple-bond character exists in the $Cr-NPh_2$ link *(183)*.

Interestingly, the reaction of $CpCr(NO)(CO)_2$ with $Me_3P=CH_2$ proceeds via ylide addition to coordinated CO, i.e.

$$CpCr(NO)(CO)_2 \xrightarrow{2\ Me_3P=CH_2} (PMe_4)[CpCr(NO)(CO)(C(O)CH=PMe_3)] \xrightarrow[-Me_3P=CH_2,\ -CO]{h\nu} CpCr(NO)(CO)(CH_2PMe_3)$$

which upon photolysis, yields the diamagnetic Cr-ylide complex *(184)*. The anionic mononitrosyls $[CpM(NO)(CO)M'R_3]^-$ (M = Mo, W; M' = Ge, Sn) are known, and these react with allyl halides to form olefin (M' = Sn) or η^1-allyl (M' = Ge) complexes *(185)*. Vinyl complexes of the form Cp*W(NO)(Cl)(CH=CHC(O)Me) have also been prepared *(171,179,186)*.

The dinitrosyl complexes $Cp'M(NO)_2Cl$ are prepared by the reaction of their $Cp'M(NO)(CO)_2$ precursors with nitrosyl chloride *(57,187,188)*.

$$Cp'M(NO)(CO)_2 + ClNO \longrightarrow Cp'M(NO)_2Cl + 2\ CO \qquad (4.11)$$

The analogous bromide and iodide complexes are then obtained by reaction of the dinitrosyl chloride complex with excess NaX (X = Br, I) in THF *(189)*. The molecular structures of the $CpCr(NO)_2Cl$ and $CpW(NO)_2Cl$ congeners reveal that they possess the familiar three-legged piano-stool geometries *(190)*, and PES studies on these compounds reveal that (i) the W-Cl bond is stronger than the Cr-Cl bond, and (ii) in going from Cr to W, backbonding of electron density to the NO group increases and electron density is withdrawn from the Cl atoms *(189,191)*. The Lewis acid $FeCl_3$ interacts with $CpCr(NO)_2Cl$ mainly via a $CrCl..FeCl_3$ interaction (see Chapter 7) *(192)*. Furthermore, a nitrosyl ligand in $CpCr(NO)_2Cl$ is expelled from the metal's coordination sphere by a range of strong electron donors (L) to produce the 17-electron CpCr(NO)Cl(L) species *(193)*.

Although the CpMo and CpW dinitrosyl chlorides both undergo *reversible* reductions to their 19-electron radical anions, the CpCr complex undergoes an *irreversible* reduction under identical experimental conditions *(194)*. Furthermore, whereas all the $Cp'M(NO)_2R$ derivatives (R = alkyl, aryl) are readily obtained by the treatment of their chloride precursors with AlR_3 reagents, only the Cr complexes can be obtained by similar reactions with Grignard reagents *(58,61,195,196)*, an observation that reflects the different chemical behavior of the M-X bonds in these complexes. Interestingly, the $Cp'M(NO)_2R$ compounds may also be obtained in high yields from the reaction of their electrophilic $[Cp'M(NO)_2]^+$ cations with $[BR_4]^-$ or SnR_4 *(57,58)*. Nevertheless, the $CpM(NO)_2R$ compounds (R = H, alkyl, aryl) also undergo reversible one-electron reductions to produce the corresponding 19-electron radical anions *(194,197)*,

$$CpM(NO)_2R \underset{}{\overset{+e^-}{\rightleftharpoons}} [CpM(NO)_2R]^{\cdot -} \qquad (4.12)$$

with a substantial delocalization of the extra electron density onto the $M(NO)_2$ fragment *(194,197)*. In contrast, these $CpM(NO)_2R$ compounds (R = halide, alkyl, aryl) exhibit irreversible oxidation behavior *(194,196)*, and the reaction of $Cp'M(NO)_2Cl$ with silver salts in the presence of Lewis basses (L) results in simple metathesis and the generation of the 18-electron $[Cp'M(NO)_2L]^+$ species *(56,58)*. These latter species may also be obtained by the reaction of $Cp'M(NO)(CO)_2$ with NO^+ to yield initially $[Cp'M(NO)_2CO]^+$, a process followed by CO displacement by other Lewis bases (M = Mo, W) *(21,59)*.

When the $Cp'M(NO)_2Cl$ complexes are reacted with silver salts of weakly coordinating anions such as BF_4^-, PF_6^-, or $N(SO_2F)_2^-$ in the absence of coordinating ligands, the $Cp'M(NO)_2Y$ compounds (Y = weakly coordinating anion) are produced *(56-58,198)*. Nevertheless, these $Cp'M(NO)_2Y$ complexes behave as functional equivalents of the 16-electron $[Cp'M(NO)_2]^+$ cations *(57)*.

Interestingly, the Cl atom in $CpCr(NO)_2Cl$ is metathesized by $[CpCo(NO)]^-$ to produce a heterobimetallic complex *(70)*.

$$CpCr(NO)_2Cl + Na[CpCo(NO)] \longrightarrow CpCo(\mu\text{-}NO)_2Cr(Cp)(NO) \quad (4.13)$$

The Cr-R bonds in $CpCr(NO)_2R$ may be cleaved by strong acids to generate $[CpCr(NO)_2]^+$ *(56)*. Curiously, formaldoxime $N(CH_2)OH$ is sufficiently activated by the $[CpCr(NO)_2]^+$ cation to effect the protolytic cleavage of $CpCr(NO)_2CH_3$ to produce a bimetallic formaldoximato complex *(199)*.

$$[CpCr(NO)_2\{N(OH)=CH_2\}]^+ \xrightarrow[-\,CH_4]{CpCr(NO)_2CH_3} [CpCr(NO)_2\{N(=CH_2)O\}Cr(NO)_2Cp]^+ \quad (4.14)$$

This reaction is of interest since free oximes are not acidic enough to effect this cleavage of the $Cr\text{-}CH_3$ bond.

Furthermore, the Cr-R bonds in these $CpCr(NO)_2R$ complexes are prone to undergo insertion by neutral electrophiles such as SO_2 and TCNE *(200)*, or cationic electrophiles such as NO^+ and ArN_2^+ *(196)* (Section 7.4.2.5). Diazomethane also inserts into the Cr-Cl bonds of $Cp'Cr(NO)_2Cl$ to result in the formation of halomethyl complexes *(201)*.

Only a few unsaturated alkyl or heteroatom-substituted groups for the $Cp'M(NO)_2R'$ complexes are known, and these include the cases where R' = propargyl *(202)*, alkenyl ketone *(179,203)*, trifluoroacetate *(204)*, *p*-toluenesulfonate *(58)*, nitrate *(56)*, nitrite *(205,206)*, η^1-Cp *(207)* and carbamoyl *(56)*.

None of the $Cp'M(NO)_3$ complexes of the Group 6 metals (which would contain one bent and two linear NO ligands) are known to date. However, the related dinitrosyl-thionitrosyl complex $[CpCr(NO)_2NS]^{2+}$ has been obtained via the reaction described in eq 4.15 *(208)*.

$$[CpCr(NO)_2NSF]AsF_6 + AsF_5 \longrightarrow [CpCr(NO)_2NS](AsF_6)_2 \quad (4.15)$$

A nitrosyl-aryldiazonium compound $CpMo(NO)(N_2Ph)Cl$ has been briefly mentioned in the literature *(209)*, but is reported to be unreactive.

4.3.4: Manganese and Rhenium

The Cp'M(NO)(CO)X complexes are known for manganese (X = I) and rhenium (X = Cl, Br, I) *(90,210,211)*. The related thiocarbonyl (MeCp)Mn(NO)(CS)I *(212)* and cyano $[\{CpMn(NO)(PPh_3)CN\}_2(\mu\text{-}Ag)]^+$ *(123)* complexes have also been prepared. The CO ligands in the Cp'Mn(NO)(CO)I compounds are replaceable by soft Lewis bases (L) to produce Cp'Mn(NO)(L)I, whereas with hard Lewis bases only the bimetallic $Cp'_2Mn_2(NO)_3I$ complexes are formed *(90)*. The CpMn(NO)(dithiolene) and related complexes are also produced when the $[CpMn(NO)(CO)_2]^+$ cation is reacted with mono- and di-anionic bidentate sulfur ligands *(213)*.

Surprisingly, alkylation of $Cp'Mn(NO)(PPh_3)I$ (Cp' = Cp, MeCp) produces the new 17-electron $Cp'Mn(NO)R_2$ dialkyl complexes (R = Me, Et, Pr) rather than the expected $Cp'Mn(NO)(PPh_3)R$ compounds *(214)*.

Unlike the surprising absence of Mn complexes of the type Cp'M(NO)(L)R (R = alkyl, aryl), many such complexes are known for rhenium.

Reduction of a CO ligand in $[CpRe(NO)(CO)_2]^+$ produces the formyl complex CpRe(NO)(CO)(CHO) or the hydroxymethyl complex $CpRe(NO)(CO)(CH_2OH)$, depending of the choice of reducing agent *(215,216)*. Many analogous formyl and acyl complexes of this type are known for rhenium *(89,217-219)*, and interestingly, the CpRe(NO)(CO)(CHO) formyl complex acts as a hydride-transfer agent to coordinated carbonyls in what are essentially transformylation reactions. For example, $[CpMn(NO)(CO)_2]^+$ is transformed to CpMn(NO)(CO)(CHO) when reacted with the Re-formyl complex *(217)*.

Reduction of the Re-formyl complex with BH_3 produces the methyl complex CpRe(NO)(CO)Me, whereas reduction with triethylborohydride produces the anionic bis-formyl complex *(215)*.

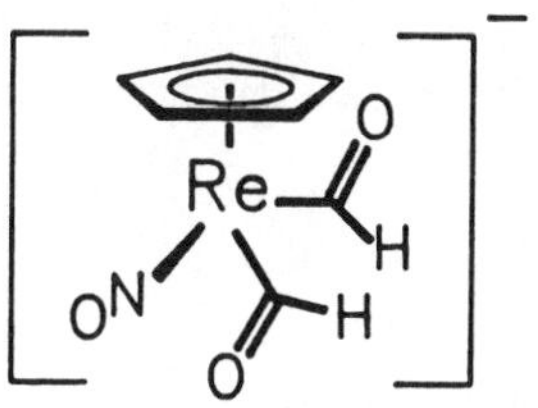

The reduction chemistry of the coordinated CO group in the $[Cp'Re(NO)(CO)L]^+$ cations has been reviewed *(215,216)*.

The oxidative cleavage of the Re-R bonds in $Cp'Re(NO)(PPh_3)R$ compounds proceeds with retention of configuration at rhenium *(220)* and involves the formation of cationic alkyl-hydride complexes as shown in Scheme 4.1 *(221)*. Stable dihydrides and alkyl hydrides of rhenium are isolable, and these include the cases where R = H and CH_2Ph. Reaction of the $Cp'Re(NO)(PPh_3)R$ compounds with trityl cation results in the formation of carbene or olefin complexes (vide infra).

Treatment of CpRe(NO)(CO)Me with trimethylphosphine does not result in CO insertion, but rather results in a hapticity change of the Cp ligand from η^5 to η^1 to produce $(\eta^1\text{-Cp})Re(NO)(PMe_3)_2Me$,

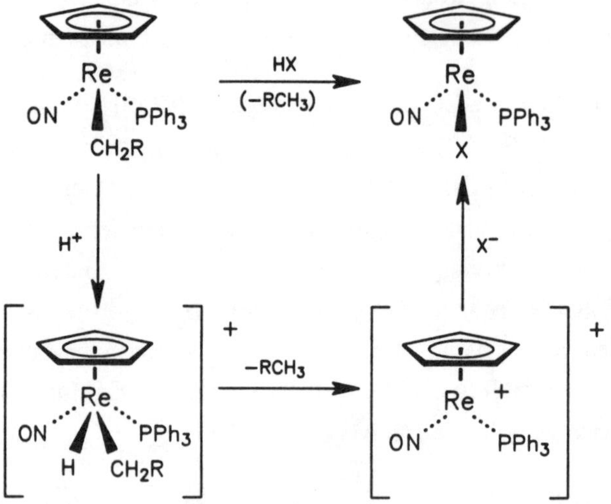

Scheme 4.1

whose solid-state structure has been determined *(12)*. However, a similar reaction with PBu_3 (at 100 °C) does indeed induce CO insertion to generate $CpRe(NO)(PBu_3)(COMe)$ as the major product and $CpRe(NO)(PBu_3)Me$ as the minor product *(12)*. Other compounds of the form $Cp'Re(NO)(PPh_3)R'$, where R' = vinyl *(222,223)*, acetylide *(224)*, allyl *(225)* are also known, as are heterobimetallic μ-carboxylate complexes of Ge, Sn and Pb *(74)*. Rhenium amide (NR_2) and phosphide (PR_2) complexes of this general form have also been synthesized and contain very basic N and P atoms, respectively *(83-85,226,227)*.

The manganese α-ketoacyl complex shown below is produced via a novel nitrite-induced carbyne-to-acyl conversion.

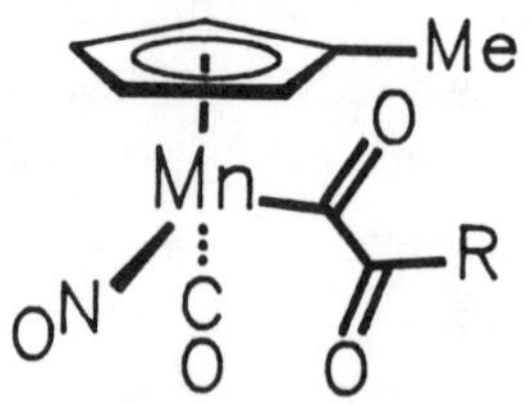

(R=Ph, tol)

It exhibits rich reactivity towards electrophiles, nucleophiles and alkynes *(228,229)*. Other $Cp'M(NO)(L)X$ complexes of manganese and rhenium, where X is carbamoyl (C(O)NHR) *(69,230)*, acyl *(231-233)*, allyl *(234)*, $C(O)PR_2$ *(235)* and $C(PR)OSiMe_3$ *(236)* have been prepared by various routes.

4.3.5: Iron and Ruthenium

The $Cp'M(NO)X_2$ dihalide complexes of iron and ruthenium serve as convenient precursors to their corresponding $Cp'M(NO)R_2$ compounds (R = alkyl, aryl) *(114,237-239)*. Methylation of the $[CpFe(NO)]_2^{\cdot -}$ radical anion yields the dimeric $[CpFe(NO)Me]_2$ complex, and thermolysis of this monoalkyl dimer yields $CpFe(NO)Me_2$ as one of the products *(240)*.

Thermolysis of the $Cp'M(NO)R_2$ complexes in the presence of phosphines leads to intramolecular insertions of the bound NO ligands (Section 7.3.3.2) *(237,238,240)*. A reductive elimination of the σ-bound alkyl or aryl groups may also occur when the $Cp'M(NO)R_2$ compounds are thermolyzed *(114,238)*, i.e.,

$$Cp'M(NO)R_2 \xrightarrow[\Delta]{} \text{"}Cp'M(NO)\text{"} + R\text{-}R \qquad (4.16)$$

The $[CpRu(NO)(PPh_3)Cl]^+$ cation has been synthesized from the reaction of $CpRu(PPh_3)_2Cl$ with nitrite *(112)*, and the electrolytic reduction of the related $(\eta^5\text{-}C_5Ph_5)Ru(NO)(L)Br]^+$ complex (L = phosphine) generates the neutral $(\eta^5\text{-}C_5Ph_5)Ru(NO)X_2$ complexes. Reduction of the acyl compounds $[(\eta^5\text{-}C_5Ph_5)Ru(NO)(L)(COMe)]^+$ yields the neutral 19-electron $(\eta^5\text{-}C_5Ph_5)$-Ru(NO)(L)(COMe) complexes, whose ESR spectra reveal extensive localization of the unpaired electron on the NO ligands of these complexes *(111)*. The related α-ketoacyl compound $[CpFe(NO)(CO)\{C(O)C(O)Me\}]^+$ has been obtained by an entirely different route *(241)*.

As mentioned earlier, reactions of diazoalkanes with the $[Cp'M(NO)]_2$ dimers of Fe and Ru produce methylidene-bridged complexes *(105-107)*. Treatment of one such compound, namely $[CpFe(NO)]_2(\mu$-

CHMe), with the trityl cation produces a bridging vinyl complex as shown *(242,243)*.

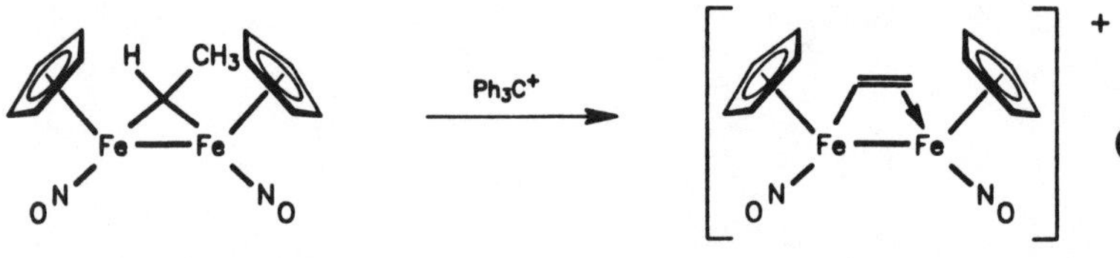

(4.17)

4.3.6: Cobalt and Rhodium

The Cp'Co(NO)X compounds are prepared from the reactions of the parent $[Cp'Co(NO)]_2$ dimers with the appropriate halogen *(244,245)*. These monohalide complexes are monomeric in solution and in the vapor phase, but are associated in the solid state via NO bridges. Curiously, the Cp*Co(NO)Cl complex undergoes spontaneous decomposition to the binuclear $[Cp^*CoCl]_2(\mu\text{-NO})(\mu\text{-Cl})$ complex, whose long Co-Co separation of 3.108 Å is indicative of the absence of a metal-metal bond *(244)*. The alkyl derivatives CpCo(NO)R are best obtained by the reaction of the nitrosylcobaltate $[CpCo(NO)]^-$ with alkyl halides, and these alkyl derivatives undergo intramolecular NO insertion reactions into their Co-R bonds *(70,246)* (Section 7.3.3.2). The related $CpCo(NO)M'X_3$ compounds, where M' = Ge or Sn, and X = halide or alkyl are also known *(70,247)*.

The rhodium complex $(\eta^5\text{-}C_5Ph_5)Rh(NO)I$ readily loses the iodide ligand to form the bimetallic $[(\eta^5\text{-}C_5Ph_5)Rh(NO)]_2$ compound *(118)*.

4.4: Organometallic Nitrosyl Hydrides

The organometallic nitrosyl hydrides known to date belong to the Group 6, Group 7 or Group 8 metals, and representative examples of these are presented in Table 4.2.

4.4.1: Molybdenum and Tungsten

Even though $CpW(NO)_2H$ contains two strong π-acceptor NO ligands, its

Table 4.2 Some Organometallic Nitrosyl Hydrides

Compound	δ_H (ppm)	ν_{NO} (cm^{-1})	Ref
$CpW(NO)_2H$	2.27	1718, 1632	*248*
$[CpW(NO)H_2]_2$	6.99, -2.05	1599	*250*
$Cp^*W(NO)H(CH_2SiMe_3)(PMe_3)$	-1.25	1545	*172*
$CpRe(NO)(CO)H$	-8.2	1722	*251*
$Cp^*Re(NO)(CO)H$	-7.59	1699	*252*
$(MeCp)Mn(NO)(PPh_3)H$	-6.48	1683	*253*
$[CpRe(NO)(PPh_3)(GePh_3)H]^+$	-1.58	1760	*254*
$[\{Cp^*Re(NO)(CO)\}_2(\mu\text{-}H)]^{+a}$	-13.54 -12.33		*252*
$(PhC_2)Os(NO)(H)(PPr_3)_2Cl$	-0.72	1795	*4b*

[a] 2 isomers

chemistry is dominated by its hydridic behavior *(248)*. [The Cp* analogue is also known *(58)*.] This hydridic behavior has been rationalized by considering that the dissociation of H^+ from $CpW(NO)_2H$ would generate the unstable conjugate base $[CpW(NO)_2]^-$, whereas the loss of H^- (in the presence of solvent) would generate the stable $[CpW(NO)_2(solv)]^+$ solvated species *(249)*. In any event, the $CpM(NO)_2H$ complexes form 1:1 adducts with the corresponding $[CpM(NO)_2]^+$ cations via a bridging hydride ligand (M = Mo, W) *(255)*.

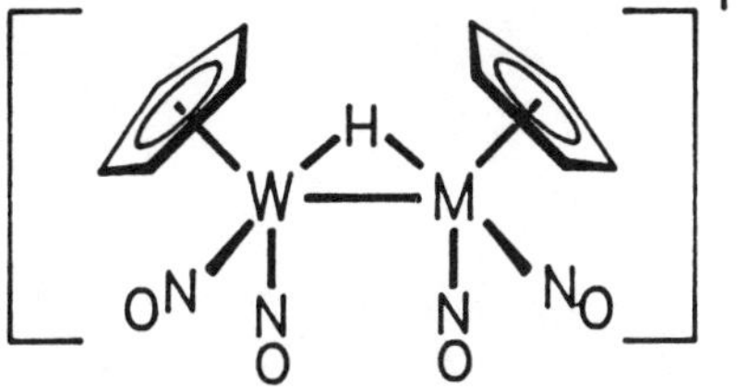

+
H
W
M
ON
N
O
N
O
NO

or with $Cr(CO)_5(THF)$ to form the $CpW(NO)_2H.Cr(CO)_5$ adduct which is held together by a W-Cr interaction *(255)*. The $CpW(NO)_2H$ complex undergoes a reversible one-electron reduction to its corresponding 19-electron radical anion, which is isolable as its cobalticinium salt *(194)*.

A series of tungsten nitrosyl hydrides $[Cp'W(NO)(X)H]_2$ and Cp'W(NO)(X)H(L) have been synthesized (X = H, I, alkyl; L = phosphine or phosphite) *(253,256)*. The dimeric hydrides contain $W(\mu\text{-}H)_2W$ bridges which are best viewed as "fused" single units held together by delocalized bonding, and are formulated as

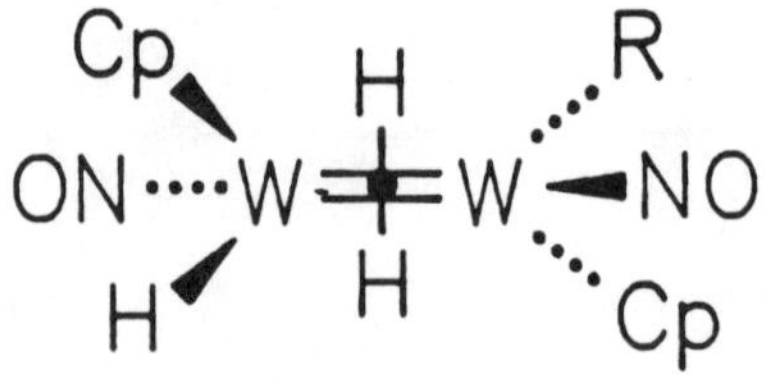

to reflect this delocalization *(250)*. Such a bridge system has been determined (with the aid of molecular orbital calculations) to be composed of an overall four-center, four-electron interaction which results from strong delocalized W-(μ-H)-W interactions *(257)*.

The alkyl hydride complexes $Cp^*W(NO)(CH_2SiMe_3)H(PMe_3)$ are thermally stable, and activate C-H bonds both intra- and intermolecularly *(172)*.

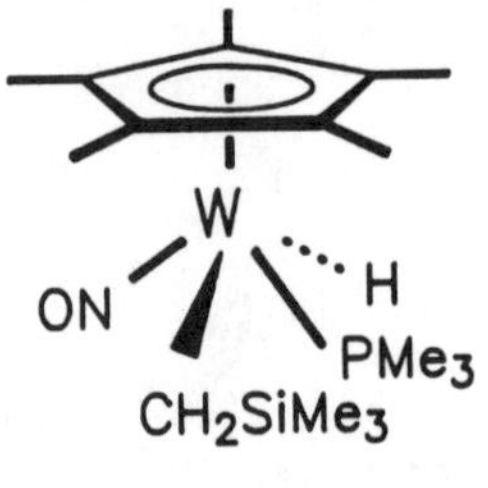

Thus, although the non-bonded separation between the Cα carbon and the hydride ligand is 3.80 Å in the solid state, thermolysis of this complex in benzene results in the elimination of $SiMe_4$ and intermolecular C-H activation of the solvent to produce $Cp^*W(NO)(Ph)H(PMe_3)$ *(172)*.

4.4.2: Manganese and Rhenium

The rhenium hydride CpRe(NO)(CO)H was first prepared by the reaction of $[CpRe(NO)(CO)_2]^+$ with triethylamine in a water-acetone mixture *(251)*. [The manganese analogue, $(MeCp)Mn(NO)(PPh_3)H$, has also been synthesized *(253)*.] Interestingly, protonation of Cp*Re(NO)(CO)H at 195 K affords the dihydrogen complex $[Cp^*Re(NO)(CO)(\eta^2\text{-}H_2)]^+$ and the dihydride complex $[Cp^*Re(NO)(CO)(H)_2]^+$ *(252)*. The ratio of these products is 93:7 at 225 K. Analogues of the dihydride complex (with PPh_3 in place of CO in the complex) have also been obtained, and related alkyl hydride complexes are also known *(221)*.

The deprotonation of $CpRe(NO)(PPh_3)H$ proceeds in two steps *(258)*:

$$CpRe(NO)(PPh_3)H \xrightarrow[-78^\circ]{BuLi} (LiCp)Re(NO)(PPh_3)H \xrightarrow{-32^\circ} Li^+[(HCp)Re(NO)(PPh_3)]^- \qquad (4.18)$$

It has been observed that the less acidic Cp proton (pK_a = 36) is abstracted kinetically (pK_a of Re-H = 26 - 30), an observation that has been substantiated by molecular orbital calculations *(259)*. Indeed, the relative rates of Cp deprotonation in various $CpRe(NO)(PPh_3)R$ complexes are in the order halide, acyl > alkyl $\geq$ silyl or hydride, and this trend correlates well with the electron-withdrawing natures of these R ligands *(259-262)*. In the chloro complexes (i.e., R = Cl) double deprotonation occurs to yield the 1,3-dilithiocyclopentadienyl ligand attached to the rhenium center *(260)*. In any event, when a solution of the $(LiCp)Re(NO)(PPh_3)H$ complex shown in the middle of eq 4.18 is warmed, an intramolecular H-migration from Re to Cp occurs *(258)*. [The trimethylsilyl group is also known to migrate from

Re to Cp in the carbonyl analogue *(263)*.] Curiously, deprotonation of the related $CpRe(NO)(PPh_3)CH_2CN$ complex occurs at the ligating carbon of the CH_2CN group *(264)*.

The secondary phosphine complex $[CpRe(NO)(CO)(PR_2H)]^+$ reacts with $Pt(PPh_3)_4$ to form the terminal hydride complex $[CpRe(NO)H(\mu\text{-}PR_2)Pt(PPh_3)_2]^+$, which converts to the bridging hydride $[CpRe(NO)(\mu\text{-}H)(\mu\text{-}PR_2)Pt(PPh_3)_2]^+$ upon standing *(265)*. The manganese analogue of the latter complex has also been obtained via a similar route *(266)*.

4.4.3: Ruthenium

Bimetallic hydrides of ruthenium have resulted from the protonation of bridging methylidene complexes (R = H, Me) as shown *(105)*.

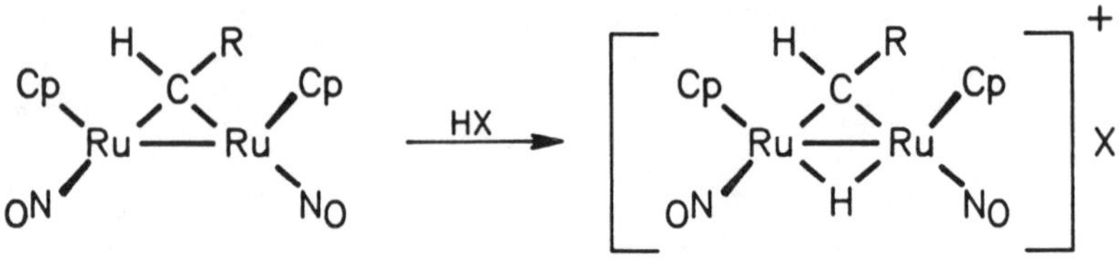

4.5: Organometallic Nitrosyl Carbenes

4.5.1: Chromium, Molybdenum and Tungsten

Mononitrosyl carbene complexes of all three Group 6 metals can be made by nucleophilic addition to the coordinated CO ligand of their dicarbonyl nitrosyl precursors *(267)*, e.g.,

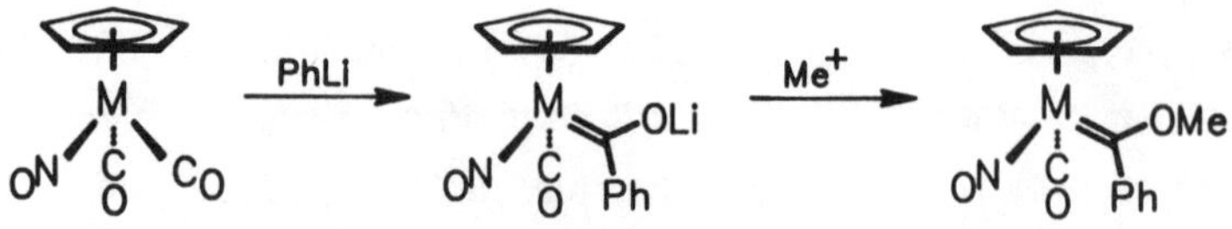

(4.19)

In the case of Mo, the intermediate carbene complex also reacts with the $[CpFe(CO)_2(THF)]^+$ cation to result in the production of a heterobimetallic bridging acyl complex as shown in Scheme 4.2 *(268)*.

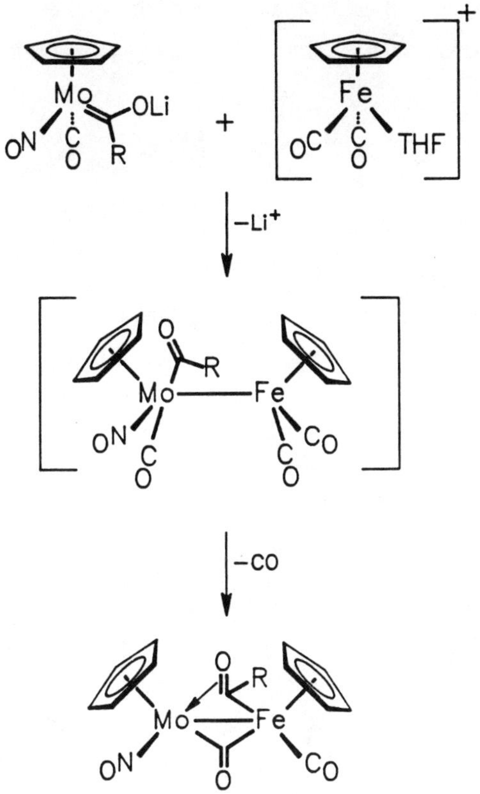

Mo, OLi, ON, C, O, R + [Fe, OC, C, O, THF]$^+$ —(−Li$^+$)→ [Mo—Fe, O, R, ON, C, O, C, O, CO] —(−CO)→ Mo—Fe, O, R, ON, O, CO

Scheme 4.2

The diphenylcarbene complex of chromium shown below has been studied by X-ray and spectroscopic methods *(269)*.

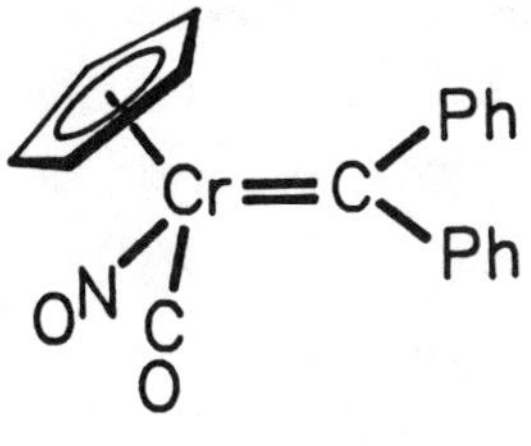

Cr=C, Ph, Ph, ON, C, O

In this complex, the $Cr{=}CPh_2$ conformation is locked in its position due to a restricted Cr-carbene rotation. Other electron-rich carbene complexes of the Group 6 metals are also known *(151,270)*.

A dinitrosyl carbene complex has been obtained by alkoxide attack on a coordinated isocyanide ligand followed by the addition of acid *(56)*.

$$[CpCr(NO)_2(CNMe)]^+ \xrightarrow[\text{(ii) } H^+]{\text{(i) } OR^-} [CpCr(NO)_2{=}C(OR)(NHMe)]^+ \qquad (4.20)$$

R = Me, Et

Dinitrosyl carbene complexes of molybdenum of the form $[(AlCl_2)_2(\mu\text{-}OR)_2Mo(NO)_2({=}CHMe)]$ are generated during the reaction of $Mo(NO)_2(OR)_2$ with $EtAlCl_2$ or $Et_4Sn/AlCl_3$ *(271)*.

4.5.2: Manganese and Rhenium

Cationic manganese carbene complexes are obtained by aziridine or oxirane addition to a coordinated CO group in $[CpMn(NO)(CO)_2]^+$ *(272)*, i.e.,

$$[M{-}C{\equiv}O]^+ + X\triangleleft \longrightarrow [M{=}\overline{CO(CH_2)_2X}]^+ \qquad (4.21)$$

(X = NH, O)

Very high stereospecifities have been observed in the hydride-abstraction reactions of the pseudooctahedral $Cp'Re(NO)(PPh_3)R$ complexes by Ph_3C^+ to form the corresponding carbene (α-H abstraction) or olefin (β-H abstraction) complexes. Extended Hückel calculations performed on the model compound $CpRe(NO)(PPhH_2)R$ reveal that in general, for R groups with a single α-substituent (e.g., CH_2Me, CH_2Ph), the preferred conformation is such that the $C\alpha$ substituent lies between the Cp and the NO ligand *(273)*.

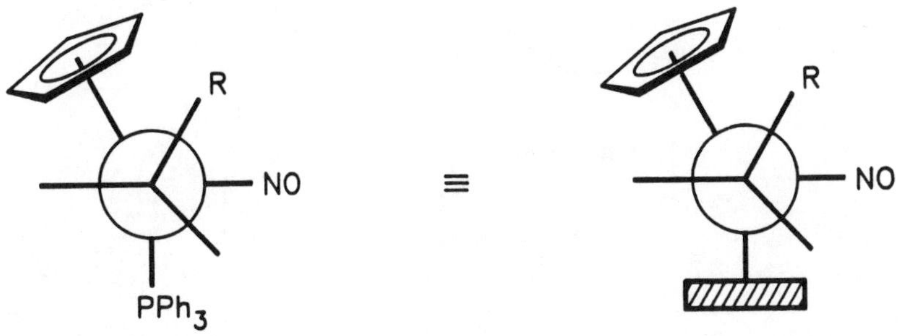

This is because one Ph group of the PPh_3 ligand lies in a plane close to (3 - 4 Å) and parallel to the plane containing the Cα-Re-NO group as shown *(273)*. For less sterically demanding substituents (e.g., Me) the less stable conformation where R lies between the Cp and PPh_3 ligands is also accessible. Due to its steric bulk, the trityl cation could only approach an α-hydrogen that is located between the Cp and the NO ligands (likely) or the Cp and PPh_3 ligands (less likely), e.g.,

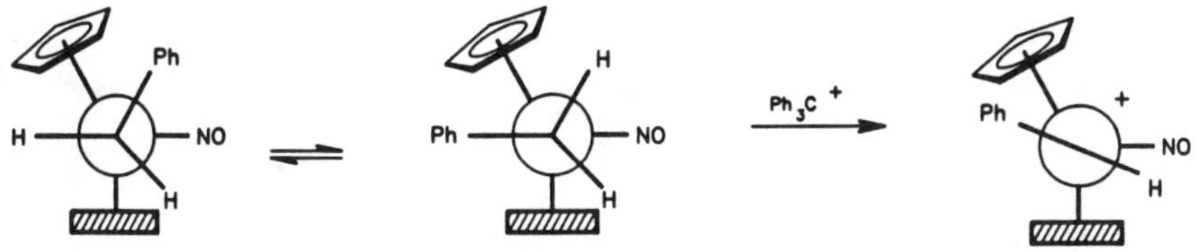

For example, when R is CH_2Ph, the *kinetic* product is that in which the Ph group is located "away" from the NO ligand *(274-276)*. When the substituent is mesityl, however, this group is locked in between the Cp and the NO ligands, and the H atom located between the Cp and the PPh_3 ligands is the one that is abstracted to give the product in which the mesityl is located "cis" to the nitrosyl.

When the R group contains two α-substituents (e.g., $CHMe_2$), the two substituents straddle the Cp ligand *(273)* as shown:

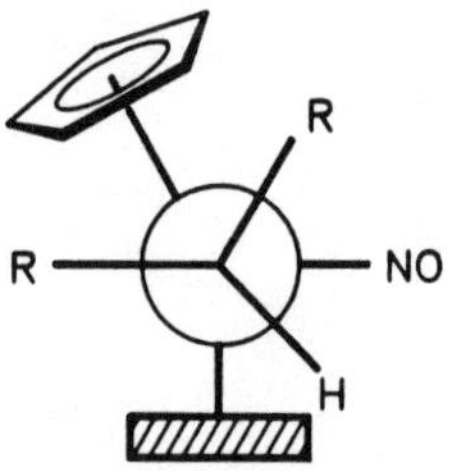

In this case, the α-H is not accessible to Ph_3C^+, thus β-H abstraction occurs to yield olefin complexes *(277)*. The product distributions observed in the reaction of trityl cation with various $CpRe(NO)(PPh_3)R$ complexes are shown in Table 4.3.

Table 4.3 Summary of Regiochemistry of H-abstraction by Trityl Cation

Re-R	α-abstraction (%)	β-abstraction (%)
$Re\text{-}CH_2CH_3$	100	0
$Re\text{-}CH_2CH_2CH_3$	100	0
$Re\text{-}CH_2CH_2Ph$	63	36
$Re\text{-}CH_2CHMe_2$	0	100
$Re\text{-}CHMe_2$	0	100

Data taken from reference 277.

A mechanistic study of such reactions reveals that the overall H-abstraction reaction probably proceeds by initial electron transfer from the Re-R complex to the trityl cation, followed by H-atom transfer *(278,279)*, i.e.,

$$Re\text{-}R + [Ph_3C]^+ \rightleftharpoons [Re\text{-}R]^{\cdot +} + Ph_3C\cdot$$

Nevertheless, the carbene complexes produced display a rich and varied chemistry as shown in Scheme 4.3. Thus, anionic or neutral bases (e.g., thioethers) coordinate to the methylene carbon to yield cationic Re-CH_2L complexes *(89,280)*.

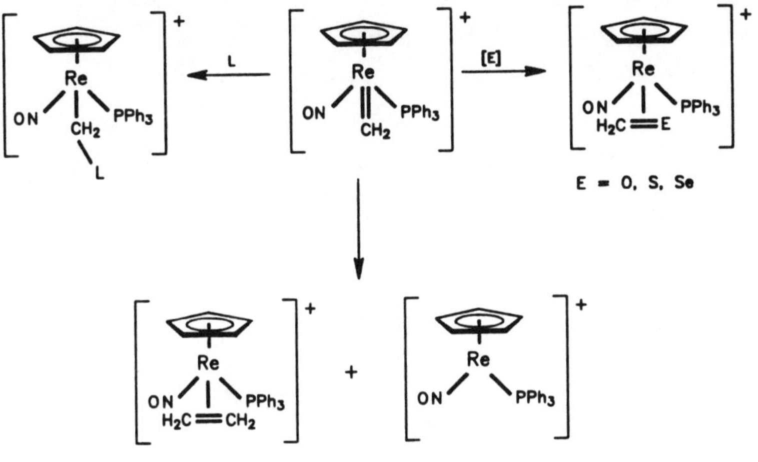

Re, ON, PPh_3, CH_2, L, [E], $H_2C=E$, E = O, S, Se, $H_2C=CH_2$

Scheme 4.3

Reactions with *O*-, *S*- or *Se*-donors (E) result in the formation of formaldehyde, thioformaldehyde and selenoformaldehyde ligands (i.e., η^2-$E=CH_2$) respectively attached to the rhenium center *(281-283)*, whereas disproportionation of the methylene to coordinated ethylene may also occur *(284)*. A remarkable enantiomer self-recognition is observed in the latter disproportionation reaction,

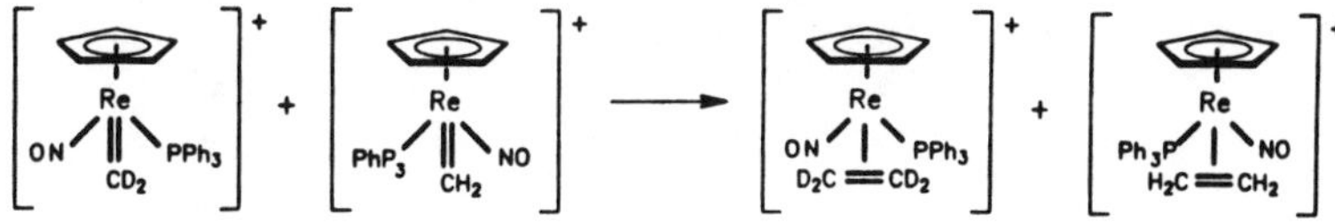

Re, ON, PPh_3, CD_2, PhP_3, NO, CH_2, $D_2C=CD_2$, Ph_3P, $H_2C=CH_2$ (4.23)

Reaction of Cp*Re(NO)(PPh_3)Me with $[Cp_2Fe]^+$ results, effectively, in the oxidative disproportionation of the coordinated methyl group to coordinated ethylene and methane *(285)*. Related germylene *(254)*, silylene *(286)* and vinylidene *(224)* complexes containing the $Cp'Re(NO)(PPh_3)$ fragment have also been prepared.

4.5.3: Iron, Ruthenium and Osmium

Electron-rich olefins may be employed as sources of the carbene group for the generation of various mono- and bis-carbene complexes of the Group 8 metals *(287-289)*. Monocarbene complexes of the form $M(NO)(=CR_2)(PPh_3)_2Cl$ (R = H, F) for ruthenium *(290)* and osmium *(4,291)* are also known. The dinitrosyl carbene complex shown below has been obtained by a suitable modification of a coordinated carbonyl ligand in $Fe(NO)_2(CO)_2$ *(292)*.

$$Fe(CO)_2(NO)_2 \xrightarrow[\text{(ii) } Et_3O^+]{\text{(i) } Ph_3Si^-} (ON)_2(OC)Fe{=}C(OEt)(SiPh_3) \qquad (4.24)$$

4.5.4: Nickel

Both PPh_3 ligands are lost from $Ni(NO)(PPh_3)_2X$ when this nitrosyl complex is reacted with electron-rich olefins *(287)*. In these reactions, the products are the bis-carbene complexes $Ni(NO)(=CR_2)_2X$, where CR_2 = $C(NRCH_2)_2$.

4.6: Olefins and Acetylenes

4.6.1: Chromium, Molybdenum and Tungsten

Photoinduced substitution of a CO ligand in $Cp'M(NO)(CO)_2$ (M = Cr, Mo) in the presence of olefins yields the corresponding $Cp'M(NO)(CO)(\eta^2$-olefin) complexes *(52,293,294)*. The $CpM(NO)(CO)(\eta^2$-olefin) complexes of Mo and W may also be obtained from their allyl precursors *(171,185,295)* as described in Section 4.7.1, and the related $CpW(NO)(L)(\eta^2$-acetylene) analogues (L = CO, PPh_3) are also obtainable by other routes *(186)*.

The only known cyclopentadienyl dinitrosyl olefin complexes of the Group 6 metals are the $[CpM(NO)_2(\eta^2\text{-cyclooctene})]^+$ cations of Mo and W *(58,255)*. Other unsaturated organic groups undergo subsequent transformations in the metals' coordination spheres (Section 7.4.2.2).

4.6.2: Rhenium

Olefins and acetylenes bind to the coordinatively unsaturated $[CpRe(NO)(PPh_3)]^+$ cation *(296-298)*. The olefin complexes undergo stereospecific deprotonations to yield the corresponding vinyl derivatives, whereas the terminal-acetylene complexes undergo deprotonations to produce the corresponding acetylide complexes or undergo a thermal rearrangement to generate the vinylidene complexes. The CpRe(NO)(L)Ph phenyl complexes undergo protonation to generate η^2-benzene compounds as shown in eq 4.25 *(299,300)*. The η^2-triphenylmethane and the η^2-cycloheptatriene analogues of the complexes where L = CO have been obtained by the reactions of CpRe(NO)(CO)H with trityl and tropylium cations, respectively *(301)*.

$$CpRe(NO)(CO)(C_6H_4R) \underset{Et_3N}{\overset{H^+}{\rightleftharpoons}} [CpRe(NO)(CO)(\eta^2\text{-}C_6H_5R)]^+ \quad (4.25)$$

(R=H, Me, CF_3)

The chloro-bridged $[Re(NO)(CO)(cyclooctene)Cl_2]_2$ complex is obtained when $[Re(NO)(CO)_2Cl_2]_2$ is reacted with cyclooctene *(302)*.

4.6.3: Iron, Ruthenium and Osmium

UV photolysis of $Fe(NO)_2(CO)_2$ dissolved in liquid xenon (and doped with 1-butene) affords $Fe(NO)_2(CO)(\eta^2\text{-1-butene})$ and $Fe(NO)_2(\eta^2\text{-1-butene})_2$, which are observable only at low temperatures *(303)*. A similar reaction employing 1,3-butadiene forms $Fe(NO)_2(CO)(\eta^2\text{-}C_4H_6)$, $Fe(NO)_2(\eta^4\text{-}C_4H_6)$ and $Fe(NO)_2(\eta^2\text{-}C_4H_6)_2$ *(304)*. Other Fe-olefin complexes are generally obtained from their allyl precursors *(305,306)* (Section 4.7.2). Tetrafluoroethylene and hexafluorobut-2-yne form adducts with $Ru(NO)L_2Cl$ (L = phosphine) to generate the corresponding olefin and acetylene complexes, respectively *(307)*. However, the Os-ethylene complex $[Os(NO)(CO)(PPh_3)_2(C_2H_4)]^+$ is obtained by halide abstraction from $Os(NO)(CO)(PPh_3)_2Cl$ with a silver salt and then further reaction with ethylene *(308)*. The Os-acetylene analogues are obtained via similar routes *(4b,309)*.

The ylide complexes $M(NO)(CH_2PPh_3)(\eta^2\text{-}C_2F_4)(PPh_3)Cl$ of ruthenium and osmium are obtained by treatment of their carbene precursors with the olefin *(290)*, i.e.

$$M(NO)(=CH_2)L_2Cl + C_2F_4 \longrightarrow M(NO)(CH_2L)(\eta^2\text{-}C_2F_4)(L)Cl \quad (4.26)$$

$$(L = PPh_3)$$

4.6.4: Cobalt, Rhodium and Iridium

Many olefin complexes of the Group 9 metals possess the general formula $M(NO)L_2$(olefin), and examples include $Co(NO)(CO)_2(\eta^2$-butene) *(303)*, $Co(NO)(CNR)_2(TCNE)$ (R = aryl) *(310)*, $[Ir_2(NO)(\mu\text{-pz})_2(COD)_2]^+$ *(311)*, and the $M(NO)(PPh_3)_2(\eta^2$-olefin) complexes for all three metals *(307,312)*. The cyclopentadienyl complexes $[CpM(NO)(\eta^2\text{-}C_2H_4)]^+$ for Co and Rh have been obtained by the reaction of the precursor $[Cp_2M_2(CO)_2(\mu\text{-}NO)]^+$ complexes with ethylene *(313)*. Analogous cyclooctene complexes are also obtained by this route. The iridium analogue $[Cp^*Ir(NO)(\eta^2\text{-}C_2H_4)]^+$ has also been obtained from the reaction of $Cp^*Ir(\eta^2\text{-}C_2H_4)_2$ with NO^+ *(127)*.

The cationic $[(\eta^4\text{-COD})M(NO)L_2]^+$ complexes of rhodium (L = $\{(NH)_2C_{10}H_6\}Rh(COD)$) and iridium (L = $(pz)_2Ir(COD)$) each contain a bent NO ligand *(314,315)*.

4.7: Allyls

4.7.1: Molybdenum and Tungsten

The electronic asymmetry induced by the substitution of the two CO ligand in $CpM(CO)_2(\eta^3\text{-allyl})$ by nitrosyl and iodide ligands is known to cause severe distortions in the allyl moiety from the symmetrical π mode in the dicarbonyl towards the unsymmetrical σ-π mode in the nitrosyl iodide complexes *(316,317)*.

Symmetrical π Unsymmetrical $\sigma-\pi$

In the latter formulation, the sigma bond forms at the terminus *cis* to the NO ligand, and the C-C bond of the allyl that has more double-bond character is situated *trans* to the NO ligand. The nitrosyl iodide complexes of Mo are obtained by reaction of iodide with the precursor $[CpMo(NO)(CO)(allyl)]^+$ complexes, a reaction that occurs with retention of configuration at the metal center *(318)*. Nucleophilic attack on these $[CpMo(NO)(CO)(\eta^3\text{-allyl})]^+$ complexes to form $CpMo(NO)(CO)(\eta^2\text{-olefin})$ occurs regioselectively; in each case, the new nucleophile-carbon bond is directed by (and forms *cis* to) the nitrosyl group *(319-322)*. Such a desirable property of these Mo-allyl cations has been used for specific transformations of coordinated allyls such as allylic alkylations *(323,324)* or carbanion addition *(319)*, and in the stereoselective formation of substituted tetrahydropyrans by the reaction of the Mo-allyl cations with aldehydes *(325,326)*. Reaction with the carbonylmetallates $[M'(CO)_5]^-$ (M' = Mn, Re) leads to σ,π-allyl-bridged complexes $CpMo(NO)(CO)(\mu\text{-allyl})M'(CO)_5$ *(326)*. A brief summary of the reactions of Mo-allyl complexes is presented on the next page *(327,330)*.

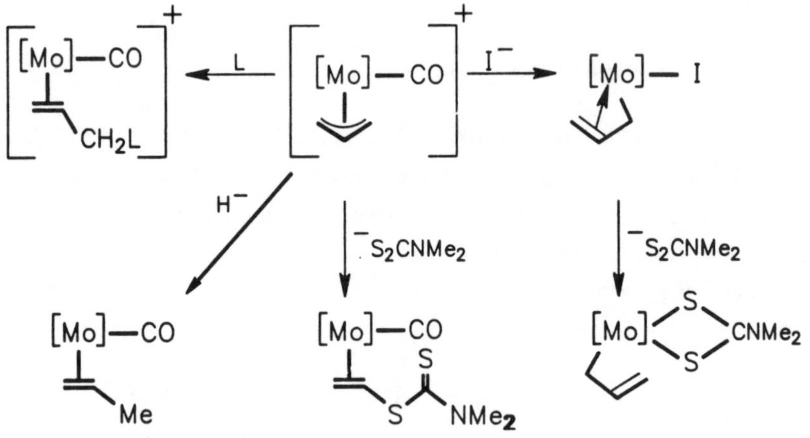

[Mo] = CpMo(NO)

4.7.2: Iron and Ruthenium

The (η^3-allyl)Fe(NO)L_2 compounds (L = CO, CNMe, phosphine) display some interesting chemistry *(17,331-333)*. For example, the reaction described by eq 4.23 results in the formation of a Zwitterionic adduct,

$$(\eta^3\text{-}C_3H_5)Fe(NO)(CO)_2 + L \longrightarrow (\eta^2\text{-}CH_2{=}CH_2CH_2L)^+Fe^-(NO)(CO)_2 \quad (4.27)$$

$$(L = PR_3)$$

via attack of the basic phosphine of the terminus of the allyl ligand *(306)*. With less basic phosphines (or phosphites) only carbonyl substitution occurs *(306)*. The general mechanism of the reaction is believed to involve the initial coordination of the incoming Lewis base (L) to form the five-coordinate (η^1-C_3H_5)Fe(NO)$(CO)_2$L compounds, which may then undergo allyl-L coupling or CO loss to yield the observed products. In some cases, CO insertion occurs into the Fe-(η^1-allyl) bond to yield but-3-enoyl complexes *(334)*. Chemical reduction, however, results in the loss of the allyl moiety from (η^3-allyl)Fe(NO)$(CO)_2$ *(335)*.

Treatment of (η^3-allyl)Fe(NO)(CO)L with NO^+ affords the cationic dinitrosyls $[(\eta^3\text{-allyl})Fe(NO)_2L]^+$, whose observed fluxionality involves slow allyl rotation at room temperature *(336)*. The reactions of these dinitrosyl allyl cations with nucleophiles are dominated by the loss of the allyl ligand, and this property has been exploited for the allylic alkylations of Fe- and Ru-complexed cyclopentadienyl or cyclooctatetraene groups *(337,338)*. The (η^3-allyl)Ru(NO)(PPh_3)$_2$ compound possesses a linear RuNO group, but addition of CO to this compound results in the bending of the nitrosyl ligand *(339)*.

The [(η^3-allyl)Fe(NO)$_2$]$_2$SnCl$_2$ compound has been obtained from the reaction of [Fe(NO)$_2$Cl]$_2$ and tetraallyltin *(340)*. This complex catalyzes the conversion of butadiene to 4-vinylcyclohexadiene (Section 7.4.2.2).

4.7.3: Iridium and Rhodium

The iridium complex [(η^3-C_3H_5)Ir(NO)(PPh_3)$_2$]$^+$ undergoes a facile linear-to-bent IrNO interconversion *(341)*. Interestingly, the linear IrNO bonding mode is found in the PF_6^- salt of this complex, whereas the bent IrNO bonding mode (bond angle of 129°) is found in the BF_4^- salt of this complex *(341)*. Both the Ir and Rh allyl complexes of the form [(η^3-allyl)M(NO)L_2]$^+$ react with CO to yield acrolein oxime *(341)*.

3.7.4: Cyclopropenyl and Related Complexes

Both the cyclopropenyl complexes (η^3-C_3R_3)Fe(NO)(CO)$_2$ and the η^3-oxocyclobutenyl complexes [η^3-C_3R_3(CO)]Fe(NO)(CO)$_2$ (R = H, alkyl, aryl) are obtained from the reaction of cyclopropenyl cations with [Fe(NO)(CO)$_3$]$^-$ *(342)*.

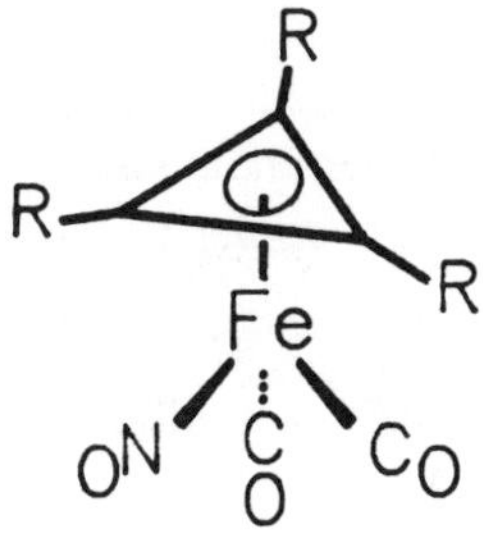

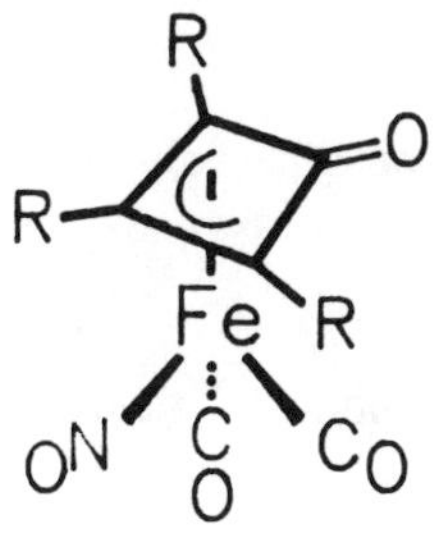

Carbonyl insertion into the η^3-cyclopropenyl ring to produce the oxocyclobutenyl derivatives is induced by reaction of the former complexes with phosphines *(343)*.

The related $[\{\eta^3\text{-}C_4H_4(PMe_3)\}Fe(NO)(CO)_2]^+$ cation is obtained by nucleophilic addition of the phosphine to the coordinated cyclobutadiene in $[(\eta^4\text{-}C_4H_4)Fe(NO)(CO)_2]^+$ *(344)*.

4.8: Dienes

4.8.1: Molybdenum

The Cp'Mo(NO)(η^4-diene) compounds are prepared by the chemical reduction of their diiodo precursors in the presence of the appropriate acyclic, conjugated diene *(295,345)*. Molecular orbital calculations of the model systems containing *cis*- and *trans*-butadiene bound to the CpMo(NO) fragment indicate that the HOMO energy of the *trans* complex is *ca.* 0.9 eV lower than that of the *cis* complex, a feature that results from the asymmetry of the bound *trans*-diene ligand *(346)*. Indeed, it appears that the *trans* complexes are the thermodynamically preferred complexes, the *cis*-complexes being only observable when the diene is 2,3-dimethylbutadiene *(295)*.

Proton abstraction from the cationic $[CpMo(NO)(CO)(\eta^3\text{-}C_8H_{13})]^+$ complex yields the η^2-cyclooctadiene complex CpMo(NO)(CO)(η^2-COD), in which only one of the double-bonds of COD coordinates to the metal center *(320)*.

4.8.2: Manganese

Excess 1,3-butadiene and $Mn(NO)(CO)_4$ react under photolytic conditions to produce $(\eta^4\text{-}C_4H_6)Mn(NO)(CO)_2$ together with $(\eta^4\text{-}C_4H_6)_2Mn(CO)$ and $Mn(NO)_3(THF)$ *(347)*. Other diene complexes of manganese are generated by nucleophilic attack on the η^5-bonded rings in $[(\eta^5\text{-}cyclohexadienyl)Mn(NO)(CO)L]^+$ (L = CO, phosphine) and related complexes *(93,95,348-351)* as shown in eq 4.28. While most nucleophiles end up in an *exo* position, hydride (and deuteride) is positioned stereospecifically *endo (94,95)*. Indirect evidence suggests the formation of an intermediate formyl complex which then converts to a metal carbonyl-hydride prior to H-migration to the ring. Hydride addition to the rhenium analogue also proceeds *endo* to yield a cyclohexadiene product *(94)*.

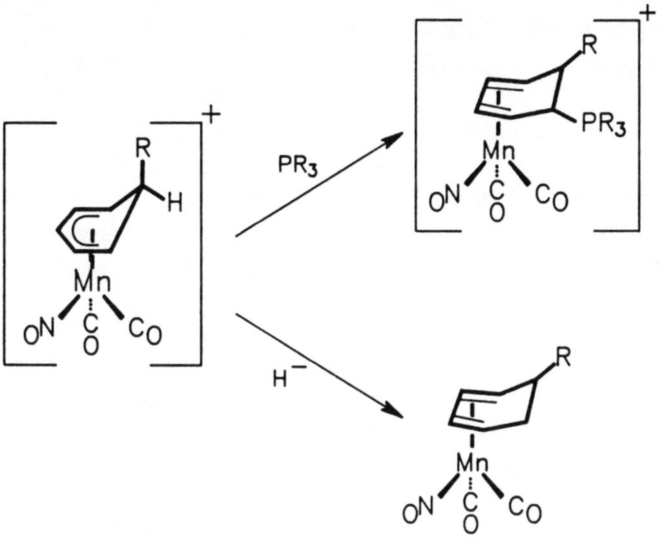

(4.28)

4.8.3: Cyclobutadiene Complexes

The only cyclobutadiene nitrosyl complexes known to date are those of iron and manganese. The iron complexes are prepared by the reaction of their tricarbonyl precursors with nitrosonium or nitronium salts *(352,353)*. These compounds react with two equivalents of neutral bases ER_3 (E= P, As) via attack at the cyclobutadiene ring and/or CO substitution *(352-354)*:

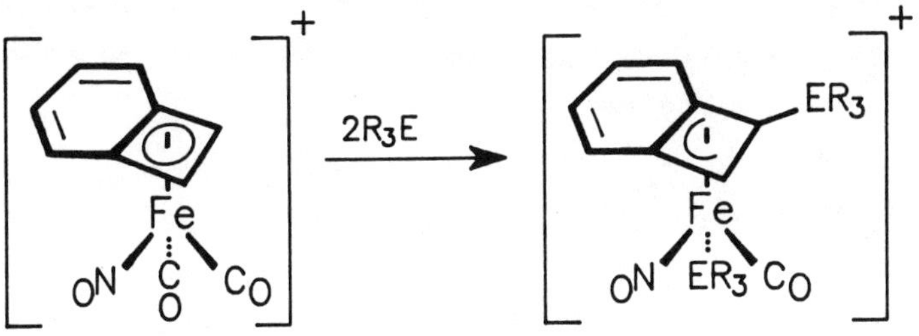

(4.29)

The reaction of $[(\eta^4\text{-}C_4Ph_4)Fe(NO)(CO)_2]^+$ with anionic nucleophiles such as Cl^-, NCO^-, NCS^-, NO_2^- or N_3^- yields dinuclear $[(\eta^4\text{-}C_4Ph_4)Fe(NO)X]_2$. With potentially chelating anions such as NO_3^- and $S_2CNEt_2^-$, the monomeric $(\eta^4\text{-}C_4Ph_4)Fe(NO)X$ complexes are produced instead *(355)*.

The manganese complex $(\eta^4\text{-}C_4Ph_4)Mn(NO)(CO)_2$ is generated by the reaction of diphenylacetylene with $Mn(NO)(CO)_4$ *(356)*.

4.9: Arenes

The only η^6-arene nitrosyl complexes known to date are those of chromium. These are usually made by the reaction of the precursor $(arene)Cr(CO)_3$ or $(arene)Cr(CO)_2(CS)$ complexes with the nitrosonium cation to produce the cationic $[(arene)Cr(NO)(CO)_2]^+$ and $[(arene)Cr(NO)(CO)(CS)]^+$ complexes, respectively *(41,42,357,358)*. The η^6-arene ligands in these complexes are normally heavily substituted. The addition of nucleophiles to the cationic dicarbonylnitrosyl complex results in the generation of the neutral cyclohexadienyl complexes via attack of the incoming nucleophiles at the bound arene rings in these complexes *(41)*.

The $(arene)Cr(NO)_2$ compound remains unknown, and is not expected to be stable *(359,360)*.

4.10: References

1. Richter-Addo, G. B.; Legzdins, P. *Chem. Rev.* **1988**, *88*, 991.
2. Mantell, D. R.; Gladfelter, W. L. *J. Organomet. Chem.* **1988**, *347*, 333.
3. Jacob, K.; Thiele, K. H. *Z. Anorg. Allg. Chem.* **1981**, *479*, 143.
4. (a) Clark, G. R.; Greene, T. R.; Roper, W. R. *Aust. J. Chem.* **1986**, *39*, 1315. (b) Werner, H.; Michenfelder, A.; Schulz, M. *Angew. Chem., Int. Ed. Engl.* **1991**, *30*, 596.
5. Reed, C. A.; Roper, W. R. *J. Chem. Soc. A* **1970**, 3054.
6. Seidel, W.; Geinitz, D. *Z. Chem.* **1975**, *15*, 71.
7. Green, M.; Osborn, R. B. L.; Rest, A. J.; Stone, F. G. A. *J. Chem. Soc. A* **1968**, 2525.
8. Reed, C. A.; Roper, W. R. *J. Chem. Soc., Dalton Trans.* **1973**, 1014.

9. Neve, F.; Ghedini, M. *Inorg. Chim. Acta* **1990**, *175*, 111.
10. Shin, S. K.; Beauchamp, J. L. *J. Am. Chem. Soc.* **1990**, *112*, 2066.
11. Yu, Y. F.; Chau, C. N.; Wojcicki, A.; Calligaris, M.; Nardin, G.; Balducci, G. *J. Am. Chem. Soc.* **1984**, *106*, 3704.
12. Casey, C. P.; Jones, W. D. *J. Am. Chem. Soc.* **1980**, *102*, 6154.
13. Casey, C. P.; Jones, W. D.; Harsy, S. G. *J. Organomet. Chem.* **1981**, 206, C38.
14. DiMauro, P. T.; Wolczanski, P. T.; Párkanyi, L.; Petach, H. H. *Organometallics* **1990**, *9*, 1097.
15. Berke, H.; Kundel, P. *Z. Naturforsch., B: Anorg. Chem. Org. Chem.* **1986**, *41B*, 527.
16. Jungbauer, A.; Behrens, H. *Z. Naturforsch., B: Anorg. Chem., Org. Chem.* **1979**, *34B*, 1641.
17. Chaudhari, F. M.; Knox, G. R.; Pauson, P. L. *J. Chem. Soc. (C)* **1967**, 2255.
18. Herberhold, M.; Trampisch, H. *Z. Naturforsch., B: Anorg. Chem., Org. Chem.* **1982**, *37B*, 614.
19. Herberhold, M.; Kremnitz, W.; Trampisch, H.; Hitam, R. B.; Rest, A. J.; Taylor, D. J. *J. Chem. Soc., Dalton Trans.* **1982**, 1261.
20. Moran, M.; Gayoso, M. *Z. Naturforsch., B: Anorg. Chem., Org. Chem.* **1981**, *36B*, 434.
21. Herberhold, M.; Klein, R.; Smith, P. D. *Angew. Chem., Int. Ed. Engl.* **1979**, *18*, 220.
22. Rehder, D.; Wenke, D. *J. Organomet. Chem.* **1988**, *348*, 205.
23. Bottomley, F.; Darkwa, J.; White, P. S. *J. Chem. Soc. Dalton Trans.* **1985**, 1435.
24. Chin, T. T.; Hoyano, J. K.; Legzdins, P.; Malito, J. T. *Inorg. Synth.* **1990**, *28*, 196.
25. Malito, J. T.; Shakir, R.; Atwood, J. L. *J. Chem. Soc., Dalton Trans.* **1980**, 1253.
26. Alt, H. G.; Frister, T.; Trapl, E. E.; Engelhardt, H. E. *J. Organomet. Chem.* **1989**, *362*, 125.
27. (a) Dryden, N. H.; Legzdins, P.; Batchelor, R. J.; Einstein, F. W. B. *Organometallics* **1991**, *10*, 2077. (b) Dryden, N. H.; Legzdins, P.; Einstein, F. W. B.; Jones, R. H. *Can. J. Chem.* **1988**, *66*, 2100.
28. Shakir, R.; Atwood, J. L. *Acta Crystallogr., Sect. B* **1981**, *B37*, 1656.

29. Herberhold, M.; Bernhagen, W. *Angew. Chem., Int. Ed. Engl.* **1976**, *15*, 617.
30. Atwood, J. L.; Shakir, R.; Malito, J. T.; Herberhold, M.; Kremnitz, W.; Bernhagen, W. P. E.; Alt, H. G. *J. Organomet. Chem.* **1979**, *165*, 65.
31. Gubitosa, G.; Brintzinger, H. H. *J. Organomet. Chem.* **1977**, *140*, 187.
32. Dötz, K. H.; Rott, J. *J. Organomet. Chem.* **1988**, *338*, C11.
33. Macomber, D. W.; Rausch, M. D. *Organometallics* **1983**, *2*, 1523.
34. Rausch, M. D.; Mintz, E. A.; Macomber, D. W. *J. Org. Chem.* **1980**, *45*, 689.
35. Rogers, R. D.; Shakir, R.; Atwood, J. L.; Macomber, D. W.; Wang, Y.-P.; Rausch, M. D. *J. Crystallogr. Spectrosc. Res.* **1988**, *18*, 767.
36. Moriarty, K. J.; Rausch, M. D. *J. Organomet. Chem.* **1989**, *370*, 75.
37. Rausch, M. D.; Wang, Y.-P. *Organometallics* **1991**, *10*, 1438.
38. Wang, Y.-P; Hwu, J.-M.; Wang, S.-L. *J. Organomet. Chem.* **1990**, *390*, 179.
39. Wang, Y.-P.; Hwu, J.-M. *J. Organomet. Chem.* **1990**, *399*, 141.
40. Rausch, M. D.; Kowalski, D. J.; Mintz, E. A. *J. Organomet. Chem.* **1988**, *342*, 201.
41. Connelly, N. G.; Kelly, R. L. *J. Chem. Soc., Dalton Trans.* **1974**, 2334.
42. Ball, D. E.; Connelly, N. G. *J. Organomet. Chem.* **1973**, *55*, C24.
43. Weber, L.; Boese, R. *Chem. Ber.* **1985**, *118*, 1545.
44. Weber, L. *Angew. Chem., Int. Ed. Engl.* **1983**, *22*, 516.
45. Weber, L. *Z. Naturforsch., B: Anorg. Chem., Org. Chem.* **1985**, *40B*, 373.
46. Geiger, W. E.; Rieger, P. H.; Tulyathan, B.; Rausch, M. D. *J. Am. Chem. Soc.* **1984**, *106*, 7000.
47. Solodovnikov, S. P.; Tumanskii, B. L.; Bubnov, N. N.; Kabachnik, M. I. *Izv. Akad. Nauk SSSR, Ser Khim.* **1986**, 2147. CA107(17):154444n.
48. Hitam, R. B.; Rest, A. J.; Herberhold, M.; Kremnitz, W. *J. Chem. Soc., Chem. Commun.* **1984**, 471.
49. Reisner, M. G.; Bernal, I.; Brunner, H.; Doppelberger, J. *J. Chem. Soc., Dalton Trans.* **1978**, 1664.
50. Brunner, H. *J. Organomet. Chem.* **1969**, *16*, 119.
51. Herberhold, M.; Alt, H. *Liebigs Ann. Chem.* **1976**, 292.
52. Herberhold, M.; Smith, P. D. *Angew. Chem., Int. Ed. Engl.* **1979**, *18*, 631.

53. Behrens, H.; Landgraf, G.; Merbach, P.; Moll, M.; Trummer, K. H. *J. Organomet. Chem.* **1983**, *253*, 217.
54. Greaves, W. W.; Angelici, R. J. *Inorg. Chem.* **1981**, *20,* 2983.
55. Hunter, A. D.; Legzdins, P. *Organometallics* **1986**, *5,* 1001.
56. Regina, F. J.; Wojcicki, A. *Inorg. Chem.* **1980**, *19,* 3803.
57. Legzdins, P.; Richter-Addo, G. B.; Einstein, F. W. B.; Jones, R. H. *Organometallics* **1990**, *9,* 431.
58. Legzdins, P.; Martin, D. T. *Organometallics* **1983**, *2,* 1785.
59. Stewart, R. P., Jr.; Moore, G. T. *Inorg. Chem.* **1975**, *14,* 2699.
60. Yu, Y. S.; Jacobson, R. A.; Angelici, R. J. *Inorg. Chem.* **1982**, *21,* 3106.
61. Kolthammer, B. W. S.; Legzdins, P.; Malito, J. T. *Inorg. Synth.* **1979**, *19*, 208.
62. Legzdins, P. *Can. Chem. News* **1990**, *42(10)*, 19.
63. Calderón, J. L.; Fontana, S.; Frauendorfer, E.; Day, V. W. *J. Organomet. Chem.* **1974**, *64*, C10.
64. Legzdins, P.; Wassink, B. *Organometallics* **1984**, *3,* 1811.
65. Bursten, B. E.; Cayton, R. H.; Gatter, M. G. *Organometallics* **1988**, *7,* 1342.
66. Connelly, N. G. *Inorg. Synth.* **1974**, *15*, 91.
67. Fischer, E. O.; Strametz, H. *Z. Naturforsch., B: Anorg. Chem., Org. Chem.* **1967**, *23B*, 278.
68. Treichel, P. M.; Stenson, J. P.; Benedict, J. J. *Inorg. Chem.* **1971**, *10,* 1183.
69. Busetto, L.; Palazzi, A.; Pietropaolo, D.; Dolcetti, G. *J. Organomet. Chem.* **1974**, *66*, 453.
70. Weiner, W. P.; Hollander, F. J.; Bergman, R. G. *J. Am. Chem. Soc.* **1984**, *106,* 7462.
71. Busetto, L.; Monari, M.; Palazzi, A.; Albano, V.; Demartin, F. *J. Chem. Soc., Dalton Trans.* **1983**, 1849.
72. Housmekerides, C. E.; Pilato, R. S.; Geofffroy, G. L.; Rheingold, A. L. *J. Chem. Soc., Chem. Commun.* **1991**, 563.
73. Pilato, R. S.; Housmekerides, C. E.; Jernakoff, P.; Rubin, D.; Geoffroy, G. L.; Rheingold, A. L. *Organometallics* **1990**, *9,* 2333.
74. Senn, D. R.; Gladysz, J. A.; Emerson, K.; Larsen, R. D. *Inorg. Chem.* **1987**, *26,* 2737.
75. Tso, C. T.; Cutler, A. R. *J. Am. Chem. Soc.* **1986**, *108,* 6069.

76. Winter, C. H.; Gladysz, J. A. *J. Organomet. Chem.* **1988**, *354*, C33.

77. Winter, C. H.; Arif, A. M.; Gladysz, J. A. *Organometallics* **1989**, *8*, 219.

78. (a) Winter, C. H.; Veal, W. R.; Garner, C. M.; Arif, A. M.; Gladysz, J. A. *J. Am. Chem. Soc.* **1989**, *111*, 4766. (b) Igau, A.; Gladysz, J. A. *Organometallics* **1991**, *10*, 2327.

79. Kulawiec, R. J.; Crabtree, R. H. *Coord. Chem. Rev.* **1990**, *99*, 89.

80. Dalton, D. M.; Fernández, J. M.; Emerson, K.; Larsen, R. D.; Arif, A. M.; Gladysz, J. A. *J. Am. Chem. Soc.* **1990**, *112*, 9198.

81. Garner, C. M.; Méndez, N. Q.; Kowalczyk, J. J.; Fernández, J. M.; Emerson, K.; Larsen, R. D.; Gladysz, J. A. *J. Am. Chem. Soc.* **1990**, *112*, 5146.

82. Garner, C. M.; Fernández, J. M.; Gladysz, J. A. *Tetrahedron Lett.* **1989**, *30*, 3931.

83. Dewey, M. A.; Bakke, J. M.; Gladysz, J. A. *Organometallics* **1990**, *9*, 1349.

84. Buhro, W. E.; Arif, A. M.; Gladysz, J. A. *Inorg. Chem.* **1989**, *28*, 3837.

85. Buhro, W. E.; Zwick, B. D.; Georgiou, S.; Hutchinson, J. P.; Gladysz, J. A. *J. Am. Chem. Soc.* **1988**, *110*, 2427.

86. Agbossou, S. K.; Smith, W. W.; Gladysz, J. A. *Chem. Ber.* **1990**, *123*, 1293.

87. (a) Agbossou, S. K.; Fernández, J. M.; Gladysz, J. A. *Inorg. Chem.* **1990**, *29*, 476. (b) Méndez, N. Q.; Arif, A. M.; Gladysz, J. A. *Organometallics* **1991**, *10*, 2199.

88. Casey, C. P.; Andrews, M. A.; McAlister, D. R.; Jones, W. D.; Harsy, S. G. *J. Mol. Catal.* **1981**, *13*, 43.

89. Tam, W.; Lin, G. Y.; Wong, W. K.; Kiel, W. A.; Wong, V. K.; Gladysz, J. A. *J. Am. Chem. Soc.* **1982**, *104*, 141.

90. Hames, B. W.; Kolthammer, B. W. S.; Legzdins, P. *Inorg. Chem.* **1981**, *20*, 650.

91. Efraty, A.; Arneri, R.; Ruda, W. A. *Inorg. Chem.* **1977**, *16*, 3124.

92. Kirchner, R. M.; Marks, T. J.; Kristoff, J. S.; Ibers, J. A. *J. Am. Chem. Soc.* **1973**, *95*, 6602.

93. Sweigart, D. A.; Chung, Y. K.; Honig, E. D.; Alavosus, T. J.; Halpin, W. A.; Williams, J. C.; Williard, P. G.; Connelly, N. G. In *Organometallic Syntheses*; King, R. B.; Eisch, J. J., Eds.; Elsevier: New York, 1988; Vol. 3, p 108.

94. Pike, R. D.; Ryan, W. J.; Lennhoff, N. S.; Epp, J. V.; Sweigart, D. A. *J. Am. Chem. Soc.* **1990**, *112*, 4798.

95. Ittel, S. D.; Whitney, J. F.; Chung, Y. K.; Williard, P. G.; Sweigart, D. A. *Organometallics* **1988**, *7*, 1323.

96. Pike, R. D.; Alavosus, T. J.; Camaioni-Neto, C. A.; Williams, J. C., Jr.; Sweigart, D. A. *Organometallics* **1989**, *8*, 2631.

97. King, R. B. *Inorg. Chem.* **1967**, *6*, 30.

98. Calderon, J. L.; Fontana, S.; Frauendorfer, E.; Day, V. E.; Stults, B. R. *Inorg. Chim. Acta* **1976**, *17*, L31.

99. Bottomley, F. *Inorg. Chem.* **1983**, *22*, 2656.

100. Lichtenberger, D. L.; Copenhaver, A. S.; Hubbard, J. L. *Polyhedron* **1990**, *9*, 1783.

101. Kubat-Martin, K. A.; Barr, M. E.; Spencer, B.; Dahl, L. F. *Organometallics* **1987**, *6*, 2570.

102. Mougenot, P.; Demuynck, J.; Benard, M. *J. Phys. Chem.* **1988**, *92*, 571.

103. Granozzi, G.; Mougenot, P.; Demuynck, J.; Benard, M. *Inorg. Chem.* **1987**, *26*, 2588.

104. Schugart, K. A.; Fenske, R. F. *J. Am. Chem. Soc.* **1986**, *108*, 5094.

105. Herrmann, W. A.; Floel, M.; Weber, C.; Hubbard, J. L.; Schaefer, A. *J. Organomet. Chem.* **1985**, *286*, 369.

106. Kalcher, W.; Herrmann, W. A.; Pahl, C.; Ziegler, M. L. *Chem. Ber.* **1984**, *117*, 69.

107. Herrmann, W. A.; Bauer, C. *Chem. Ber.* **1982**, *115*, 14.

108. Conroy-Lewis, F. M.; Simpson, S. J. *J. Organomet. Chem.* **1987**, *322*, 221.

109. Treichel, P. M.; Molzahn, D. C.; Wagner, K. P. *J. Organomet. Chem.* **1979**, *174*, 191.

110. Szczepura, L. F.; Takeuchi, K. J. *Inorg. Chem.* **1990**, *29*, 1772.

111. Connelly, N. G.; Loyns, A. C.; Manners, I.; Mercer, D. L.; Richardson, K. E.; Rieger, P. H. *J. Chem. Soc., Dalton Trans.* **1990**, 2451.

112. Conroy-Lewis, F. M.; Redhouse, A. D.; Simpson, S. J. *J. Organomet. Chem.* **1990**, *399*, 307.

113. Ashok, R. F. N.; Gupta, M.; Arulsamy, K. S.; Agarwala, U. C. *Can. J. Chem.* **1985**, *63*, 963.

114. Chang, J.; Bergman, R. G. *J. Am. Chem. Soc.* **1987**, *109*, 4298.

115. Berg, D. J.; Andersen, R. A. *J. Am. Chem. Soc.* **1988**, *110*, 4849.

116. Pilloni, G.; Zecchin, S.; Casarin, M.; Granozzi, G. *Organometallics* **1987**, *6*, 597.
117. Connelly, N. G.; Payne, J. D.; Geiger, W. E. *J. Chem. Soc., Dalton Trans.* **1983**, 295.
118. Connelly, N. G.; Raven, S. J.; Geiger, W. E. *J. Chem. Soc., Dalton Trans.* **1987**, 467.
119. Wochner, F.; Keller, E.; Brintzinger, H. H. *J. Organomet. Chem.* **1982**, *236*, 267.
120. Clamp, S.; Connelly, N. G.; Howard, J. A. K.; Manners, I.; Payne, J. D.; Geiger, W. E. *J. Chem. Soc., Dalton Trans.* **1984**, 1659.
121. Herrmann, W. A.; Bernal, I. *Angew. Chem., Int. Ed. Engl.* **1977**, *16*, 172.
122. Schugart, K. A.; Fenske, R. F. *J. Am. Chem. Soc.* **1986**, *108*, 5100.
123. Connelly, N. G.; Lucy, A. R.; Galas, A. M. R. *J. Chem. Soc., Chem. Commun.* **1981**, 43.
124. Faraone, F.; Pietropaolo, R.; Troilo, G. G.; Piraino, P. *Inorg. Chim. Acta* **1973**, *7*, 729.
125. Faraone, F.; Bruno, G.; Tresoldi, G.; Faraone, G.; Bombieri, G. *J. Chem. Soc. Dalton Trans.* **1981**, 1651.
126. Dimas, P. A.; Shapley, J. R. *J. Organomet. Chem.* **1982**, *228*, C12.
127. Einstein, F. W. B.; Yan, X.; Sutton, D. *J. Chem. Soc., Chem. Commun.* **1990**, 1466.
128. Cox, A. P.; Randell, J.; Legon, A. C. *Chem. Phys. Lett.* **1988**, *153*, 253.
129. Shirokii, V. L.; Sutormin, A. B.; Maier, N. A.; Ol'dekop, Y. A. *Zh. Obshch. Khim.* **1984**, *54*, 1436.
130. Piper, T. S.; Cotton, F. A.; Wilkinson, G. *J. Inorg. Nucl. Chem.* **1955**, *1*, 165.
131. King, R. B. *Organometallic Syntheses*; Academic: New York, 1965; Vol. 1.
132. Fischer, E. O.; Vogler, A. *Z. Naturforsch., B: Anorg. Chem., Org. Chem.* **1963**, *18B*, 771.
133. Fischer, E. O.; Schuster-Woldan, H. *Z. Naturforsch., B: Anorg. Chem., Org. Chem.* **1964**, *19B*, 766.
134. Kokulich, S. G.; Rund, J. V.; Pauley, D. J.; Bumgarner, R. E. *J. Am. Chem. Soc.* **1988**, *110*, 7356.
135. Boehm, M. C. *Z. Naturforsch.* **1981**, *36A*, 1361.
136. Müller, J.; Goll, W. *Chem. Ber.* **1973**, *106*, 1129.

137. Burnier, R. C.; Freiser, B. S. *Inorg. Chem.* **1979**, *18*, 906.
138. Georgiou, S.; Wight, C. A. *J. Chem Phys.* **1988**, *88*, 7418.
139. Georgiou, S.; Wight, C. A. *Chem. Phys. Lett.* **1986**, *132*, 511.
140. Jack, T. R.; May, C. J.; Powell, J. *J. Am. Chem. Soc.* **1977**, *99*, 4707.
141. Welcker, P. S.; Todd, L. J. *Inorg. Chem.* **1970**, *9*, 286.
142. Middleton, A. R.; Wilkinson, G. *J. Chem. Soc., Dalton Trans.* **1980**, 1888.
143. Bottomley, F.; Darkwa, J.; White, P. S. *Organometallics* **1985**, *4*, 961.
144. Herring, F. G.; Legzdins, P.; Richter-Addo, G. B. *Organometallics* **1989**, *8*, 1485.
145. Gomez-Sal, P.; de Jesús, E.; Michiels, W.; Royo, P.; de Miguel, A. V.; Martinez-Carrera, S. *J. Chem. Soc., Dalton Trans.* **1990**, 2445.
146. Seddon, D.; Kita, W. G.; Bray, J.; McCleverty, J. A. *Inorg. Synth.* **1976**, *16*, 24.
147. Legzdins, P.; Martin, D. T.; Nurse, C. R. *Inorg. Chem.* **1980**, *19*, 1560.
148. Christensen, N. J.; Hunter, A. D.; Legzdins, P.; Sanchez, L. *Inorg. Chem.* **1987**, *26*, 3344.
149. McCleverty, J. A.; Seddon, D. *J. Chem. Soc., Dalton Trans.* **1972**, 2526.
150. Kita, W. G.; McCleverty, J. A.; Patel, B.; Williams, J. *J. Organomet. Chem.* **1974**, *74*, C9.
151. McCleverty, J. A.; Williams, J. *Transition Met. Chem.* **1976**, *1*, 288.
152. James, T. A.; McCleverty, J. A. *J. Chem. Soc. (A)* **1971**, 1596.
153. Frisch, P. D.; Hunt, M. M.; Kita, W. G.; McCleverty, J. A.; Rae, A. E.; Seddon, D.; Swann, D.; Williams, J. *J. Chem. Soc., Dalton Trans.* **1979**, 1819.
154. James, T. A.; McCleverty, J. A. *J. Chem. Soc. (A)* **1971**, 1068.
155. Bonnesen, P. V.; Yau, P. K. L.; Hersh, W. H. *Organometallics* **1987**, *6*, 1587.
156. Legzdins, P.; Nurse, C. R. *Inorg. Chem.* **1982**, *21*, 3110.
157. Hunt, M. M.; McCleverty, J. A. *J. Chem. Soc., Dalton Trans.* **1978**, 480.
158. Koridze, A. A.; Yanovskii, A. I.; Slovokhotov, Y. I.; Andrianov, V. G.; Struchkov, Y. T. *Koord. Khim.* **1982**, *8*, 541.
159. King, R. B. *Inorg. Chem.* **1968**, *7*, 90.
160. Calderon, J. L.; Cotton, F. A.; Legzdins, P. *J. Am. Chem. Soc.* **1969**, *91*, 2528.
161. Calderon, J. L.; Cotton, F. A. *J. Organomet. Chem.* **1971**, *30*, 377.

162. Hunt, M. M.; Kita, W. G.; McCleverty, J. A. *J. Chem. Soc., Dalton Trans.* **1978**, 474.

163. Pfeiffer, E.; Vrieze, K.; McCleverty, J. A. *J. Organomet. Chem.* **1979**, *174*, 183.

164. Legzdins, P.; Rettig, S. J.; Sanchez, L. *Organometallics* **1988**, *7*, 2394.

165. Legzdins, P.; Jones, R. H.; Phillips, E. C.; Yee, V. C.; Trotter, J.; Einstein, F. W. B. *Organometallics* **1991**, *10*, 986.

166. de Jesús, E.; de Miguel, A. V.; Royo, P.; Lanfredi, A. M. M.; Tiripicchio, A. *J. Chem. Soc., Dalton Trans.* **1990**, 2779.

167. Alegre, B.; de Jesus, E.; de Miguel, A. V.; Royo, P.; Lanfredi, A. M. M.; Tiripichio, A. *J. Chem. Soc., Dalton Trans.* **1988**, 819.

168. Legzdins, P.; Rettig, S. J.; Sanchez, L. *J. Am. Chem. Soc.* **1985**, *107*, 1411.

169. Bursten, B. E.; Cayton, R. H. *Organometallics* **1987**, *6*, 2004.

170. (a) Dryden, N. H.; Legzdins, P.; Trotter, J.; Yee, V. C. *Organometallics* **1991**, *10*, 2857. (b) Dryden, N. H.; Legzdins, P.; Phillips, E. C.; Trotter, J.; Yee, V. C. *Organometallics* **1990**, *9*, 882.

171. Alt, H.; Hayen, H. I.; Klein, H. P.; Thewalt, U. *Angew. Chem., Int. Ed. Engl.* **1984**, *23*, 809.

172. Legzdins, P.; Martin, J. T.; Einstein, F. W. B.; Jones, R. H. *Organometallics* **1987**, *6*, 1826.

173. Legzdins, P.; Phillips, E. C.; Sanchez, L. *Organometallics* **1989**, *8*, 940.

174. Gomez-Sal, P.; de Jesus, E.; Royo, P.; de Miguel, A. V.; Martinez-Carrera, S.; Garcia-Blanco, S. *J. Organomet. Chem.* **1988**, *353*, 191.

175. Legzdins, P.; Sanchez, L. *J. Am. Chem. Soc.* **1985**, *107*, 5525.

176. Ashby, M. T.; Enemark, J. H. *J. Am. Chem. Soc.* **1986**, *108*, 730.

177. Clark, G. R.; Hall, D.; Marsden, K. *J. Organomet. Chem.* **1979**, *177*, 411.

178. McCleverty, J. A.; Seddon, D. *J. Chem. Soc., Dalton Trans.* **1972**, 2588.

179. Alt, H. G.; Freytag, U.; Herberhold, M.; Hayen, H. I. *J. Organomet. Chem.* **1987**, *336*, 361.

180. Rott, J.; Guggolz, E.; Rettenmeier, A.; Ziegler, M. L. *Z. Naturforsch., B: Anorg. Chem., Org. Chem.* **1982**, *37B*, 13.

181. (a) Legzdins, P.; Nurse, C. R. *Inorg. Chem.* **1985**, *24*, 327. (b) Herring, F. G.; Legzdins, P.; McNeil, W. S.; Shaw, M. J. *J. Am. Chem Soc.* **1991**, *113*, 7049.

182. Hardy, A. D. U.; Sim, G. A. *Acta Crystallogr., Sect. B* **1979**, *B35*, 1463.
183. Sim, G. A.; Woodhouse, D. I.; Knox, G. R. *J. Chem. Soc., Dalton Trans.* **1979**, 83.
184. Malisch, W.; Blau, H.; Weickert, P.; Griessmann, K.-H. *Z. Naturforsch., B: Anorg. Chem., Org. Chem.* **1983**, *38B*, 711.
185. Carre, F.; Colomer, E.; Corriu, R. J. P.; Vioux, A. *Organometallics* **1984**, *3*, 970.
186. Alt, H.; Hayen, H. I. *Angew. Chem., Int. Ed. Engl.* **1983**, *22*, 1008.
187. Hoyano, J. K.; Legzdins, P.; Malito, J. T. *Inorg. Synth.* **1978**, *18*, 126.
188. Legzdins, P.; Malito, J. T. *Inorg. Chem.* **1975**, *14*, 1875.
189. Morris-Sherwood, B. J.; Kolthammer, B. W. S.; Hall, M. B. *Inorg. Chem.* **1981**, *20*, 2771.
190. Greenhough, T. J.; Kolthammer, B. W. S.; Legzdins, P.; Trotter, J. *Acta Crystallogr., Sect. B* **1980**, *B36*, 795.
191. Chen, H. W.; Jolly, W. L.; Xiang, S. F.; Legzdins, P. *Inorg. Chem.* **1981**, *20*, 1779.
192. Pankowski, M.; Bigorgne, M.; Chauvin, Y. *J. Organomet. Chem.* **1976**, *110*, 331.
193. Fischer, E. O.; Strametz, H. *J. Organomet. Chem.* **1967**, *10*, 323.
194. Legzdins, P.; Wassink, B. *Organometallics* **1988**, *7*, 482.
195. Hoyano, J. K.; Legzdins, P.; Malito, J. T. *J. Chem. Soc., Dalton Trans.* **1975**, 1022.
196. Legzdins, P.; Richter-Addo, G. B.; Wassink, B.; Einstein, F. W. B.; Jones, R. H.; Willis, A. C. *J. Am. Chem. Soc.* **1989**, *111*, 2097.
197. Legzdins, P.; Wassink, B.; Einstein, F. W. B.; Jones, R. H. *Organometallics* **1988**, *7*, 477.
198. Froböse, R.; Mews, R.; Glemser, O. *Z. Naturforsch., B: Anorg. Chem., Org. Chem.* **1976**, *31B*, 1497.
199. Legzdins, P.; Richter-Addo, G. B.; Einstein, F. W. B.; Jones, R. H. *Organometallics* **1987**, *6*, 1807.
200. Wojcicki, A. *Adv. Organomet. Chem.* **1974**, *12*, 31.
201. Hubbard, J. L.; McVicar, W. K. *Organometallics* **1990**, *9*, 2683.
202. Leung, T. W.; Christoph, G. G.; Gallucci, J.; Wojcicki, A. *Organometallics* **1986**, *5*, 846.
203. Alt, H. G.; Hayen, H. I. *J. Organomet. Chem.* **1986**, *316*, 105.
204. Alt, H. G.; Hayen, H. I. *J. Organomet. Chem.* **1986**, *316*, 301.

205. Eremenko, I. L.; Pasynskii, A. A.; Kalinnikov, V. T.; Struchkov, Y. T.; Aleksandrov, G. G. *Inorg. Chim. Acta* **1981**, *52*, 107.
206. Alt, H. G.; Hayen, H. I. *J. Organomet. Chem.* **1986**, *315*, 337.
207. Hames, B. W.; Legzdins, P.; Martin, D. T. *Inorg. Chem.* **1978**, *17*, 3644.
208. Beran, K.; Steinke, G.; Mews, R. *Chem. Ber.* **1989**, *122*, 1613.
209. Deane, M.; Lalor, F. J. *J. Organomet. Chem.* **1973**, *57*, C61.
210. Lichtenberger, D. L.; Rai-Chaudhuri, A.; Seidel, M. J.; Gladysz, J. A.; Agbossou, S. K.; Igau, A.; Winter, C. H. *Organometallics* **1991**, *10*, 1355.
211. Kolthammer, B. W. S.; Legzdins, P. *Inorg. Chem.* **1979**, *18*, 889.
212. Potenza, J. A.; Johnson, R.; Rudich, S.; Efraty, A. *Acta Crystallogr., Sect. B* **1980**, *B36*, 1933.
213. McCleverty, J. A.; James, T. A.; Wharton, E. J.; Winscom, C. J. *J. Chem. Soc., Chem. Commun.* **1968**, 933.
214. Becalska, A.; Hill, R. H. *J. Chem. Soc., Chem. Commun.* **1989**, 1626.
215. Casey, C. P.; Neumann, S. M.; Andrews, M. A.; McAlister, D. R. *Pure Appl. Chem.* **1980**, *52*, 625.
216. Casey, C. P.; Andrews, M. A.; McAlister, D. R.; Rinz, J. E. *J. Am. Chem. Soc.* **1980**, *102*, 1927.
217. Tam, W.; Lin, G.-Y.; Gladysz, J. A. *Organometallics* **1982**, *1*, 525.
218. Bodner, G. S.; Patton, A. T.; Smith, D. E.; Georgiou, S.; Tam, W.; Wong, W.-K.; Strouse, C. E.; Gladysz, J. A. *Organometallics* **1987**, *6*, 1954.
219. Sheridan, J. B.; Geoffroy, G. L.; Rheingold, A. L. *J. Am. Chem. Soc.* **1987**, *109*, 1584.
220. Merrifield, J. H.; Fernandez, J. M.; Buhro, W. E.; Gladysz, J. A. *Inorg. Chem.* **1984**, *23*, 4022.
221. Fernandez, J. M.; Gladysz, J. A. *Organometallics* **1989**, *8*, 207.
222. Bodner, G. S.; Smith, D. E.; Hatton, W. G.; Heah, P. C.; Georgiou, S.; Rheingold, A. L.; Geib, S. J.; Hutchinson, J. P.; Gladysz, J. A. *J. Am. Chem. Soc.* **1987**, *109*, 7688.
223. Smith, D,. E.; Gladysz, J. A. *Organometallics* **1985**, *4*, 1480.
224. Senn, D. R.; Wong, A.; Patton, A. T.; Marsi, M.; Strouse, C. E.; Gladysz, J. A. *J. Am. Chem. Soc.* **1988**, *110*, 6096.
225. Coe, J. V.; Snodgrass, J. T.; Freidhoff, C. B.; McHugh, K. M.; Bowen, K. H. *J. Chem Phys.* **1987**, *87*, 4302.

226. Dewey, M. A.; Arif, A. M.; Gladysz, J. A. *J. Chem. Soc., Chem. Commun.* **1991**, 712.
227. Dewey, M. A.; Gladysz, J. A. *Organometallics* **1990**, *9*, 1351.
228. Bassner, S. L.; Sheridan, J. B.; Kelly, C.; Geoffroy, G. L. *Organometallics* **1989**, *8*, 2121.
229. Stufkens, D. J.; Sheridan, J. B.; Geoffroy, G. L. *Inorg. Chem.* **1990**, *29*, 4347.
230. Messer, D.; Landgraf, G.; Behrens, H. *J. Organomet. Chem.* **1979**, *172*, 349.
231. King, R. B.; Bisnette, M. B.; Fronzaglia, A. *J. Organomet. Chem.* **1965**, *4*, 256.
232. Brunner, H.; Langer, M. *J. Organomet. Chem.* **1973**, *54*, 221.
233. Blau, H.; Malisch, W.; Weickert, P. *Chem. Ber.* **1982**, *115*, 1488.
234. Alt, H.; Engelhardt, H. E. *J. Organomet. Chem.* **1988**, *346*, 211.
235. Weber, L.; Reizig, K. *Angew. Chem., Int. Ed. Engl.* **1985**, *24*, 53.
236. Weber, L.; Reizig, K.; Boese, R.; Polk, M. *Organometallics* **1986**, *5*, 1098.
237. Chang, J.; Seidler, M. D.; Bergman, R. G. *J. Am. Chem. Soc.* **1989**, *111*, 3258.
238. Diel, B. N. *J. Organomet. Chem.* **1985**, *284*, 257.
239. Efraty, A.; Elbaze, G. *J. Organomet. Chem.* **1984**, *260*, 331.
240. Seidler, M. D.; Bergman, R. G. *Organometallics* **1983**, *2*, 1897.
241. Sheridan, J. B.; Han, S. H.; Geoffroy, G. L. *J. Am. Chem. Soc.* **1987**, *109*, 8097.
242. Casey, C. P.; Roddick, D. M. *Organometallics* **1986**, *5*, 436.
243. Bursten, B. E.; Cayton, R. H. *J. Am. Chem. Soc.* **1987**, *109*, 6053.
244. Strauss, R. C.; Keller, E.; Brintzinger, H. H. *J. Organomet. Chem.* **1988**, *340*, 249.
245. Brunner, H.; Wachsmann, H. *J. Organomet. Chem.* **1968**, *15*, 409.
246. Weiner, W. P.; Bergman, R. G. *J. Am. Chem. Soc.* **1983**, *105*, 3922.
247. Brunner, H.; Loskot, S. *Z. Naturforsch., B: Anorg. Chem., Org. Chem.* **1973**, *28B*, 314.
248. Legzdins, P.; Martin, D. T. *Inorg. Chem.* **1979**, *18*, 1250.
249. Bursten, B. E.; Gatter, M. G. *J. Am. Chem. Soc.* **1984**, *106*, 2554.
250. Legzdins, P.; Martin, J. T.; Einstein, F. W. B.; Willis, A. C. *J. Am. Chem. Soc.* **1986**, *108*, 7971.

251. Collman, J. P.; McDevitt, J. T.; Leidner, C. R.; Yee, G. T.; Torrance, J. B.; Lettle, W. A. *J. Am. Chem. Soc.* **1987**, *109*, 4606.
251. Stewart, R. P.; Okamoto, N.; Graham, W. A. G. *J. Organomet. Chem.* **1972**, *41*, C32.
252. Chinn, M. S.; Heinekey, D. M.; Payne, N. G.; Solfield, C. D. *Organometallics* **1989**, *8*, 1824.
253. Hames, B. W.; Legzdins, P.; Oxley, J. C. *Inorg. Chem.* **1980**, *19*, 1565.
254. Lee, K. E.; Arif, A. M.; Gladysz, J. A. *Organometallics* **1991**, *10*, 751.
255. Hames, B. W.; Legzdins, P. *Organometallics* **1982**, *1*, 116.
256. Legzdins, P.; Martin, J. T.; Oxley, J. C. *Organometallics* **1985**, *4*, 1263.
257. Bursten, B. E.; Cayton, R. H. *Organometallics* **1988**, *7*, 1349.
258. Crocco, G. L.; Gladysz, J. A. *J. Am. Chem. Soc.* **1988**, *110*, 6110.
259. Milletti, M. C.; Fenske, R. F. *Organometallics* **1989**, *8*, 420.
260. Crocco, G. L.; Gladysz, J. A. *Chem. Ber.* **1988**, *121*, 375.
261. Heah, P. C.; Patton, A. T.; Gladysz, J. A. *J. Am. Chem. Soc.* **1986**, *108*, 1185.
262. Crocco, G. L.; Young, C. S.; Lee, K. E.; Gladysz, J. A. *Organometallics* **1988**, *7*, 2158.
263. Labanova, I. A.; Zdanovich, V. I.; Kolobova, N. E. *Metalloorg. Khim.* **1988**, *1*, 1176. CA111(21):194987f.
264. Crocco, G. L.; Gladysz, J. A. *J. Am. Chem. Soc.* **1985**, *107*, 4103.
265. Powell, J.; Sawyer, J. F.; Stainer, M. V. R. *Inorg. Chem.* **1989**, *28*, 4461.
266. Powell, J.; Fuchs, E.; Gregg, M. R.; Phillips, J.; Stainer, M. V. R. *Organometallics* **1990**, *9*, 387.
267. Fischer, E. O. *Pure Appl. Chem.* **1970**, *24*, 407.
268. Bonnesen, P. V.; Baker, A. T.; Hersh, W. H. *J. Am. Chem. Soc.* **1986**, *108*, 8304.
269. Herrmann, W. A.; Hubbard, J. L.; Bernal, I.; Korp, J. D.; Haymore, B. L.; Hillhouse, G. L. *Inorg. Chem.* **1984**, *23*, 2978.
270. Lappert, M. F.; Pye, P. L.; McLaughlin, G. M. *J. Chem. Soc., Dalton Trans.* **1977**, 1272.
271. Keller, A. *J. Organomet. Chem.* **1991**, *407*, 237. ibid. *415*, 97.
272. Singh, M. M.; Angelici, R. J. *Inorg. Chem.* **1984**, *23*, 2691.
273. Seeman, J. I.; Davies, S. G. *J. Am. Chem. Soc.* **1985**, *107*, 6522.
274. Kiel, W. A.; Buhro, W. E.; Gladysz, J. A. *Organometallics* **1984**, *3*, 879.

275. Kiel, W. A.; Lin, G. Y.; Constable, A. G.; McCormick, F. B.; Strouse, C. E.; Eisenstein, O.; Gladysz, J. A. *J. Am. Chem. Soc.* **1982**, *104*, 4865.

276. Kiel, W. A.; Lin, G.-Y.; Gladysz, J. A. *J. Am. Chem. Soc.* **1980**, *102*, 3299.

277. Kiel, W. A.; Lin, G.-Y.; Bodner, G. S.; Gladysz, J. A. *J. Am. Chem. Soc.* **1983**, *105*, 4958.

278. Bodner, G. S.; Gladysz, J. A.; Nielsen, M. F.; Parker, V. D. *J. Am. Chem. Soc.* **1987**, *109*, 1757.

279. Bodner, G. S.; Gladysz, J. A.; Nielsen, M. F.; Parker, V. D. *Organometallics* **1987**, *6*, 1628.

280. McCormick, F. B.; Gleason, W. B.; Zhao, X.; Heah, P. C.; Gladysz, J. A. *Organometallics* **1986**, *5*, 1778.

281. Buhro, W. E.; Georgiou, S.; Fernandez, J. M.; Patton, A. T.; Strouse, C. E.; Gladysz, J. A. *Organometallics* **1986**, *5*, 956.

282. Buhro, W. E.; Etter, M. C.; Georgiou, S.; Gladysz, J. A.; McCormick, F. B. *Organometallics* **1987**, *6*, 1150.

283. McCormick, F. B. *Organometallics* **1984**, *3*, 1924.

284. Merrifield, J. H.; Lin, G. Y.; Kiel, W. A.; Gladysz, J. A. *J. Am. Chem. Soc.* **1983**, *105*, 5811.

285. Tilset, M.; Bodner, G. S.; Senn, D. R.; Gladysz, J. A.; Parker, V. D. *J. Am. Chem. Soc.* **1987**, *109*, 7551.

286. Lee, K. E.; Arif, A. M.; Gladysz, J. A. *Chem. Ber.* **1991**, *124*, 309.

287. Lappert, M. F.; Pye, P. L. *J. Chem. Soc., Dalton Trans.* **1978**, 837.

288. Hitchcock, P. B.; Lappert, M. F.; Pye, P. L.; Thomas, S. *J. Chem. Soc., Dalton Trans.* **1979**, 1929.

289. Lappert, M. F.; MacQuitty, J. J.; Pye, P. L. *J. Chem. Soc., Dalton Trans.* **1981**, 1583.

290. Burrell, A. K.; Clark, G. R.; Rickard, C. E. F.; Roper, W. R.; Wright, A. H. *J. Chem. Soc., Dalton Trans.* **1991**, 609.

291. Herberhold, M.; Hill, A. F. *J. Organomet. Chem.* **1986**, *309*, C29.

292. Fischer, E. O.; Stadler, P. Z. *Naturforsch., B: Anorg. Chem., Org. Chem.* **1981**, *36B*, 781.

293. Doetz, K. H.; Lyon, C.; Rott, J. *J. Organomet. Chem.* **1988**, *345*, 117.

294. Herberhold, M.; Alt, H.; Kreiter, C. G. *Liebigs Ann. Chem.* **1976**, 300.

295. Christensen, N. J.; Hunter, A. D.; Legzdins, P. *Organometallics* **1989**, *8*, 930.

296. Kowalczyk, J. J.; Arif, A. M.; Gladysz, J. A. *Organometallics* **1991**, *10*, 1079.

297. Bodner, G. S.; Peng, T.-S.; Arif, A. M.; Gladysz, J. A. *Organometallics* **1990**, *9*, 1191.

298. Peng, T.-S.; Gladysz, J. A. *Organometallics* **1990**, *9*, 2884.

299. Agbossou, S. K.; Bodner, G. S.; Patton, A. T.; Gladysz, J. A. *Organometallics* **1990**, *9*, 1184.

300. Sweet, J. R.; Graham, W. A. G. *J. Am. Chem. Soc.* **1983**, *105*, 305.

301. Sweet, J. R.; Graham, W. A. G. *Organometallics* **1983**, *2*, 135.

302. Trovati, A.; Uguagliati, P.; Zingales, F. *Inorg. Chem.* **1971**, *10*, 851.

303. Gadd, G. E.; Poliakoff, M.; Turner, J. J. *Inorg. Chem.* **1986**, *25*, 3604.

304. Gadd, G. E.; Poliakoff, M.; Turner, J. J. *Organometallics* **1987**, *6*, 391.

305. Cardaci, G. *J. Organomet. Chem.* **1980**, *202*, C81.

306. Cardaci, G. *J. Chem. Soc., Dalton Trans.* **1984**, 815.

307. Clemens, J.; Green, M.; Stone, F. G. A. *J. Chem. Soc., Dalton Trans.* **1973**, 375.

308. Segal, J. A.; Johnson, B. F. G. *J. Chem. Soc., Dalton Trans.* **1975**, 677.

309. Segal, J. A.; Johnson, B. F. G. *J. Chem. Soc., Dalton Trans.* **1975**, 1990.

310. Kawakami, K.; Ishii, K.; Tanaka, T. *Bull. Chem. Soc. Jpn.* **1975**, *48*, 1051.

311. Fjeldsted, D. O. K.; Stobart, S. R.; Zaworotko, M. J. *J. Am. Chem. Soc.* **1985**, *107*, 8258.

312. Monica, G. L.; Navazio, G.; Sandrini, P. Cenini, S. *J. Organomet. Chem.* **1971**, *31*, 89.

313. Clamp, S.; Connelly, N. G.; Payne, J. D. *J. Chem. Soc., Chem. Commun.* **1981**, 897.

314. Connelly, N. G.; Loyns, A. C.; Fernandez, M. J.; Modrego, J.; Oro, L. A. *J. Chem. Soc., Dalton Trans.* **1989**, 683.

315. Blohm, M. L.; Gladfelter, W. L. *Organometallics* **1985**, *4*, 45.

316. Faller, J. W.; Chodosh, D. F.; Katahira, D. *J. Organomet. Chem.* **1980**, *187*, 227.

317. Greenhough, T. J.; Legzdins, P.; Martin, D. T.; Trotter, J. *Inorg. Chem.* **1979**, *18*, 3268.

318. Faller, J. W.; Shvo, Y.; Chao, K.; Murray, H. H. *J. Organomet. Chem.* **1982**, *226*, 251.

319. Van Arsdale, W. E.; Winter, R. E. K.; Kochi, J. K. *Organometallics* **1986**, *5*, 645.
320. Faller, J. W.; Chao, K. H.; Murray, H. H. *Organometallics* **1984**, *3*, 1231.
321. Faller, J. W.; chao, K. H. *Organometallics* **1984**, *3*, 927.
322. Benamou, C.; Benaim, J. *J. Organomet. Chem.* **1985**, *280*, 377.
323. Faller, J. W.; Chao, K. H. *J. Am. Chem. Soc.* **1983**, *105*, 3893.
324. Faller, J. W.; Linebarrier, D. *Organometallics* **1988**, *7*, 1670.
325. Faller, J. W.; Linebarrier, D. L. *Organometallics* **1990**, *9*, 3182.
326. Müller, H.-J.; Nagel, U.; Beck, W. *Organometallics* **1987**, *6*, 193.
327. McCleverty, J. A.; Murray, A. J. *Transition Met. Chem. (Weinheim)* **1979**, *4(5)*, 273.
328. McCleverty, J. A.; Murray, A. J. *J. Chem. Soc., Dalton Trans.* **1979**, 1424.
329. Bailey, N. A.; Kita, W. G.; McCleverty, J. A.; Murray, A. J.; Mann, B. E.; Walker, N. W. J. *J. Chem. Soc., Chem. Commun.* **1974**, 592.
330. Baxter, J. S.; Gren, M.; Lee, T. V. *J. Chem. Soc., Chem. Commun.* **1989**, 1595.
331. Eberhardt, U.; Mattern, G. *Chem. Ber.* **1988**, *121*, 1531.
332. Perugini, D.; Innorta, G.; Torroni, S.; Foffani, A. *Inorg. Chim. Acta* **1988**, *146*, 223.
333. Moll, M.; Behrens, H.; Seibold, H.-J.; Merbach, P. *Z. Naturforsch., B: Anorg. Chem., Org. Chem.* **1983**, *38B*, 409.
334. Cardaci, G.; Foffani, A. *J. Chem. Soc., Dalton Trans.* **1974**, 1808.
335. Cardaci, G. *J. Organomet. Chem.* **1983**, *244*, 153.
336. Baker, P. K.; Clamp, S.; Connelly, N. G.; Murray, M.; Sheridan, J. B. *J. Chem. Soc., Dalton Trans.* **1986**, 459.
337. Connelly, N. G.; Gilbert, M.; Orpen, A. G.; Sheridan, J. B. *J. Chem. Soc., Dalton Trans.* **1990**, 1291.
338. Connelly, N. G.; Gilbert, M. *J. Chem. Soc., Dalton Trans.* **1990**, 373.
339. Schoonover, M. W.; Eisenberg, R. *J. Am. Chem. Soc.* **1977**, *99*, 8371.
340. Maxfield, P. L. *Inorg. Nucl. Chem. Lett.* **1970**, *6*, 707.
341. Schoonover, M. W.; Baker, E. C.; Eisenberg, R. *J. Am. Chem. Soc.* **1979**, *101*, 1880.
342. Hughes, R. P.; Lambert, J. M. J.; Whitman, D. W.; Hubbard, J. L.; Henry, W. P.; Rheingold, A. L. *Organometallics* **1986**, *5*, 789.

343. Hughes, R. P.; Lambert, J. M. J.; Hubbard, J. L. *Organometallics* **1986,** *5,* 797.

344. Potenza, J. A.; Johnson, R.; Williams, D.; Toby, B. H.; Lalacentte, R. A.; Efraty, A. *Acta Crystallogr., Sect. B* **1981,** *B37,* 442.

345. Hunter, A. D.; Legzdins, P.; Nurse, C. R.; Einstein, F. W. B.; Willis, A. C. *J. Am. Chem. Soc.* **1985,** *107,* 1791.

346. Hunter, A. D.; Legzdins, P.; Einstein, F. W. B.; Willis, A. C.; Bursten, B. E.; Gatter, M. G. *J. Am. Chem. Soc.* **1986,** *108,* 3843.

347. Herberhold, M.; Razavi, A. *Angew. Chem., Int. Ed. Engl.* **1975,** *14,* 351.

348. Pearson, A. J.; Bruhn, P. R.; Richards, I. C. *Isr. J. Chem.* **1984,** *24(2),* 93.

349. Honig, E. D.; Sweigart, D. A. *J. Chem. Soc., Chem. Commun.* **1986,** 691.

350. Chung, Y. K.; Sweigart, D. A.; Connelly, N. G.; Sheridan, J. B. *J. Am. Chem. Soc.* **1985,** *107,* 2388.

351. Hyeon, T. H.; Chung, Y. K. *J. Organomet. Chem.* **1989,** *372,* C12.

352. Efraty, A.; Liebman, D.; Sikora, J.; Denney, D. Z. *Inorg. Chem.* **1976,** *15,* 886.

353. Efraty, A.; Bystrek, R.; Geaman, J. A.; Sandhu, S. S., Jr.; Huang, M. H. A.; Herber, R. H. *Inorg. Chem.* **1974,** *13,* 1269.

354. Efraty, A. *Chem. Rev.* **1977,** *77,* 691.

355. Lalor, F. J.; Brookes, L. H.; Ferguson, G.; Parvez, M. *J. Chem. Soc., Dalton Trans.* **1984,** 245.

356. Rausch, M. D.; Edwards, B. H.; Atwood, J. L.; Rogers, R. *Organometallics* **1982,** *1,* 1567.

357. Mailvaganam, B.; Frampton, C. S.; Top, S.; Sayer, B. G.; McGlinchey, M. J. *J. Am. Chem. Soc.* **1991,** *113,* 1177.

358. Connelly, N. G.; Demidowicz, Z.; Kelly, R. L. *J. Chem. Soc., Dalton Trans.* **1975,** 2335.

359. Bursten, B. E.; Gatter, M. G.; Goldberg, K. I. *Polyhedron* **1990,** *9,* 2001.

360. Middleton, R.; Hull, J. R.; Simpson, S. R.; Tomlinson, C. H.; Timms, P. L. *J. Chem. Soc., Dalton Trans.* **1973,** 120.

5

Nitrosyl Clusters

5.1: The Interaction of NO with Metal Surfaces and Supports

5.1.1: The Interaction of NO with Metal Surfaces

The interaction of NO with metal surfaces has been studied with a view to understanding how surfaces of metals (or metal oxides) assist in the catalytic decomposition (eq 5.1) or reduction (eq 5.2) of NO *(1)*.

$$NO \longrightarrow 1/2\ N_2 + 1/2\ O_2 \qquad (5.1)$$

$$NO + CO\ (\text{or}\ H_2) \longrightarrow 1/2\ N_2 + CO_2\ (\text{or}\ H_2O) \qquad (5.2)$$

However, much of the work done on the interaction of NO with metals has been performed on well-defined single crystal surfaces, and obviously, these differ from the surfaces of real catalysts. For example, the interaction of NO on Al(100) *(2)*, Si(100) *(3)*, W(100) and (110) *(4)*, Re(001) *(5)*, Pt(111) *(6-8)*, Ag(100) and (111) *(7,9)*, Mo(100) and (111) *(10)*, Pd(100) and (111) *(11,12)* have been studied. Most of these interactions appear to result in the formation of adsorbed nitrosyl species (step *a*),

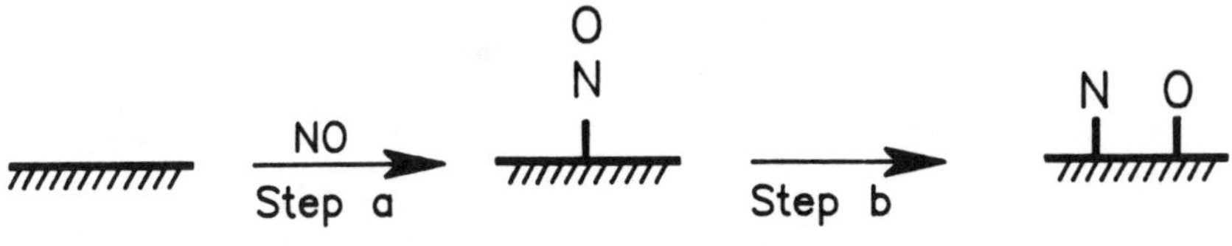

and many such nitrosyl species have even been detected (or postulated as intermediate species) for metals which do not traditionally form nitrosyl complexes, such as magnesium *(13)*, tin (in SnO_2) *(14)*, silicon *(3)*, aluminum *(2)* or even silver *(15,16)*.

More interesting is the dissociative adsorption of NO to form a nitrogen overlayer (or nitride) (N_{ads}) and oxide (O_{ads}) on the metal surface (step *b*) to eventually result in the formation of N_2 and O_2. Alternatively, CO_2 (or even isocyanate species) and H_2O may form in the presence of reductants such as CO and H_2, respectively.

$$N_2 + O_2 (N_2O)$$

$$\uparrow$$

$$\text{OCN}_{(surface)} + CO_2 \xleftarrow{2CO} \text{N}\ \text{O}_{(surface)} \xrightarrow{H_2} \text{NH}_x\ \text{OH}_{x\,(surface)} + y\ H_2O$$

Such dissociative adsorption of NO to N_2 and O_2 has been observed on Ir, Mo, Al, Pt, Pd, Rh, Ru, Ni, Ti, Zn, or even the Pt-Rh(100) alloy *(10,17-22)*.

In many respects, the bonding of NO to a metal surface mimics the bonding of NO to a semi-infinite array of metal atoms. Theoretical studies on the Group 10 metals indicate that the MNO link may either be linear or bent depending on the metal and its electronic configuration *(23-25)*.

Atom	Electronic configuration	Predicted MNO geometry
Ni	d^9s^1	linear
	d^{10}	linear or bent
Pd	d^9s^1	bent
	d^{10}	linear
Pt	d^9s^1	bent
	d^{10}	linear

Furthermore, the electroadsorption of NO on a smooth Pd electrode in acidic solution results in the generation of what is believed to be a bent PdNO species *(26)*. Experimentally, however, isolation of these metal atoms or imbedding these atoms (or ions) onto inactive supports such as inorganic oxides should therefore serve as models for the study of the interaction of NO with such atoms (or ions).

5.1.2: Metal Supports

5.1.2.1: Silica (SiO_2)

The addition of NO to metals supported on silica results in the formation of mononitrosyl and/or dinitrosyl species, e.g.,

$$\text{M (surface)} \xrightarrow{2\ \text{NO}} \text{M(NO)}_2\ \text{(surface)} \qquad (5.3)$$

M=Cr,Ni

Much work has been done on Cr-SiO_2 *(27-30)*, where the newly-formed dinitrosyl -$Cr(NO)_2$ species may also add Lewis bases to form -$Cr(NO)_2L$ derivatives *(28,30)*. Similar work on Ni-SiO_2 also results in the formation of surface dinitrosyl species which then decompose with the formation of surface nitrides and oxides *(31,32)*. In the presence of CO, surface isocyanate species are formed.

5.1.2.2: Alumina (Al_2O_3)

Surface dinitrosyl entities are also formed when NO gas is reacted with metals supported on alumina. Such metals include Rh *(33,35)*, Cr, Mo, and W *(34)* and Co *(36,37)*. For the case in which the supported Co-alumina compound is formed by reaction of $Co(NO)(CO)_3$ with alumina, cobalt-isocyanate species as well as cobalt-dinitrosyl (or trinitrosyl) groups are also formed *(36-38)*. Similarities in the IR spectra of $[Mo(NO)_2(CH_3CN)_4]BF_4$, supported on alumina or as a silica wafer, and NO adsorbed on reduced MoO_3-alumina provide additional evidence that a $-Mo(NO)_2$ species is formed in the latter *(39)*. Likewise, NO adsorption on chromia-Al_2O_3 or even chromia-$AlPO_4$ results in the formation of dinitrosyls *(40)*.

5.1.2.3: Other metal oxides

Many metal oxides such as Fe_2O_3, Cu_2O, NiO, Co_3O_4, V_2O_5 and MnO all effect NO decomposition to N_2, and the use of such oxides as catalysts for NO_x decomposition for auto exhausts is well documented *(41-44)* (Section 7.4). Many perovskite-type catalysts are also known to effect NO decomposition to N_2 *(45-46)*. The simultaneous reduction of NO and SO_2 with H_2S occurs in a process catalyzed by MoO_3-MgO to give N_2, H_2O and S *(47)*.

Rapid adsorption of NO into BaO-CuO mixed oxides generates nitrite (not nitrosyl) species, and in the presence of oxygen, nitrates are produced instead *(48)*.

5.1.2.4: Carbon supports

NO decomposes on a Co-La_2O_3-Pt system supported on active carbon without added reductant to generate N_2 and CO_2 *(49)*. As expected, carbon-supported "Ru-NO" reacts with hydrogen to yield ammonia *(50)*.

5.1.2.5: Aluminosilicates (zeolites)

Zeolites are crystalline aluminosilicates with cavities or tunnels of definite size present in fibrous crystals or in a three-dimensional array of SiO_4^- and AlO_4^- tetrahedra *(51)*. They may be classified according to their pore size (Table 5.2).

Table 5.2 Zeolite Classification

small pore	medium pore	large pore
A	ZSM-5	Faujasite
Erionite	ZSM-11	X-/Y-zeolite
Chabazite	ZSM-22	mordenite
	ZSM-23	L
	ZSM-48	ZSM-4
	silicalite	ZSM-12
		Z

Data from reference 51.

For example, small pore A-zeolites possess cavities of diameter 4.1 Å, whereas large pore Y-zeolites possess cavities of diameter 7.4 Å, and the medium pore (pentasil) zeolites have tubular diameter of 5.5 Å x 5.6 Å *(51)*. Their variable acidity, shape-selective properties, and thermal stability have rendered them particularly useful as catalysts for various organic transformations *(51,52)*. Combined with the fact that both the Si and Al atoms in zeolites may be replaced by transition elements, they may be made to serve as excellent "supports" for such metallic elements in their reactions with NO.

a) Formation of Nitrosyl Species

When Cr-X or Cr-Y zeolites are reacted with NO, geminal Cr-dinitrosyl species form, an observation that has been confirmed by ^{15}NO labelling studies *(53)*. A model for the orientation of the two NO ligands on Cr in Cr-X zeolite has been proposed on the basis of IR and ESR data and molecular orbital calculations, i.e.,

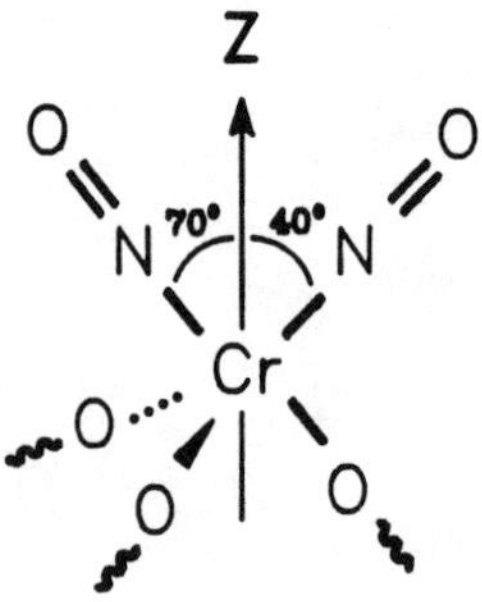

In this model, the HOMO is *ca.* 50 % metal in character and contains a significant contribution from the ligands. In addition to the dinitrosyl species, some surface nitrite is also observed, which may arise either from the disproportionation of NO, or from the direct reaction of NO with oxide ions in the lattice. Typical IR frequencies of these species are listed in Table 5.3.

Table 5.3 IR Stretching Frequencies (cm^{-1}) of Adsorbed Species on Cr-X and Cr-Y Zeolites

species	Cr-Y	Cr-X
$[Cr(^{14}NO)_2]^{3+}$	1900, 1775	1895, 1770
$[Cr(^{15}NO)_2]^{3+}$	1870, 1745	1865, 1740
$^{14}NO_2^-$		1650,1260
$^{15}NO_2^-$		1620, 1235
$^{15}NO_3^-$		1370

Data from reference 53.

Similar dinitrosyl species are formed in Co or Rh zeolites (X and Y) by reaction with NO *(54,55)* and, interestingly, the reaction of the $Co(NO)_2$

complex in zeolite Y with oxygen affords nitro and nitrito complexes which oxidize 2-propanol to acetone, thereby regenerating the $Co(NO)_2$ moiety *(56)*.

The $Co(NO)(CO)_3$ complex is physisorbed within the confines of Na-Y zeolite at room temperature, probably via the interaction between the nitrosyl oxygen and the Lewis acid sites of the zeolite *(57)*. Subsequent decomposition of the complex in the zeolite (either X or Y) results in the formation of dinitrosyl or trinitrosyl entities *(38,58)*. A similar decomposition of $Fe(NO)_2(CO)_2$ in zeolite-Y (with added NO) is believed to result in the formation of an $Fe(NO)_3$ complex *(59)*.

Ni-A zeolite also forms a nitrosyl complex whose form depends on the concentration of Ni in the Ni_xNa-A zeolite *(60)*. Interestingly, while Ca-Y and Na-Y zeolites form adsorbed N_xO_y species with N_2O and NO_2 as the final gaseous products, H-Y zeolite is inactive *(61)* and again underlines the importance of metal-substituted zeolites in NO decomposition. Ag-Y zeolite reacts with NO to produce Ag(NO) and $Ag_2(\mu\text{-}NO)$ species *(62)*.

b) Catalytic Decomposition of NO

The Cu-ZSM-5 zeolite is an effective catalyst for NO decomposition to N_2 and O_2 *(63,64)*. The catalytic activity of this Cu(II)-exchanged zeolite is very high. The catalytic activity of the related Co(III)-exchanged zeolite is very low, and no catalytic activity is observed to the Fe(III), Zn(II) and Mg(II)-exchanged zeolites *(64)*.

In the presence of ammonia, Co-Y zeolites reduce NO to N_2 and N_2O, possibly via $[L_nCo(NO)_2]^{2+}$ complexes (L= amine) *(65,66)*. The related $[Ru(O_{zeolite})_3(NH_3)_x(NO)]$ complex (x = 1, 2) is formed by the oxidation of a bound amine ligand under conditions of thermolysis *(67,68)*.

Indeed, many catalysts have been employed for the catalytic reduction of NO by NH_3 in the presence or absence of oxygen. Such processes may be described by the following reactions:

$$4\,NH_3 + 6\,NO \longrightarrow 5\,N_2 + 6\,H_2O \qquad (5.4)$$

$$4\ NH_3 + 4\ NO + O_2 \longrightarrow 4\ N_2 + 6\ H_2O \qquad (5.5)$$

$$8\ NH_3 + 12\ NO + 5\ O_2 \longrightarrow 10\ N_2O + 12\ H_2O \qquad (5.6)$$

Thus, while current technology utilizes catalysts made up of vanadium and titanium oxides, there appears to be an upsurge in research utilizing metal-doped zeolites. Of course, the catalytic activity of these zeolites depends markedly of the nature of the metal, with the later transition metals (i.e. more electron-rich metals) apparently being the more reactive. For example, Pt-mordenite exhibits very high (100%) catalytic activity for the reduction of NO by NH_3, Cu-mordenite exhibits only moderate activity, and V-modernite exhibits no significant activity *(69)*. Also, Ir-X zeolite is found to be more effective than Ir-alumina or Ir-silica for the reduction of NO by CO to give CO_2 *(70)*.

Finally, whereas Rh metal effects the decomposition of NO to yield N_2 as the major product, Rh-Y zeolite effects the decomposition of NO in the presence of CO to yield N_2O and CO_2 *(71)*, i.e.,

$$[Y\text{-}Rh(NO)(CO)_2]^+ + NO \longrightarrow [Y\text{-}Rh(CO)]^+ + N_2O + CO_2 \qquad (5.7)$$

The formation of $[Y\text{-}Rh(NO)(CO)_2]^+$ has been invoked in the latter reaction (it may be noted that this complex is analogous to the Co-amine species described earlier). Nevertheless, both surface $[Rh(CO)_2]^+$ and $[Rh(NO)_2]^+$ species are also detectable spectroscopically during the course of this reaction *(71)*.

5.2: Metal Nitrosyl Clusters

In the previous section, the interaction of NO with metal surfaces and with metal-doped silica and alumina was detailed. In this section, the interactions of NO with clusters of metal atoms will be considered. In many ways, metal clusters bear a qualitative similarity to finite metal surfaces, with the added (and often desirable!) advantage that heteropolymetallic surfaces may be predictably constructed for the selective binding and activation of small molecules such as NO.

M M M — homotrimetallic   M' M M'' — heterotrimetallic

The primary interest in cluster-bound NO derives from the need to explore conditions necessary for the rupture of the N-O bond of the bound nitrosyl group. Most of the work done on metal nitrosyl clusters has been somewhat limited to the syntheses and characterization of these polymetallic molecules, and on the reduction chemistry of the bound NO group ultimately converting it to the NOH, NH and NH_2 groups *(72)* (Sections 7.2 and 7.3).

As will be shown in this section, all forms of NO binding may be found in metal nitrosyl clusters. Thus, both the terminal (linear or bent) and bridging (doubly, triply and quadruply) forms of MNO bonding may be found in these cluster systems. This section of metal nitrosyl clusters will be limited to complexes containing three or more metals, with at least one metal-metal bond present in the trimetallic core. Also, compounds which are simple derivatives of their monometallic analogues, e.g. $L_xM'[Fe(NO)(CO)_2L']_{4-x}$ where M' = Pb *(81)*, Hg *(82)*, Ge *(83)* or even Pt *(84)*, will not be considered. Furthermore, reactions in the gas phase resulting in complexes such as $Co_{2+x}NO^+$ (x = 0, 1) *(85,86)* or $Co_x(CO)_y(NO)_z^+$ (x,y,z = 5 - 8) *(87)* will not be discussed.

5.2.1: Syntheses of Metal Nitrosyl Clusters

In general, the synthetic routes to monometallic and bimetallic nitrosyl complexes (Section 2.1) should be applicable to the syntheses of metal nitrosyl clusters. The methods that have been used to date for this purpose include the use of NO gas *(73)*, NO^+ *(74)*, $PPN(NO_2)$ *(75)*, or the condensation of metal nitrosyls *(76,77)*. Examples of each are shown below:

$$Os_3(CO)_{12} + 2\,NO \longrightarrow Os_3(CO)_9(NO)_2 + 3\,CO \qquad (5.8)$$

$$[Re_7C(CO)_{21}]^{3-} + NO^+ \longrightarrow [Re_7C(CO)_{20}(NO)_2]^- + CO \qquad (5.9)$$

$$Os_6(CO)_{18} + PPN(NO_2) \longrightarrow [Os_6(CO)_{17}(NO)]^- + CO_2 \qquad (5.10)$$

$$[CpFe(NO)]_2 + CpCo(C_2H_4)_2 \longrightarrow Cp_3FeCo_2(NH)(NO) \qquad (5.12)$$

The latter method is especially applicable for the syntheses of heteropolymetallic nitrosyl clusters. Other preparations have employed other nitrosylating agents such as amyl nitrite *(78)* and nitrous acid *(79)*, but perhaps a unique synthetic route is the deinsertion of NO from an alkanenitrile oxide ligand to form a tricobalt nitrosyl complex as shown in eq 5.13 *(80)*.

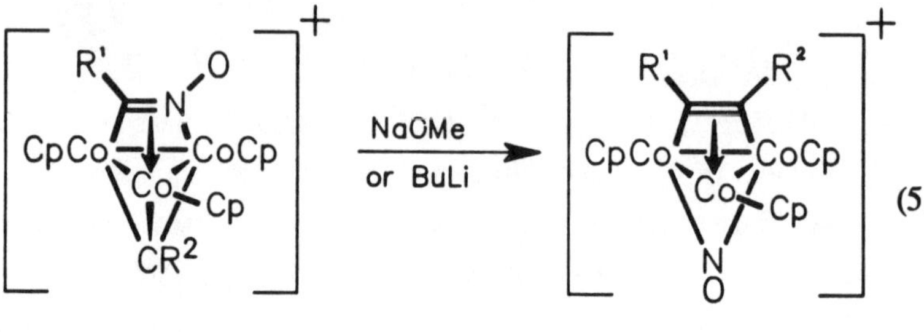

(5.13)

5.2.2: Trimetallic Clusters

5.2.2.1: Molybdenum

The trimetallic cation $[\{CpMo(NO)(OH)\}_3(\mu_3\text{-}O)]^+$ is formed when $Cp_2Mo(NO)I$ is reacted with silver salts in aqueous acetone *(90)*. A similar reaction employing $CpMo(NO)I_2$ results in the formation of the analogous $[\{CpMo(NO)I\}_3(\mu_3\text{-}O)]^+$. The spectroscopic properties of these Mo_3 clusters are in accord with a triangular arrangement of three "CpMo(NO)X" units with Mo-Mo interactions and a triply-bridging oxygen atom.

5.2.2.2: Manganese

The molecular structure of $Cp_3Mn_3(NO)_4$ has been determined to consist of an equilateral triangle of manganese atoms doubly-bridged by three NO ligands along the edges of the metal triangle below the plane *(88)*, i.e.,

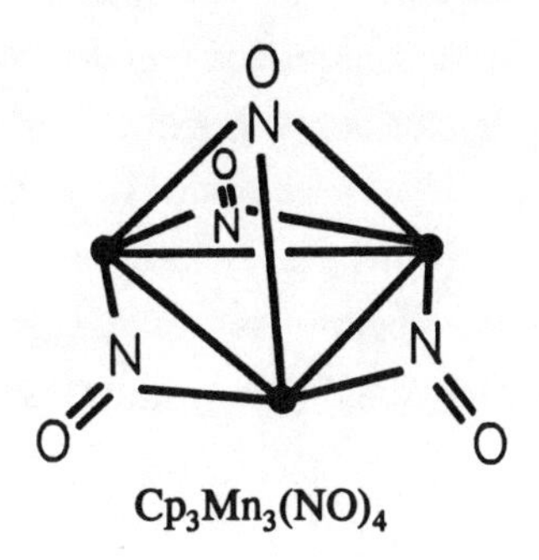

$Cp_3Mn_3(NO)_4$

Protonation of the triply-bridging NO in the methylcyclopentadienyl analogue affords a μ_3-NOH species *(89)* (Section 7.2.3).

5.2.2.3: Iron, Ruthenium and Osmium

Terminal nitrosyl ligands are present in the $Fe_3(CO)_8(NO)(NPPh_3)$ *(91)*, $[Fe_3(CO)_8(NO)(NH)]^-$ *(92)* and $Fe_2Co(CO)_8S(NO)$ *(79)* complexes. However, both μ_2-NO and μ_3-NO ligands are present in the trimetallic cluster obtained when $(MeCp)Mn(CO)_2$ is condensed with the $[CpFe(NO)]_2$ dimer *(93)*.

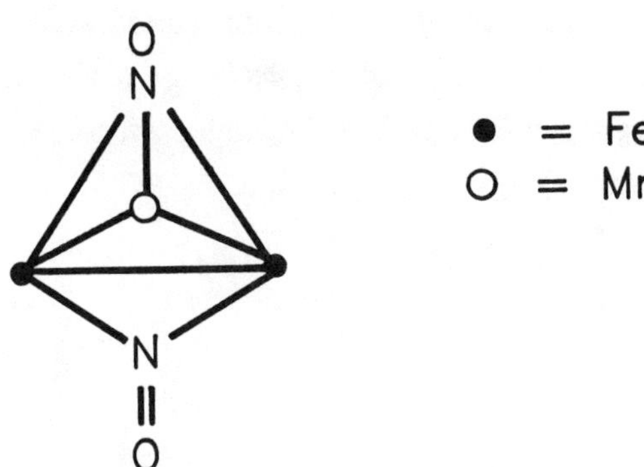

$(MeCp)Cp_2MnFe_2(CO)_2(\mu\text{-}NO)(\mu_3\text{-}NO)$

This cluster is isoelectronic with $Cp_3Mn_3(\mu\text{-}NO)_3(\mu_3\text{-}NO)$. Interestingly, the $MnFe_2$ cluster undergoes a reversible one-electron oxidation as well as a reversible one-electron reduction. In the monoanion, the added electron occupies a nondegenerate HOMO which is primarily antibonding between the two Fe atoms. Consistent with this fact is that the Fe-Fe distance increases from 2.441 Å in the neutral complex to 2.605 Å in the monoanion *(93)*. Other Fe-containing clusters with triply-bridging NO ligands include $[Cp^*Fe(CO)]_3NO$ (which is isoelectronic with $Cp_3Mn(NO)_4$) *(94)* and $Cp_3FeCo_2(NO)(NH)$ *(77)* (which is isoelectronic with $Cp_3Co_3(\mu_3\text{-}NO)_2$). The Fe_2W cluster shown below is unique in that it contains an asymmetrically-bridging NO ligand (R = *p*-tolyl) *(95)*.

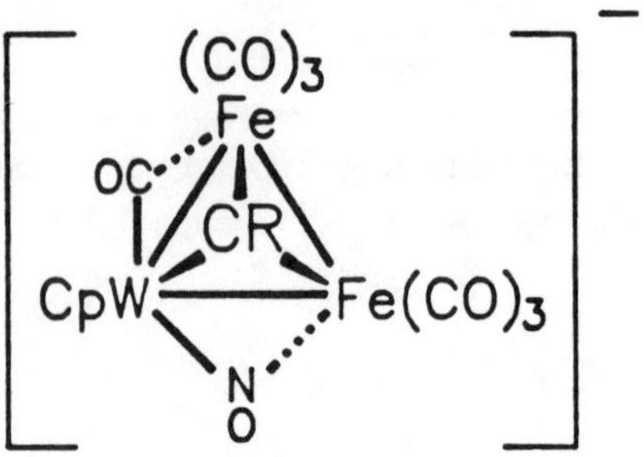

(CO)3
Fe
OC
CR
CpW
Fe(CO)3
N
O

The $[Ru_3(CO)_{10}(NO)]^-$ monoanion decomposes upon heating to yield the nitride, $[Ru_5N(CO)_{14}]^-$, whereas upon photolysis in the presence of CO, it produces the isocyanate $[Ru_3(CO)_{11}(NCO)]^-$ *(96)*. Interestingly, reaction of $[Ru_3(CO)_{10}(NO)]^-$ with strong acids results in the protonation of the nitrosyl oxygen to yield a μ_3-NOH species *(97)*(Scction 7.2.3). This then undergoes an O-H to Ru-H tautomerization in the presence of $PPN(CF_3SO_3)$ to afford $HRu_3(CO)_{10}(NO)$, i.e.,

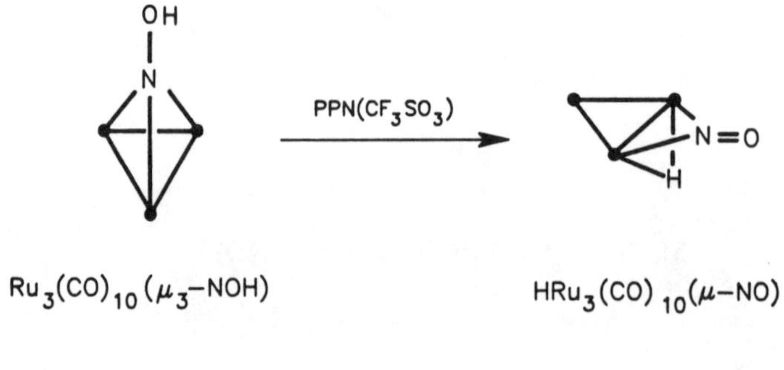

OH
N
$PPN(CF_3SO_3)$
N=O
H

(5.14)

$Ru_3(CO)_{10}(\mu_3\text{-}NOH)$ $HRu_3(CO)_{10}(\mu\text{-}NO)$

Other derivatives of this hydride are also known *(98)*, as is a Hg-bridged derivative of the form $[Ru_3(NO)(CO)_{10}]_2Hg$, which is obtained from the reaction of $[Ru_3(NO)(CO)_{10}]^-$ with $HgCl_2$ *(99)*.

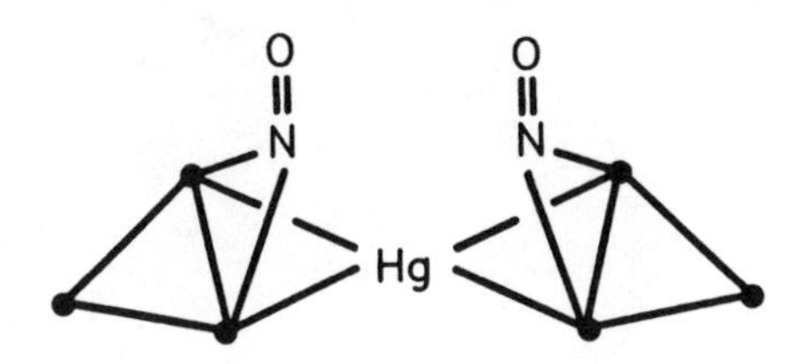

Reaction of the hydride, $HRu_3(CO)_{10}(NO)$, with H_2 results in the formal reduction of the N-O bond to generate N, NH and NH_2 species *(100,101)*, i.e.,

$$HRu_3(CO)_{10}(NO) \longrightarrow H_2Ru_3(NH)(CO)_9 + HRu_3(NH_2)(CO)_{10} + H_4Ru_4(CO)_{12} + H_3Ru_4(NH_2)(CO)_{12} + HRu_4N(CO)_{12} \quad (5.15)$$

possibly via the intermediacy of trihydride species.

Reduction of the bound NO in $Ru_3(CO)_{10}(\mu\text{-}NO)_2$ is achieved by its reaction with CO to produce the $Ru_3(CO)_{12}$, $Ru_3(CO)_{10}(NO)(NCO)$, $Ru_4N(CO)_{12}(NO)$ and $Ru_4N(CO)_{12}(NCO)$ complexes *(102)*.

Like its Ru analogue, $[Os_3(CO)_{10}(NO)]^-$ (which contains a doubly bridging nitrosyl ligand *(103)*) may be derivatized to its corresponding $HOs_3(CO)_{10}(NO)$ or $(PEt_3Au)Os_3(CO)_{10}(NO)$ complexes *(98,104)*. The structure of the latter compound is analogous to that of $HRu_3(CO)_{10}(NO)$ discussed previously. The μ-NO ligand in $HOs_3(CO)_{10}(NO)$ undergoes reduction by H_2 to yield $H_2Os_3(NH)(CO)_9$, $H_4Os_3(NH)(CO)_8$, $HOs_3(NH_2)(CO)_{10}$ and $H_4Os_4(CO)_{12}$ *(105)*.

The dinitrosyl complexes $Os_3(CO)_8(NO)_2L$ (L = CO, $P(OMe)_3$) *(73,106)* possess single $Os(NO)_2$ dinitrosyl units. In contrast, the $Os_3(CO)_9(NO)_2L'$ (L' = NMe_3, py) compounds (like the related $Ru_3(CO)_{10}(NO)_2$ *(107)*), possess $M_2(\mu\text{-}NO)_2$ fragments that have very little or no direct M-M interactions, as shown schematically *(108,109)*.

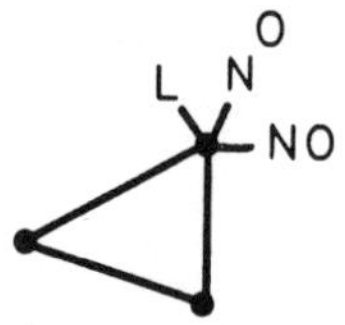

$L = CO, P(OMe)_3$

$Os_3(CO)_8(NO)_2L$

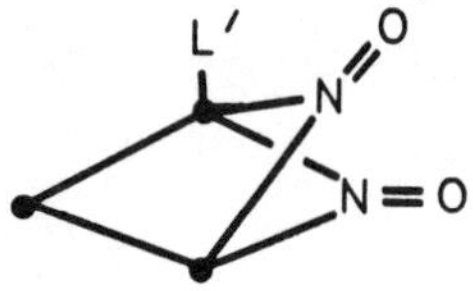

$L' = NMe_3$, py

$Os_3(CO)_9(NO)_2L'$

5.2.2.4: Cobalt and Rhodium

The $Cp'_3M_3(NO)_2$ complexes for Co *(110)* and Rh *(111)* are known to contain two μ_3-NO groups on opposite faces of the triangular array of metal atoms. The Co complex (Cp' = Cp) undergoes a reversible one-electron oxidation to its monocation, and also undergoes a one-electron reduction to its monoanion *(110)*.

The cobalt mononitrosyl clusters $[RCCo_3(CO)_7(NO)]^-$ (R = Me, Ph, COOH, Fc) contain terminal, linear CoNO groups *(112)*. In contrast, the $[Cp'_3Co_3(NO)(NH)]^n$ compounds (Cp' = Cp or MeCp; n = -1, 0, +1) contain a μ_3-NO group and a μ_3-NH group situated on opposite faces of the Co_3 triangle *(113,114)*. Reduction of the monocation (n = +1) with cobaltocene produces, as expected, the neutral Co_3 cluster. A comparison of the solid state structures of the cationic and neutral complexes reveals a Co_3 distortion that results in the transformation of a nearly equilateral Co_3 triangle to an isosceles triangle with one long and two short Co-Co bonds *(115)*.

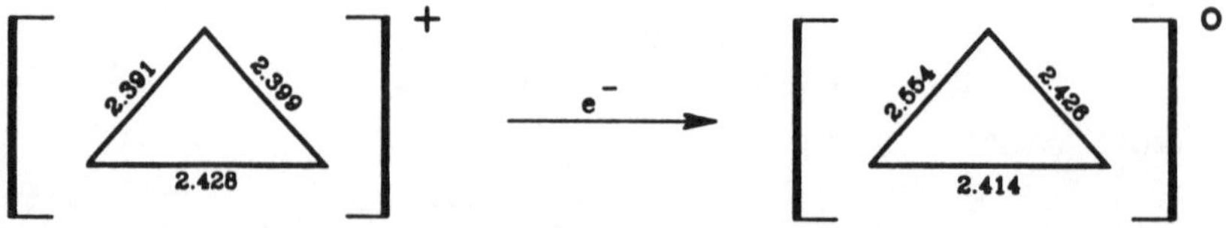

The fact that the mean Co_3-NH and Co_3-NO distances are not significantly different in the cationic and neutral complexes is consistent with the current view that the HOMO containing the added electron (in the neutral compound) is largely composed of in-plane Co_3 *d* atomic orbitals with very little contribution from the tricapped ligands *(115)*.

5.2.3: Tetrametallic Clusters

5.2.3.1: Molybdenum

Condensation of $Co_2(CO)_8$ with $Cp_2Mo_2(CO)(L)$ (L = phenylenebis(tert-butylphosphido)) generates a Co_2Mo_2 tetrametallic cluster containing the rare μ_4-η^2-NO bonding mode *(116)* (Section 2.2.1.1).

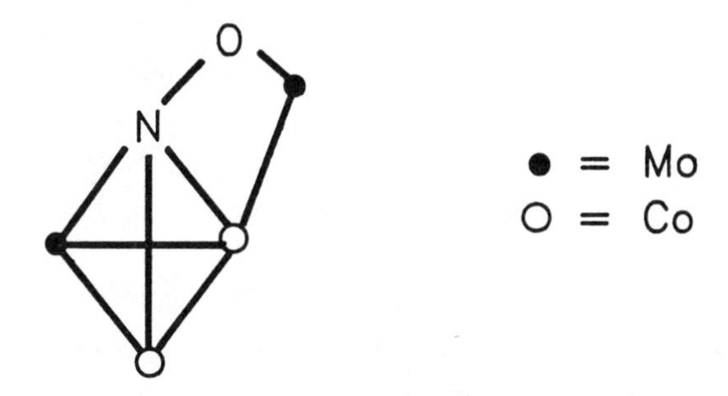

The bonding mode of NO in this Mo_2Co_2 tetrametallic cluster may be related to the (usually unobserved) intermediates of N-O dissociation on a metal surface, i.e.,

O
N
⟹ N—O ⟹ N O (5.16)

5.2.3.2: Iron, Ruthenium and Osmium

The $[Fe(NO)(CO)_3]^-$ anion condenses with $Ru_3(CO)_{12}$ to form $[FeRu_3(CO)_{12}NO]^-$, which possesses a tetrahedral $FeRu_3$ core with a linear FeNO bond *(76)*. This arrangement of the $M_4(NO)$ unit is similar to that of

● = Ru
O = Fe

$[FeRu_3(CO)_{12}(NO)]^-$ $H_3Os_4(CO)_{12}(NO)$

$H_3Os_4(CO)_{11}(NO)$ *(117)* and $Fe_4N(CO)_{11}(NO)$ *(118)*. In contrast, "butterfly" M_4 units exist in $Ru_4N(CO)_{12}(NO)$ and in $H_3Os_4(CO)_{12}(NO)$, with the NO groups in these complexes bridging the wing-tips of the butterfly units *(102)*. The $(MeCp)_4Fe_2Co_2(\mu_3\text{-}NO)$ cluster has also been prepared *(131)*.

5.2.3.3: Platinum and Palladium

The $M_4(O_2CMe)_6(NO)_2$ clusters for palladium *(119)* and platinum *(120)* contain rectangular M_4 cores with two NO groups bridging opposite edges of the rectangle as shown:

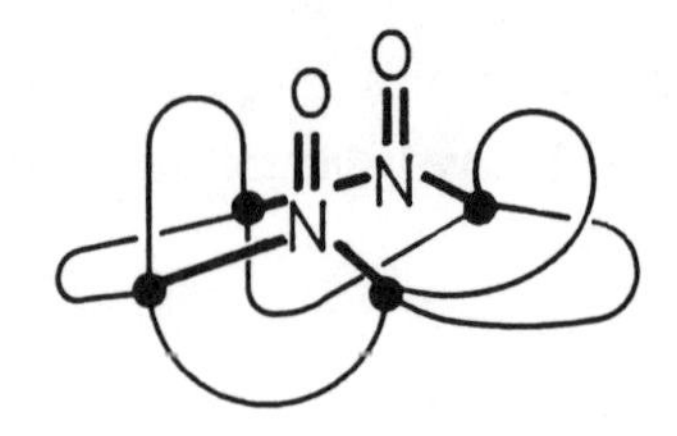

5.2.4: Pentametallic Clusters

5.2.4.1: Iron and Ruthenium

The $[M_5C(CO)_{13}(NO)]^-$ clusters for iron and ruthenium possess square pyramidal arrangements of metal atoms. In the iron complex, the NO ligand occupies an axial position on a basal metal atom *(92)*,

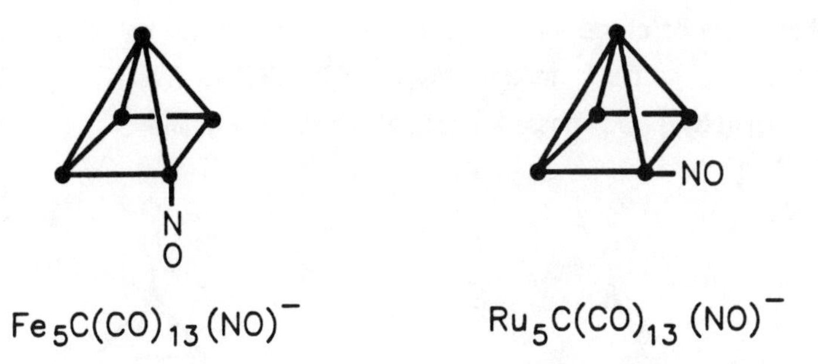

$Fe_5C(CO)_{13}(NO)^-$ $Ru_5C(CO)_{13}(NO)^-$

whereas for ruthenium, it occupies an equatorial position *(121)*. This ruthenium compound reacts with protons to form $HRu_5C(CO)_{13}(NO)$ or with $[AuPR_3]^+$ (R = Et, Ph) to form the analogous $(R_3P)AuRu_5C(CO)_{13}(NO)$ compound, which exits as isomers differing in the mode of bonding of the $AuPR_3$ ligand (i.e., μ_2 vs μ_3) *(121)*.

5.2.5 Hexametallic Clusters

5.2.5.1: Rhenium

The Re_6 cluster anion $[\{H_3Re_3(CO)_{10}\}_2(\mu_4\text{-}\eta^2\text{-NO})]^-$ contains a side-on bonded NO ligand which bridges four rhenium atoms *(122)* (Section 2.2.1.1), i.e.,

$Re(CO)_4$
H H
$(CO)_3Re$ — H — $Re(CO)_3$
(O)N=O(N)
$(CO)_3Re$ — H — $Re(CO)_3$
H H
$Re(CO)_4$
$[\ ]^-$

5.2.5.2: Iron, Ruthenium and Osmium

The hexametallic clusters of the Group 8 metals possess an octahedral metal framework. The mononitrosyls $[Fe_6C(CO)_{15}(NO)]^-$ *(123)* and $[Os_6(CO)_{17}(NO)]^-$ *(75)* possess terminal, linear NO ligands.

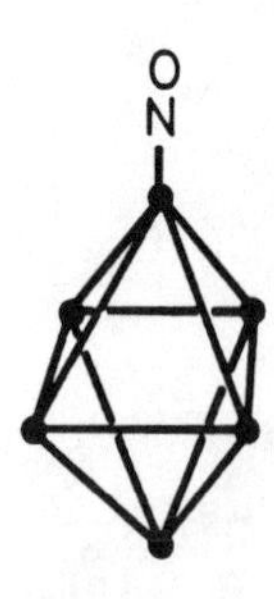

$[Fe_6C(CO)_{15}(NO)]^-$
$[Os_6(CO)_{17}(NO)]^-$

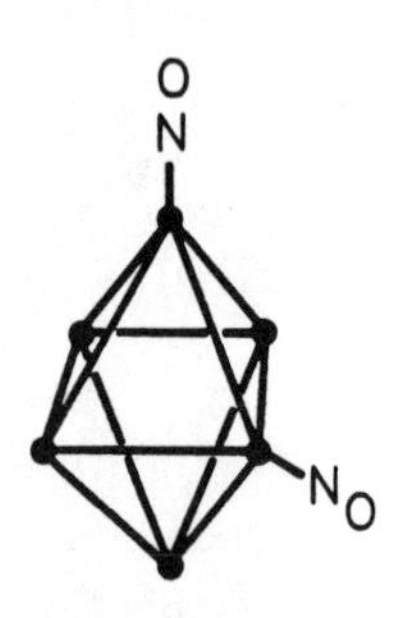

$[Fe_6C(CO)_{13}(NO)_2]^{2-}$

The $[Ru_6C(CO)_{15}NO]^-$ compound (possessing a terminal NO ligand) reacts with H^+ to produce the hydride, $HRu_6C(CO)_{15}NO$ *(124)*. The analogous $(PPh_3)AuRu_6C(CO)_{15}NO$ compound possesses a doubly-bridging $AuPPh_3$ ligand whose mode of attachment may almost be described as approaching that of triply-bridging *(124)*.

Interestingly, the dinitrosyl dianion, $[Fe_6C(CO)_{13}(NO)_2]^{2-}$, contains

$Ru_6C(CO)_{14}(NO)_2$

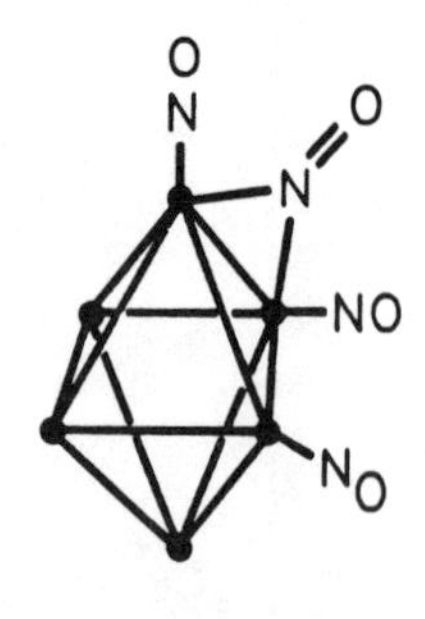

$Fe_6C(CO)_{11}(NO)_4$

two distinct and adjacent FeNO groups *(92)*, whereas the $Ru_6C(CO)_{14}(NO)_2$ compound possesses two RuNO groups on opposite vertices of the carbido-centered Ru_6 octahedron *(125)*. The tetranitrosyl hexairon cluster $Fe_6C(CO)_{11}(NO)_4$ (formed by the reaction of $[Fe_6C(CO)_{15}(NO)]^-$ with excess NO^+) possesses three terminal NO ligands on adjacent Fe atoms of one triangular face of the Fe_6 octahedron, with the fourth NO ligand bridging one edge of this face *(123)*.

5.2.5.3: Iridium

The crystal structure of $[Ir_6(CO)_{15}(NO)]^-$ has not been determined, but is believed (based on IR evidence) to contain an octahedral array of Ir atoms and a bent (terminal) NO ligand (ν_{NO} 1457 cm^{-1}; $\nu_{^{15}NO}$ 1430 cm^{-1}) *(126)*. Addition of nitrite to this anionic cluster results in the formation of N_2O, possibly via the intermediacy of the dinitrosyl $[Ir_6(CO)_{14}(NO)_2]^-$ species. The reactions of nitrite with the $Co_4(CO)_{12}$ and $Rh_6(CO)_{16}$ carbonyl precursors only result in the formation of $[Co_6N(CO)_{15}]^-$ and $[Rh_6N(CO)_{15}]^-$ which contain interstitial N atoms *(126,127)*.

5.2.5.4: Palladium

The $Pd_6(O_2CMe)_8(\mu\text{-}NO)_2$ compound is unusual in that it contains two distinct $Pd_2(\mu\text{-}NO)$ frameworks which are linked to the other two Pd atoms only by acetate bridges *(119)*.

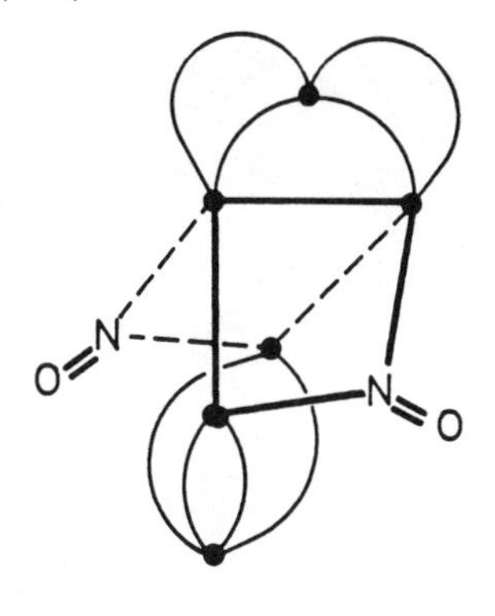

$Pd_6(O_2CMe)_8(NO)_2$

5.2.6: Heptametallic Clusters

5.2.6.1: Rhenium

Despite the fact that $[Re_7C(CO)_{21}]^{3-}$ exhibits two chemically reversible oxidations at +0.08 V and +0.42 V (vs Ag/AgCl), it reacts with NO^+ to yield the nitrosyl cluster $[Re_7C(CO)_{20}(NO)]^{2-}$ and CO *(74)*. This nitrosyl cluster also exhibits two chemically reversible oxidations, but at +0.62 V and +0.96 V instead. The observation that this nitrosyl dianion is more difficult to oxidize ($E_{1/2}$ = +0.62 V) than its carbonyl dianion precursor ($E_{1/2}$ = +0.42 V) is consistent with the presence of the strong π-acceptor NO ligand in the former complex.

5.2.7: Decametallic Clusters

5.2.7.1: Osmium

Terminal, linear NO ligands exist in $[Os_{10}C(CO)_{22}(NO)I]^{2-}$ *(128)* and in $[Os_{10}C(CO)_{23}(NO)]^-$ *(129)*, whereas in $[Os_{10}C(CO)_{24}(NO)]^-$, the wing-tip bridging mode of a butterfly arrangement of four Os atoms pertains *(129,130)*.

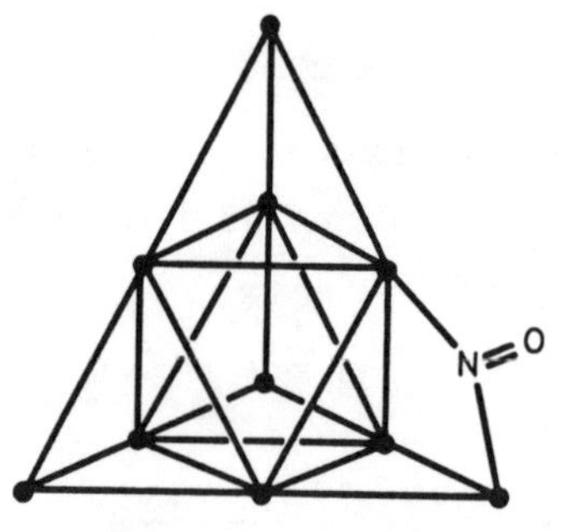

$[Os_{10}C(CO)_{24}(NO)]^-$

5.2.8: Structural Consequences of μ_3-NO Bonding

In this section, some data is presented to demonstrate how the MNO bond parameters change with the different NO bonding modes. Structural data for terminal-linear, terminal-bent, and doubly-bridging MNO geometries have already been presented and discussed in Section 2.2. Typical structural parameters for metal clusters with triply-bridging groups are now presented in Table 5.4.

Table 5.4 Triply-Bridging NO Complexes $M_3(\mu_3\text{-NO})$[a]

Compound	M-N (Å)	N-O (Å)	M-N-M (deg)	M-N-O (deg)	Ref
$[Cp^*Fe(CO)]_3(\mu_3\text{-NO})$	1.888	1.269	84.1	129.3	*94*
$[(MeCp)Cp_2MnFe_2(CO)(NO)(\mu_3\text{-NO})$	1.895	1.254	83.43	129.6	*93*
$[CpMn(NO)]_3(\mu_3\text{-NO})$	1.929	1.247	81.0	131.4	*88*
$[(MeCp)_3Co_3(NH)(\mu_3\text{-NO})]^+$	1.869	1.249	80.13	131.9	*113*
$(MeCp)_3Co_3(NH)(\mu_3\text{-NO})$	1.876	1.230	82.17	130.57	*115*
$Cp_3Co_3(\mu_3\text{-NO})_2$	1.843				*110*
$(MeCp)_4Co_2Fe_2(\mu_3\text{-NO})_2$	1.862	1.258	81.9	130.8	*131*

[a] All numbers quoted are calculated averages.

Average bond distances and angles for all the different forms of metal-nitrosyl bonding are also displayed in Table 5.6. As may be noted in Table 5.6, an increase in M-N bond distances occurs in going from the (terminal) linear mode to a bridging mode. [Also, a decrease in the MNO bond angles occurs.] However, the most important trend is in the increase in N-O bond lengths in going from linear terminal to doubly-, triply- and quadruply bridged NO bonding modes. The doubly-bridged N-O bond length of 1.22 Å is close to that of an NO double-bond (*ca.* 1.20 Å). The increase of this bond length to *ca.* 1.25 Å in triply-bonded species, and a further increase to 1.30 Å in quadruply-bridged (η^2) species is suggestive of the N-O bond approaching single-bond character (*ca.* 1.45 Å).

Table 5.6 Bond Parameters for Various NO Bonding Modes

Bonding mode	M-N (Å)	N-O (Å)	M-N-O (deg)
MNO terminal (linear)	1.70	1.17	176
μ-NO doubly-bridged			
(metal-metal bond)	1.91	1.22	134
(no metal-metal bond)	2.07		
μ_3-NO triply-bridging	1.89	1.25	131
μ_4-η^2-NO	2.08	1.30	122

The observed trends noted are important in designing metal nitrosyl clusters that could serve to cleave the N-O bond both stoichiometrically and catalytically. Furthermore, these trends are useful in correlating the experimental observations that (i) on occasion, metal nitrosyl complexes may lose their nitrosyl oxygens, (ii) metal nitrosyls do sometimes form oxo, nitrido or derivative species thereof under thermal, photolytic or reducing conditions, and (iii) the catalytic reduction (or disproportionation) of NO over metal surfaces occurs via the formation of intermediate nitrosyl species.

5.3: References

1. Roberts, M. W. *Chem. Soc. Rev.* **1977**, *6*, 373.
2. Pashutski, A.; Folman, M. *Surface Sci.* **1989**, *216*, 395.
3. Ekwelundu, E. C.; Ignatiev, A. *Surface Sci.* **1989**, *215*, 91.

4. Kioka, T.; Kawana, A.; Miki, H.; Sugai, S.; Kawasaki, K. *Surface Sci.* **1987**, *182*, 28.
5. Tatarenko, S.; Alnot, M.; Ducros, R. *Surface Sci.* **1985**, *163*, 249.
6. Hayden, B. E. *Surface Sci.* **1983**, *131*, 419.
7. Tenner, M. G.; Kuipers, E. W.; Langhout, W. Y.; Kleyn, A. W.; Nicolasen, G.; Stolte, S. *Surface Sci.* **1990**, *236*, 151.
8. Ormerod, R. M.; Lambert, R. M. *Surface Sci.* **1990**, *225*, L20.
9. Rodriguez, J. A. *Surface Sci.* **1990**, *226*, 101.
10. Kioka, T. *Surface Sci.* **1989**, *222*, 140.
11. Bertolo, M.; Jacobi, K. *Surface Sci.* **1990**, *236*, 143.
12. Yamada, T.; Matsuo, I.; Nakamura, J.; Xie, M.; Hirano, H.; Matsumoto, Y.; Tanaka, K.-I. *Surface Sci.* **1990**, *231*, 304.
13. Copperthwaite, R. G.; Carley, A. F.; Roberts, M. W. *Surface Sci.* **1986**, *165*, L1.
14. Tamaki, J.; Nagaishi, M.; Teraoka, Y.; Miura, N.; Yanazoe, N.; Moriya, K.; Nakamura, Y. *Surface Sci.* **1989**, *221*, 183.
15. Kasai, H.; Okija, A. *Surface Sci.* **1990**, *225*, L33.
16. Edamoto, K.; Maehama, S.; Miyazaki, E.; Miyahara, T.; Kato, H. *Surface Sci.* **1988**, *204*, L739.
17. Cornish, J. C. L.; Avery, N. R. *Surface Sci.* **1990**, *235*, 209.
18. Wickham, D. T.; Banse, B. A.; Koel, B. E. *Surface Sci.* **1989**, *223*, 82.
19. Carley, A. F.; Roberts, J. C.; Roberts, M. W. *Surface Sci.* **1990**, *225*, L39.
20. Nieuwenhuys, B. E. *Surface Sci.* **1983**, *126*, 307.
21. Yamada, T.; Hirano, H.; Tanaka, K.-I.; Siera, J.; Nieuwenhuys, B. E. *Surface Sci.* **1990**, *226*, 1.
22. Au, C. T.; Roberts, M. W.; Zhu, A. R. *J. Chem. Soc., Chem. Commun.* **1984**, 737.
23. Bauschlicher, C. W., Jr.; Bagus, P. S. *J. Chem Phys.* **1984**, *80*, 944.
24. Clark, D. T.; Cromarty, B. J.; Sgamellotti, A. *Chem. Phys. Lett.* **1979**, *68*, 420.
25. Smith, G. W.; Carter, E. A. *J. Phys. Chem.* **1991**, *95*, 2327.
26. Solomun, T. *J. Electroanal. Chem. Interfacial Electrochem.* **1986**, *199*, 443.
27. Ghiotti, G.; Garrone, E.; Gatta, G. D.; Fubini, B.; Giamello, E. *J. Catal.* **1983**, *80*, 249.

28. Garrone, E.; Ghiotti, G.; Giamello, E.; Fubini, B. *J. Chem. Soc., Faraday Trans. 1* **1981**, *77*, 2613.
29. Zecchina, A.; Garrone, E.; Morterra, A.; Coluccia, S. *J. Phys. Chem.* **1975**, *79*, 978.
30. Garrone, E.; Ghiotti, G.; Coluccia, S.; Zecchina, A. *J. Phys. Chem.* **1975**, *79*, 984.
31. Morrow, B. A.; Moran, L. E. *J. Catal.* **1980**, *62*, 294.
32. Morrow, B. A.; Sont, W. N.; Onge, A. St. *J. Catal.* **1980**, *62*, 304.
33. Anderson, J. A.; Millar, G. J.; Rochester, C. H. *J. Chem. Soc., Faraday Trans.* **1990**, *86*, 571.
34. Kazusaka, A.; Howe, R. F. *J. Catal.* **1980**, *63*, 447.
35. Alikina, G. M.; Davydov, A. A.; Sazonova, I. S.; Popovskii, V. V. *Kinet. Katal.* **1987**, *28*, 655. (Engl. Trans. p 571.)
36. Mauge, F.; Vallet, A.; Bachelier, J.; Duchet, J. C.; Lavalley, J. C. *Catal. Lett.* **1989**, *2*, 57.
37. Roustan, J. L.; Lijour, Y.; Morrow, B. A. *Inorg. Chem.* **1987**, *26*, 2509.
38. Newcomb, T. P.; Gopal, P. G.; Watters, K. L. *Inorg. Chem.* **1987**, *26*, 809.
39. Rosen, R. P.; Segawa, K.-I.; Millman, W. S.; Hall, W. K. *J. Catal.* **1984**, *90*, 368.
40. Conway, S. J.; Falconer, J. W.; Rochester, C. H. *J. Chem. Soc., Faraday Trans.* **1990**, *86*, 577.
41. Dwyer, F. G. *Catal. Rev.* **1972**, *6*, 261.
42. Moser, W. R. In *The Catalytic Chemistry of Nitrogen Oxides*; Klimisch, R. L.; Larson, J. G., Eds.; Plenum: New York, 1975; pp 33-43.
43. Wei, J. *Adv. Catal.* **1975**, *24*, 57.
44. Hightower, J. W. In *The Catalytic Chemistry of Nitrogen Oxides*; Klimisch, R. L.; Larson, J. G., Eds.; Plenum: New York, 1975; pp 63-93.
45. Voorhoeve, R. J. H.; Remeika, J. P.; Trimble, L. E. In *The Catalytic Chemistry of Nitrogen Oxides*; Klimisch, R. L.; Larson, J. G., Eds.; Plenum: New York, 1975; pp 215-231.
46. Yasuda, H.; Mizuno, N.; Misono, M. *J. Chem. Soc., Chem. Commun.* **1990**, 1094. *ibid* 1552.
47. Iizuka, T.; Ichikawa, T.; Egashira, S.; Tanabe, K. *Chem. Lett.* **1985**, 273.
48. Machida, Yasuoka, K. Eguchi, K.; Arai, H. *J. Chem. Soc., Chem. Commun.* **1990**, 1165.

49. Inui, T.; Otowa, T.; Takegami, Y. *J. Chem. Soc., Chem. Commun.* **1980**, 94.
50. Mahmood, T.; Williams, J. O.; Miles, R.; McNicol, B. D. *J. Catal.* **1981**, *72*, 218.
51. Hölderich, W.; Hesse, M.; Näumann, F. *Angew. Chem., Int. Ed. Engl.* **1988**, *27*, 226.
52. Perot, G.; Guisnet, M. *J. Mol. Catal.* **1990**, *61*, 173.
53. Pearce, J. R.; Sherwood, D. E.; Hall, M. B.; Lunsford, J. H. *J. Phys. Chem.* **1980**, *84*, 3215.
54. Praliaud, H.; Coudurier, G. F.; Taarit, Y. B. *J. Chem. Soc., Faraday Trans. 1* **1978**, *74*, 3000.
55. Atanasova, V. D.; Shvets, V. A.; Kazanskii, V. B. *Kinet. Katal.* **1979**, *20*, 518.
56. Park, S. E.; Lunsford, J. H. *Inorg. Chem.* **1987**, *26*, 1993.
57. Ungar, R. K.; Baird, M. C. *J. Chem. Soc., Chem. Commun.* **1986**, 643.
58. Morrow, B. A.; Baraton, M. I.; Lijour, Y.; Roustan, J. L. *Spectrochim. Acta* **1987**, *43A*, 1583.
59. Morrow, B. A.; Baraton, M. I.; Roustan, J. L. *J. Am. Chem. Soc.* **1987**, *109*, 7541.
60. Hennebert, P.; Hemidy, J. F.; Cornet, D. *J. Chem. Soc., Faraday Trans. 1* **1980**, *76*, 952.
61. Chao, C.-C.; Lunsford, J. H. *J. Am. Chem. Soc.* **1971**, *93*, 71.
62. Chao, C.-C.; Lunsford, J. H. *J. Phys. Chem.* **1974**, *78*, 1174.
63. Li, Y. Hall, W. K. *J. Phys. Chem.* **1990**, *94*, 6145.
64. Iwamoto, M.; Furukawa, H.; Mine, Y.; Uemura, F.; Mikuriya, S.; Kagawa, S. *J. Chem. Soc., Chem. Commun.* **1986**, 1272.
65. Naito, S. *J. Chem. Soc., Chem. Commun.* **1979**, 1101.
66. Lunsford, J. H. *Rev. Inorg. Chem.* **1987**, *9*, pp 26-35.
67. Pearce, J. R.; Gustafson, B. L.; Lunsford, J. H. *Inorg. Chem.* **1981**, *20*, 2957.
68. Verdonck, J. J.; Schoonheydt, R. A.; Jacobs, P. A. *J. Phys. Chem.* **1981**, *85*, 2393.
69. Haas, J.; Steinwandel, J.; Plog, C. In *Zeolites as Catalysts, Sorbents and Detergent Builders*; Karge, H. G.; Weilkamp, J., Eds.; Elsevier: Amsterdam, 1989; pp 337-346.

70. Myrdal, R.; Kolboe, S In *Zeolites as Catalysts, Sorbents and Detergent Builders*; Karge, H. G.; Weilkamp, J., Eds.; Elsevier: Amsterdam, 1989; pp 327-336.
71. Iizuka, T.; Lunsford, J. H. *J. Mol. Catal.* **1980**, *8*, 391.
72. Gladfelter, W. L. *Adv. Organomet. Chem.* **1985**, *24*, 41.
73. Bhaduri, S.; Johnson, B. F. G.; Lewis, J.; Watson, D. J.; Zuccaro, C. *J. Chem. Soc., Dalton Trans.* **1979**, 557.
74. Hayward, C. M. T.; Folkers, J. P.; Shapley, J. R. *Inorg. Chem.* **1988**, *27*, 3685.
75. Mantell, D. R.; Gladfelter, W. L. *New J. Chem.* **1988**, *12*, 487.
76. Fjare, D. E.; Gladfelter, W. L. *J. Am. Chem. Soc.* **1984**, *106*, 4799.
77. Mueller, J.; de Oliveira, G. M.; Sonn, I. *J. Organomet. Chem.* **1988**, *340*, C15.
78. Dawes, H. M.; Hursthouse, M. B.; Del Paggio, A. A.; Muetterties, E. L.; Parkins, A. W. *Polyhedron* **1985**, *4*, 379.
79. Fischer, K.; Deck, W.; Schwarz, M.; Vahrenkamp, H. *Chem. Ber.* **1985**, *118*, 4946.
80. Goldhaber, A.; Vollhardt, K. P. C.; Walborsky, E. C.; Wolfgruber, M. *J. Am. Chem. Soc.* **1986**, *108*, 516.
81. Hackett, P.; Manning, A. R. *Polyhedron* **1982**, *1*, 45.
82. Chu, C. T.-W.; Lo, F. Y.-K.; Dahl, L. F. *J. Am. Chem. Soc.* **1982**, *104*, 3409.
83. Motyl, K. M. *Diss. Abstr. Int. B* **1984**, *44*, 2420. CA100(19):156798k.
84. Braunstein, P.; Predieri, G.; Lahoz, F. J.; Tiripicchio, A. *J. Organomet. Chem.* **1985**, *288*, C13.
85. Gord, J. R.; Freiser, B. S. *J. Am. Chem. Soc.* **1989**, *111*, 3754.
86. Jacobson, D. B. *J. Am. Chem. Soc.* **1987**, *109*, 6851.
87. Fredeen, D. A.; Russell, D. H. *J. Am. Chem. Soc.* **1986**, *108*, 1860.
88. Elder, R. C. *Inorg. Chem.* **1974**, *13*, 1037.
89. Legzdins, P.; Nurse, C. R.; Rettig, S. J. *J. Am. Chem. Soc.* **1983**, *105*, 3727.
90. Legzdins, P.; Nurse, C. R. *Inorg. Chem.* **1982**, *21*, 3110.
91. Eberhardt, U.; Mattern, G.; Schiller, G. *Chem. Ber.* **1988**, *121*, 1525.
92. Gourdon, A.; Jeannin, Y. *Organometallics* **1986**, *5*, 2406.
93. Kubat-Martin, K. A.; Spencer, B.; Dahl, L. F. *Organometallics* **1987**, *6*, 2580.

94. Mueller, J.; Sonn, I.; Akhnoukh, T. *J. Organomet. Chem.* **1989**, *367*, 133.
95. Delgado, E.; Jeffery, J. C.; Simmons, N. D.; Stone, F. G. A. *J. Chem. Soc., Dalton Trans.* **1986**, 869.
96. Stevens, R. E.; Fjare, D. E.; Gladfelter, W. L. *J. Organomet. Chem.* **1988**, *347*, 373.
97. Stevens, R. E.; Guettler, R. D.; Gladfelter, W. L. *Inorg. Chem.* **1990**, *29*, 451.
98. Johnson, B. F. G.; Raithby, P. R.; Zuccaro, C. *J. Chem. Soc., Dalton Trans.* **1980**, 99.
99. Gomez-Sal, M. P.; Johnson, B. F. G.; Lewis, J.; Raithby, P. R.; Syed-Mustaffa, S. N.; Azman, B. *J. Organomet. Chem.* **1984**, *272*, C21.
100. Smieja, J. A.; Stevens, R. E.; Fjare, D. E.; Gladfelter, W. L. *Inorg. Chem.* **1985**, *24*, 3206.
101. Johnson, B. F. G.; Lewis, J.; Mace, J. M. *J. Chem. Soc., Chem. Commun.* **1984**, 186.
102. Attard, J. P.; Johnson, B. F. G.; Lewis, J.; Mace, J. M.; Raithby, P. R. *J. Chem. Soc., Chem. Commun.* **1985**, 1526.
103. Johnson, B. F. G.; Lewis, J.; Mace, J. M.; Raithby, P. R.; Stevens, R. E.; Gladfelter, W. L. *Inorg. Chem.* **1984**, *23*, 1600.
104. Burgess, K.; Johnson, B. F. G.; Lewis, J. *J. Chem. Soc., Dalton Trans.* **1983**, 1179.
105. Smieja, J. A.; Gladfelter, W. L. *J. Organomet. Chem.* **1985**, *297*, 349.
106. Rivera, A. V.; Sheldrick, G. M. *Acta Crystallogr., Sect. B* **1978**, *B34*, 3372.
107. Norton, J. R.; Collman, J. P.; Dolcetti, G.; Robinson, W. T. *Inorg. Chem.* **1972**, *11*, 382.
108. Bellard, S.; Raithby, P. R. *Acta Crystallogr., Sect. B* **1980**, *B36*, 705.
109. Johnson, B. F. G.; Lewis, J.; Raithby, P. R.; Zuccaro, C. *J. Chem. Soc., Chem. Commun.* **1979**, 916.
110. Kubat-Martin, K. A.; Rae, A. D.; Dahl, L. F. *Organometallics* **1985**, *4*, 2221.
111. Dimas, P. A.; Lawson, R. J.; Shapley, J. R. *Inorg. Chem.* **1981**, *20*, 281.
112. Colbran, S. B.; Robinson, B. H.; Simpson, J. *J. Organomet. Chem.* **1984**, *265*, 199.

113. Bedard, R. L.; Rae, A. D.; Dahl, L. F. *J. Am. Chem. Soc.* **1986**, *108*, 5924.

114. Bedard, R. L.; Dahl, L. F. *J. Am. Chem. Soc.* **1986**, *108*, 5933.

115. Bedard, R. L.; Dahl, L. F. *J. Am. Chem. Soc.* **1986**, *108*, 5942.

116. Kyba, E. P.; Kerby, M. C.; Kashyap, R. P.; Mountzouris, J. A.; Davis, R. E. *J. Am. Chem. Soc.* **1990**, *112*, 905.

117. Puga, J.; Sanchez-Delgado, R.; Braga, D. *Inorg. Chem.* **1985**, *24*, 3971.

118. Fjare, D. E.; Gladfelter, W. L. *Inorg. Chem.* **1981**, *20*, 3533.

119. Chiesa, A.; Ugo, R.; Sironi, A.; Yatsimirski, A. *J. Chem. Soc., Chem. Commun.* **1990**, 350.

120. de Meester, P.; Skapski, A. C. *J. Chem. Soc., Dalton Trans.* **1973**, 1194.

121. Henrick, K.; Johnson, B. F. G.; Lewis, J.; Mace, J.; McPartlin, M.; Morris, J. *J. Chem. Soc., Chem. Commun.* **1985**, 1617.

122. Beringhelli, T.; Ciani, G.; D'Alfonso, G.; Molinari, H.; Sironi, A.; Freni, M. *J. Chem. Soc., Chem. Commun.* **1984**, 1327.

123. Gourdon, A.; Jeannin, Y. *J. Organomet. Chem.* **1985**, *282*, C39.

124. Johnson, B. F. G.; Lewis, J.; Nelson, W. J. H.; Puga, J.; McPartlin, M.; Sironi, A. *J. Organomet. Chem.* **1983**, *253*, C5.

125. Johnson, B. F. G.; Lewis, J.; Nelson, W. J. H.; Puga, J.; Raithby, P. R.; Braga, D.; McPartlin, M.; Clegg, W. *J. Organomet. Chem.* **1983**, *243*, C13.

126. Stevens, R. E.; Liu, P. C. C.; Gladfelter, W. L. *J. Organomet. Chem.* **1985**, *287*, 133.

127. Martinengo, S.; Ciani, G.; Sironi, A.; Heaton, B. T.; Mason, J. *J. Am. Chem. Soc.* **1979**, *101*, 7095.

128. Goudsmit, R. J.; Jackson, P. F.; Johnson, B. F. G.; Lewis, J.; Nelson, W. J. H.; Puga, J.; Vargas, M. D.; Braga, D.; Henrick, K. *J. Chem. Soc., Dalton Trans.* **1985**, 1795.

129. Braga, D.; Henrick, K.; Johnson, B. F. G.; Lewis, J.; McPartlin, M.; Melson, W. J. H.; Puga, J. *J. Chem. Soc. Chem. Commun.* **1982**, 1083.

130. Johnson, B. F. G.; Lewis, J.; Nelson, W. J. H.; Puga, J.; Braga, D.; Henrick, K.; McPartlin, M. *J. Organomet. Chem.* **1984**, *266*, 173.

131. Müller, J.; Sonn, I.; Akhnoukh, T. *J. Organomet. Chem.* **1991**, *414*, 381.

6

Bioinorganic Chemistry of Nitric Oxide

The role of nitric oxide in the human body is centered around its interaction with various heme units such as those of *hemoglobin* (Hb) and *myoglobin* (Mb). The heme unit is basically a porphyrin ring that binds a metallic ion via the four nitrogen atoms of the ring. For example, the heme unit of Hb is shown in Figure 6.1 *(1)*.

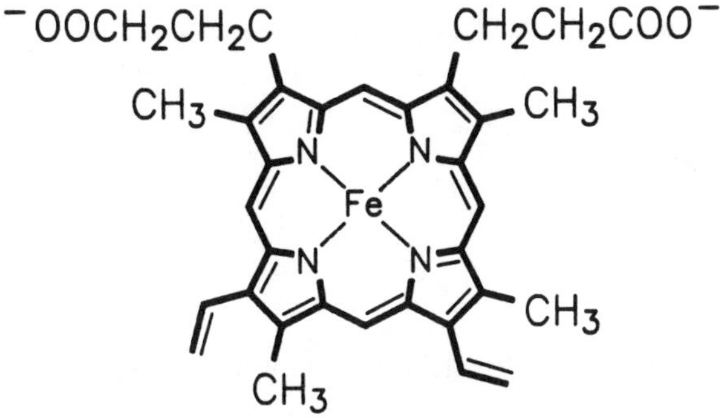

Figure 6.1. The heme group of hemoglobin.

Since a basic understanding of the porphyrin-NO interaction is essential to an understanding of the role of NO in mammals, a brief overview of the interactions of NO with various synthetic metalloporphyrins (some of which are shown in Figure 6.2.) is presented in the first part of this chapter. Later, the interactions of NO with Hb, Mb and other molecules of biological significance will be discussed.

Figure 6.2. Some synthetic porphyrin complexes [(**A**) R = -$CH=CH_2$; iron protoporphyrin dimethyl ester, (PPDME)Fe: (**B**) iron tetraphenylporphyrin, (TPP)Fe: (**C**) iron octaethylporphyrin, (OEP)Fe.]

6.1: Nitrosyl Metalloporphyrins

6.1.1: Mononitrosyl Porphyrin Complexes

There are three basic kinds of metalloporphyrin complexes, each differing in axial ligand coordination, i.e.,

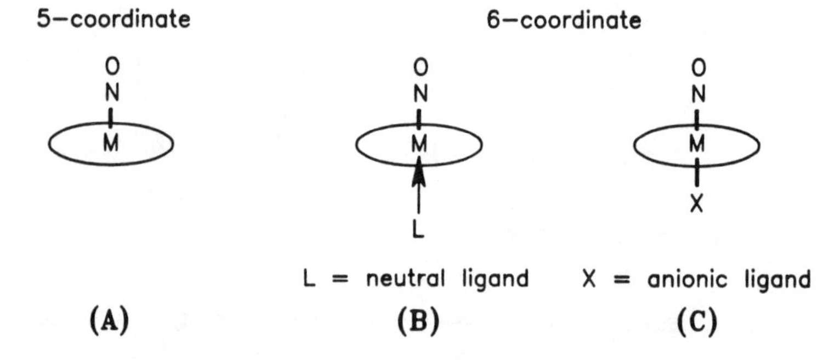

Each kind of complex will now be considered in turn.

6.1.1.1: Five-Coordinate Complexes and their Reactions with Neutral Lewis Bases

The simplest nitrosyl porphyrins are, of course, the five-coordinate complexes (structure **A**) which are essentially square pyramidal in geometry around the metal atom. A number of these mononitrosyl complexes of the Group 6 through to Group 9 metals are now known, and these may be represented generally by the formula (P)M(NO), where P is the porphyrin dianion core. The MNO bond geometry is dependent on the nature of the metal, and not on the nature of the porphyrin. For example, the tetraphenylporphyrin (TPP) and tetratolylporphyrin (TTP) mononitrosyl complexes of the Group 6 metals (Cr and Mo) are essentially $\{MNO\}^5$ derivatives and contain linear MNO linkages *(2,3)*. Similarly, the Group 7 (P)Mn(NO) complexes are essentially $\{MNO\}^6$ derivatives, and they also possess linear MNO geometries. In contrast, the Group 8 (P)Fe(NO) compounds possess bent MNO geometries, since these compounds are formally $\{MNO\}^7$ species. The Group 9 analogues also possess bent MNO linkages. This comparison of MNO geometries is illustrated below for the (TPP)M(NO) compounds of manganese *(4)*, iron *(5)* and cobalt *(6)*.

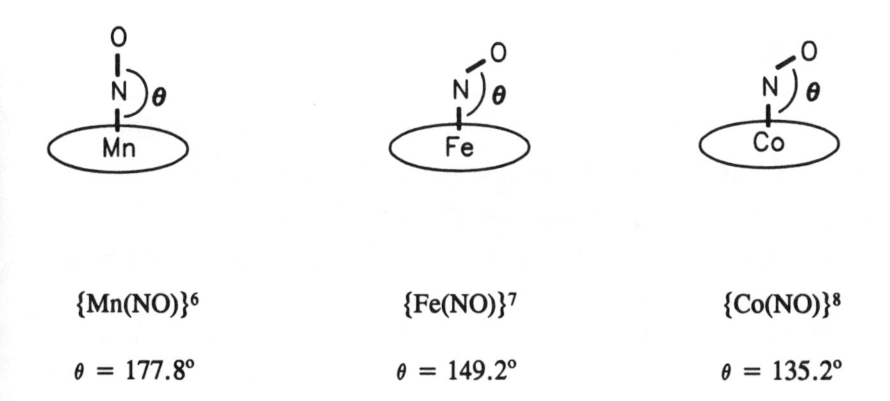

Other five-coordinate complexes of iron have been reported in the literature *(7-24)* and they all contain bent MNO geometries. However, the cationic $[(OEP)Fe(NO)]^+$ complex is formally $\{FeNO\}^6$ and should, therefore, possess a linear MNO bond geometry. That this is so is seen in its solid state structure shown in Figure 6.3, in which the compound forms a remarkable π-π dimer with an interplanar separation of only 3.36 Å (FeNO bond angle of 176.9°) *(19)*.

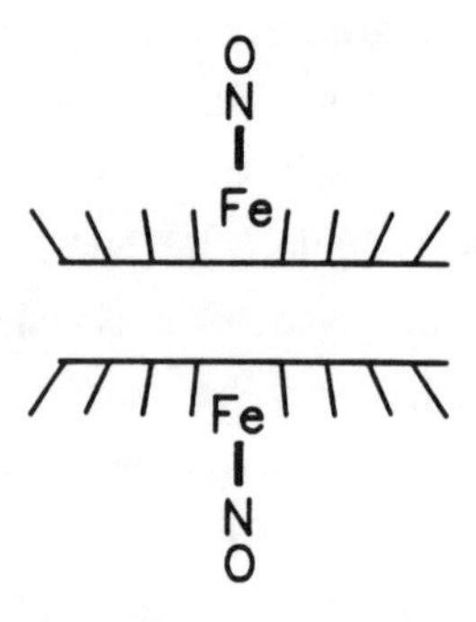

Figure 6.3. Schematic drawing of the molecular structure of $[(OEP)Fe(NO)]^+$.

Other five-coordinate porphyrin complexes of cobalt (P = TPP, TMP, OEP *(6,7,21,25-31)*) and rhodium (P = TPP, OEP *(32)*) have also been synthesized.

Since the sixth coordination site in these (P)M(NO) complexes is vacant, it is not surprising that these nitrosylporphyrin complexes coordinate neutral bases (including solvent molecules) in this axial position to generate complexes of structure **B** *(17,20,33-35,45)*. Most of the studies on axial ligand coordination to mononitrosylporphyrin complexes have employed the (P)Fe(NO) species, since the results have always been taken to be of direct relevance to various biological reactions of Hb and Mb. For instance, extensive studies have been carried out on the coordination of various *N*-donors (e.g., imidazoles, amines, pyridine *(13,36-41)*), *O*-donors (e.g., formamides, sulfoxides, ethers *(42,43)*) and *S*-donors (e.g., thioether, thiols *(42,43)*) to the (PPIXDME)Fe(NO) complex, where PPIXDME is the protoporphyrin IX dimethylesterato group. The studies involving the *N*-donors are important with respect to drawing analogies of the model complexes with Hb and Mb, whereas the studies involving the *O*- and *S*-donors as sixth axial ligands are important for drawing analogies with the cytochromes *c* and abnormal Hb, respectively *(43)*.

In all cases, increasing the basicity of the sixth axial ligand (which is *trans* to NO) lowers the ν_{NO} of the complex accordingly, an observation entirely consistent with an increased electron density in the FeNO link. Indeed, increasing the basicity of the *N*-donors in reaction 6.1 also increases the equilibrium constant for this reaction.

$$\text{(P)Fe(NO)} + \text{base} \rightleftharpoons \text{(P)Fe(NO)(base)} \qquad (6.1)$$

Since the coordination of neutral bases to (P)Fe(NO) does not formally alter its $\{FeNO\}^7$ configuration, the FeNO link would still be expected to be bent in these six-coordinate species. However, the cationic six-coordinate $[(TPP)Fe(NO)(H_2O)]^+$ complex is formally $\{FeNO\}^6$, and contains a linear FeNO group *(19)*.

Increasing the basicity of the sixth axial ligand in the (P)Fe(NO)L compounds appears to have a direct correlation with the observed g values in the ESR spectra of these complexes *(39)*. However, it is also known that the ESR parameters of NO-heme complexes containing a *trans* axial ligand are dependent on (i) the basicity and σ-bonding ability of the axial ligand, and (ii) the magnitude of the steric interaction of the axial ligand with the porphyrin core *(44)*. Thus, for systems with sterically unhindered *N*-donor axial ligands, the g_{iso} values decrease with increasing basicity of the ligand *(36,39)*. Indeed, increasing the energy of the d_{z^2} orbital would also be expected to induce the observed decrease in the ESR g values towards the free electron value *(38)*.

One final point to note about the coordination of neutral Lewis bases to the five-coordinate (P)M(NO) complexes is that not all of these ligands coordinate in the sixth axial position. For example, in the (P)Mo(NO)(MeOH) complex, the MeOH ligand is located *cis* to the NO ligand (as determined by X-ray crystallography) *(3)*. Also, it is known that the extent of binding of sixth axial molecules to the five-coordinate (P)M(NO) compounds is dependent on the polarity of the medium *(16)*. For example, in the presence of sodium dodecyl sulfate (SDS, an anionic detergent that solubulizes porphyrins that would otherwise be insoluble in aqueous solution), the (P)Fe(NO) compound is surrounded by the hydrophobic portion of SDS. This lowers the local polarity of the environment around the Fe atom, thereby lowering the coordination of donor-molecules such as DMSO (or imidazole). In such a case, the six-coordinate (P)Fe(NO)L complex then approaches five-coordination.

6.1.1.2: Six-Coordinate Complexes Containing an Anionic Sixth Axial Ligand

Most of the compounds having structure **C** contain iron as the central metal. The (P)Fe(NO)X halides (X = Cl, Br, I) are known for the cases where the porphyrin is TPP, OEP, or TMP (tetrakis(trimethylphenyl)porphyrinato) and the ν_{NO}s of these complexes for a given porphyrin ligand decrease in the order Cl > Br > I, a trend that is consistent with the increase in the basicity of the halide *(9)*. A number of six-coordinate alkyl and aryl complexes of the form (P)Fe(NO)R (R = Me, Bu, Ph, C_6H_4Me, C_6H_4OMe, C_6F_4H) are also known for the cases where P is OEP or TPP *(46-48)*, and these are generally synthesized by the reaction outlined in eq 6.2.

$$(P)Fe(R) + NO \longrightarrow (P)Fe(NO)R \qquad (6.2)$$

The observed ν_{NO}s of these complexes also correlate well with the electron-donating properties of the alkyl or aryl groups (i.e., the more electron-donating the R group is, the lower the ν_{NO} of the complex). These compounds are assumed to contain a *trans*-(ON)Fe(R) arrangement and are formulated as $\{FeNO\}^6$ species. Thus, they would be expected to possess linear FeNO linkages regardless of the natures of the R or X groups. The (P)M(NO)X compounds of ruthenium (P = TPP; X = Cl, Br) and osmium (P = OEP; X = F, OMe) have also been proposed to possess *trans* arrangements of (ON)M(X) groups *(49,50)*.

6.1.2: Dinitrosyl Porphyrin Complexes

A number of porphyrin complexes of iron *(10,51)*, ruthenium *(52)*, osmium (49), and cobalt and rhodium *(32)* bind two molecules of NO via the reaction (P = TPP, OEP, PPIXDME):

$$(P)M(NO) + NO \longrightarrow (P)M(NO)_2 \qquad (6.3)$$

On the basis of spectral data, the NO ligands are assumed to be situated in a *trans* arrangement about the metal centers. Interestingly, the only dinitrosyl porphyrin complex that has been characterized by X-ray crystallography is $(TPP)Mo(NO)_2$, and the molecular structure of this complex unambiguously reveals that the two NO ligands are mutually *cis* *(3)*. Furthermore, the Mo atom lies 0.28 Å out of the mean plane of the porphyrin core towards the NO ligands, and the $Mo(NO)_2$ group adopts an *attracto* conformation, i.e. the (ON)-Mo-(NO) bond angle (78.4°) is larger than the O-Mo-O bond angle (60.0°).

6.1.3: Reactivity

Nitrosyl complexes of metalloporphyrins undergo reactions that are characteristic of other non-porphyrin-containing metal nitrosyl complexes (Chapter 7). Thus, nitrosyl exchange and nitrosyl loss have been observed for some (P)M(NO) complexes *(7,53)*.

$$(TPP)Fe(^{14}NO) + {}^{15}NO \rightleftharpoons (P)Fe(^{15}NO) + {}^{14}NO \qquad (6.4)$$

$$(TPP)M(NO) + h\nu \longrightarrow (TPP)M + NO \qquad (6.5)$$

$$(M = Mn, Fe, Co)$$

Indeed, the general reactions exemplified by eqs 6.4 and 6.5 have been employed in the nitrosyl-transfer reactions between several neutral and cationic porphyrin complexes of iron and cobalt *(21,54)* as shown in eqs 2.46 and 2.47.

As mentioned in the previous section, the five-coordinate (P)M(NO) complexes may also coordinate NO to form the corresponding dinitrosyls. Occasionally, the five-coordinate (P)M(NO) complexes react with external sources of NO in a manner that is exemplified by eq 6.6 *(53)*.

$$(TPP)Fe(NO) + 3\,NO \longrightarrow (TPP)Fe(NO)(NO_2) + N_2O \qquad (6.6)$$

Furthermore, NO coordinates to (TPP)Co anchored to imidazole on SiO_2 *(31)* and also to (TPP)Co supported on TiO_2 *(28)* to generate excellent catalysts for NO conversion. The former precursor complex, namely

(TPP)Co(Im)-SiO_2, reacts with NO and H_2 to generate N_2, N_2O and NH_3, whereas the latter (TPP)Co-TiO_2 compound reacts with NO and CO to produce N_2 and N_2O.

6.1.4: Electrochemistry

The interpretation of the electrochemistry of metalloporphyrins poses a real challenge, since the redox changes in these complexes may occur not only at the metal centers, but also at the porphyrin π system *(14,55)*. It is not always easy to distinguish between the two, since solvent-metalloporphyrin interactions markedly affect the observed half-wave potentials as well, and since electrochemically-induced ligand-based reactions also occur somewhat frequently in these complexes *(14,56)*. However, the sagacious choice of macrocycle allows for the control of the redox site (e.g., metal vs macrocycle) *(26)*.

The (P)M(NO) complexes undergo multiple (and visually impressive!) reversible oxidation/reduction processes, i.e.,

$$[M']^{2+} \rightleftharpoons [M']^{+} \rightleftharpoons [M'] \rightleftharpoons [M']^{-} \rightleftharpoons [M']^{2-} \quad (6.7)$$

$$[M'] = (P)M(NO)$$

A number of monocations and dications of chromium *(2)*, manganese (27), iron *(9,10,21)* and cobalt *(25,27,29)* have been electrogenerated in this manner. The (P)Fe(NO) complexes (P = TPP, OEP) form stable monocations, but they transform to $[(P)Fe(NO)_2]^+$ in the presence of added NO gas *(10,46,47)*. On the other hand, the monocationic complexes of chromium and manganese are unstable with respect to the loss of their NO ligands *(2,27)*. It has been clearly demonstrated that the oxidations of the (P)Co(NO) complexes (P = porphyrin, chlorin, isobacteriochlorin) are porphyrin-ring centered, resulting in the formation of porphyrin π-cation radicals *(25,29)*.

The reductions of the (TPP)M(NO) complexes of chromium *(2)*, manganese *(27)*, iron *(9,10,57)*, and cobalt *(29)* result in the generation of their monoanionic and dianionic derivatives. The anionic complexes of iron also lose NO in the presence of a potential sixth axial ligand *(8)*. However, it has been observed that the chemically generated $[(P)Fe(NO)]^-$ anions (P = TPP, OEP) are reoxidized to their neutral precursors by phenols *(57)*. The anionic complexes of cobalt are also unstable and readily lose NO. Curiously, whereas the reductions of the manganese analogues are likely to be metal-centered, the one- and two-electron reductions of (TPP)Cr(NO) are certainly ring-centered *(58)*.

6.1.5: ESR Spectra of Nitrosyl Ferrous Heme Complexes

When NO binds to the diamagnetic (P)Fe moiety, a $\{FeNO\}^7$ species which contains an unpaired electron (formally supplied by the paramagnetic NO) results.

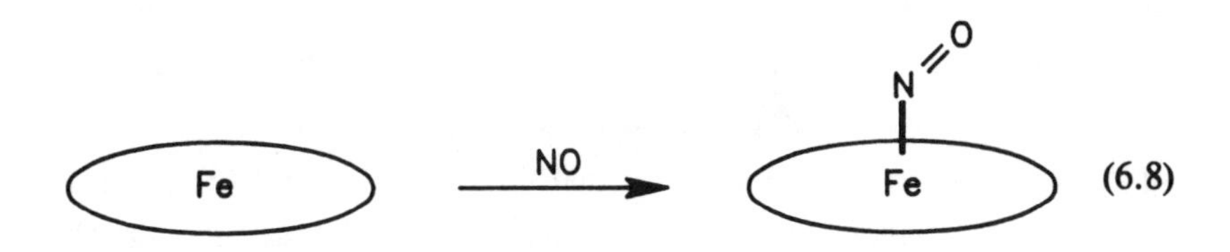

(6.8)

ESR spectroscopy is useful to some degree in characterizing the distribution of unpaired electron density in the FeNO group. In a simplistic sense, a three-line pattern would be expected to result from a five-coordinate (P)Fe(NO) complex, due to the interaction of the unpaired electron with the nitrosyl nitrogen (I = 3/2). In the presence of an *N*-donor sixth axial ligand (e.g., imidazole) this three-line pattern would be expected to convert to a nine-line pattern (actually a triplet-of-triplets) due to the interaction of the unpaired electron with the two axial nitrogen centers *(59-61)*. Unfortunately, the interpretation of the ESR spectra of various NO-heme complexes has had a long history of problems, and frequently there has been a need to invoke multiple "(heme)FeNO" species in the same mixture to account for the ESR spectra observed *(62)*. In a recent calculation, however, two distinct low-energy electronic states corresponding to the formal assignment of the unpaired electron to either NO or Fe atomic orbitals were found to be nearly degenerate, and the relative energies of these electronic states also found to be sensitive to minor changes in the axial ligand bond distances and angles *(63)*. Indeed, it was proposed that the ESR behavior of various NO-heme proteins could be understood in terms of a thermal or electronic mixing of these two electronic states derived from the same species *(63)*.

6.2: Bioinorganic Chemistry of NO

Nitric oxide has been employed as a ligand in bioinorganic chemistry for a number of reasons. As mentioned in Chapter 1, the bent metal-nitrosyl link has traditionally been thought to contain the nitroside (NO^-) group. Since this group is isoelectronic with dioxygen, it may be expected that the binding properties of both NO and O_2 should be similar. Indeed, it is such reasoning that has led to the extensive use of NO as a ligand to probe the structure and function of many heme and non-heme proteins. Thus, in simplistic terms, it may be expected that the binding of NO to biologically important compounds such as Hb, Mb and the cytochromes should mimic the binding of O_2 to these compounds. While a detailed description of the binding and transport of O_2 in the body is not the focus of this chapter, it is perhaps important to briefly review the general steps involved in this process. Oxygen in the lungs is taken up by Hb in the blood stream and is transported to the tissues where it is taken up by Mb. The oxygen is stored by the Mb until it is ultimately consumed by the *cytochromes* in the reduction of O_2 to H_2O.

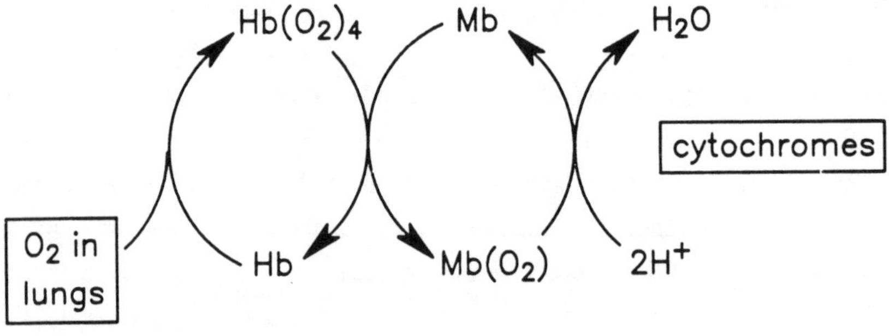

The heme group (an Fe atom ligated by the four N atoms of a porphyrin ring) is an essential unit of Hb, Mb, cytochromes and even enzymes such as catalase and peroxidase *(1)*. The heme group of Hb is shown in Figure 6.1. Since the porphyrin core is dianionic, the Fe atom is assigned the Fe^{2+} (ferrous) oxidation state. However, the ferric form (Fe^{3+}) may also occur and is referred to by the prefix *met* (e.g., metHb). However, the oxidation of Fe^{2+} to Fe^{3+} is normally prevented by the presence of polypeptide chains around the heme groups. The contact between the polypeptide chains and the Fe center is not by the heme group, but is primarily via an axial ligand (e.g., histidine) in the fifth coordination site. The binding of a substrate then occurs in the sixth coordination site as shown below.

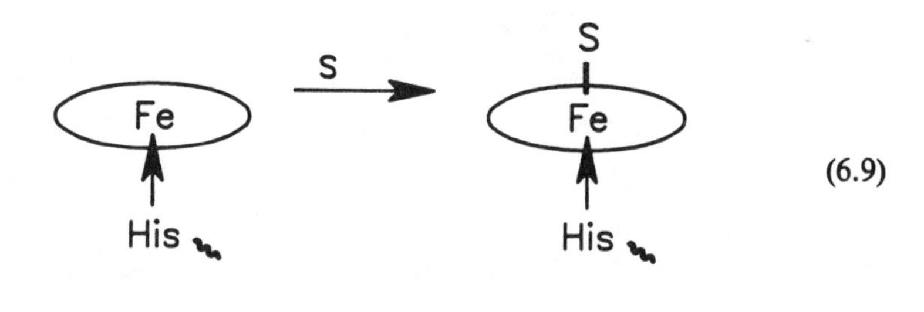

(6.9)

S = substrate

The five-coordinate form is normally referred to as being in a tensed (*T*) state, whereas the six-coordinate form is normally referred to as being in a relaxed (*R*) state. Thus, the $T \longrightarrow R$ transition involves the binding of substrate.

Each molecule of Hb contains four heme groups. Each heme group has associated with it a protein chain, and generally, each Hb molecule consists of two identical α and two identical β chains (i.e. $(\alpha)_2(\beta)_2$). The Hb molecule is thus effectively "tetrameric" and has a large molecular diameter of *ca.* 64 Å and a molecular weight of 64,650 *(1)*. However, Hb is restricted to the red blood cells and is normally found either without any ligated O_2 (deoxy form) or ligated with four molecules of O_2 (oxy form, $Hb(O_2)_4$). After O_2 binds to Hb to produce $Hb(O_2)_1$, a built-in feedback mechanism progressively speeds up the sequential binding of O_2 to ultimately produce $Hb(O_2)_4$. Under normal physiological conditions, however, one molecule of Hb binds, on average, two or three molecules of

O_2. The presence of certain organic phosphates or other reagents (protonic or anionic) influences the extent of O_2-binding to Hb, thus in effect producing $T \rightarrow R$ or $R \rightarrow T$ transitions. For example, the organophosphate 2,3-diphosphoglycerate (DPG), regulates O_2 binding in mammals via the effective decrease of O_2-affinity of Hb, i.e.,

$$Hb(O_2)_4 + DPG \longrightarrow Hb(DPG) + 4\ O_2$$

thus inducing an oxy-deoxy (or $R \rightarrow T$) transition. Other phosphates such as inositol hexaphosphate (IHP), bind to Hb more strongly than does DPG and induce similar transitions.

Hemoglobins from vertebrates differ from one another, therefore the results of studies on the Hb of one species (e.g., sperm whale) cannot be used to develop a complete understanding of the Hb of another species (e.g., humans). Slight differences may even be present in the same species. For example, the Hb of adult humans differs from the Hb of human newborns *(64)*. Also, some Hbs are unique: *earthworm erythrocruorin* is an extracellular giant Hb and all available data indicate that only two or three of its polypeptide chains contain heme (its NO adduct has been studied) *(65)*.

Myoglobin contains one heme group and is monomeric. It is found mainly in muscle tissues and its primary role is to store O_2 until it is required. However, the ultimate consumers of O_2 are the *cytochromes*, which are the final enzymes in the respiratory chain. The cytochromes are membrane proteins that contain heme groups and are essentially electron carriers by virtue of their ability to shuttle their iron atoms between two oxidation states:

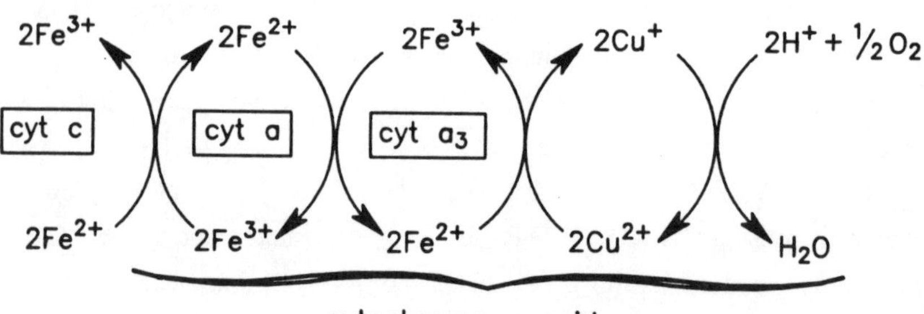

The net result of this sequence of unidirectional electron transfer is the oxidation of cytochrome *c* (extreme left) and the reduction of molecular O_2 to H_2O (extreme right) *(66)*. While cytochrome *c* oxidase actually consists of two cytochromes (cytochromes *a* and a_3), it also contains protein-bound copper. Thus, while the current view is that it is actually cytochrome a_3 that binds O_2 (via the heme group), the role of copper in the reduction of H_2O is still very uncertain *(67)*.

In the following sections, the binding of nitric oxide to Hb, Mb and the cytochromes are presented and discussed.

6.2.1: Nitrosylhemoglobin and Nitrosylmyoglobin

6.2.1.1: Syntheses and Reactivity of Nitrosylhemoglobin and Nitrosylmyoglobin

The syntheses of the nitrosyl derivatives of Hb and Mb parallel the general preparative routes to metal nitrosyls discussed previously in Chapter 2. Specifically, the methods employed include the reaction of Hb or Mb with NO gas or its solutions *(59,68-78)*, trioxodinitrate ($HN_2O_3^-$) *(79-82)*, hydroxylamine *(83)*, nitrite *(68,84-86)* or the transfer of coordinated NO from other metal nitrosyls *(87)*. The affinity of Hb for NO is on the order of a thousand times higher than for CO, and indeed, Hb and Mb form stable adducts with NO in the absence of oxygen *(75)*. The stoichiometry of the reaction is as expected: one NO group binds per each heme group at the sixth axial position. Thus, Mb reacts with NO to form MbNO, whereas Hb reacts with four molecules of NO to form the fully ligated $Hb(NO)_4$ (sometimes referred to as $(\alpha^{NO})_2(\beta^{NO})_2$). In practice, however, the extent of ligation of Hb by NO may be anywhere from $Hb(NO)_{1\text{-}4}$, and for convenience, nitrosylhemoglobin is normally just simply written as HbNO. Whether or not HbNO displays a cooperative binding of NO is still in doubt *(77,88-90)*.

Ironically, most of the characterization studies of HbNO (a $\{FeNO\}^7$ species) have involved the use of ESR spectroscopy *(60,78,89,91-105)*, although it has been noted that this technique may not be a reliable tool for the study of the coordination geometry of nitrosyl hemes *(106,107)* (vide supra). Other spectroscopic techniques used are Resonance Raman *(106,108)*, NMR *(109-112)*, and ^{14}N- and ^{1}H-ENDOR *(113,114)*. As with other metal nitrosyls, the most reliable tool for the determination of the coordination geometry of the FeNO group is X-ray crystallography, and the observed FeNO bond angle of 145° in horse heart Hb *(115)* is in accord with the prediction of a bent NO ligand in these $\{FeNO\}^7$ species.

The addition of IHP to solutions of human HbNO or MbNO results in the severe weakening of the Fe-His bond *(69,116-129)*, i.e.

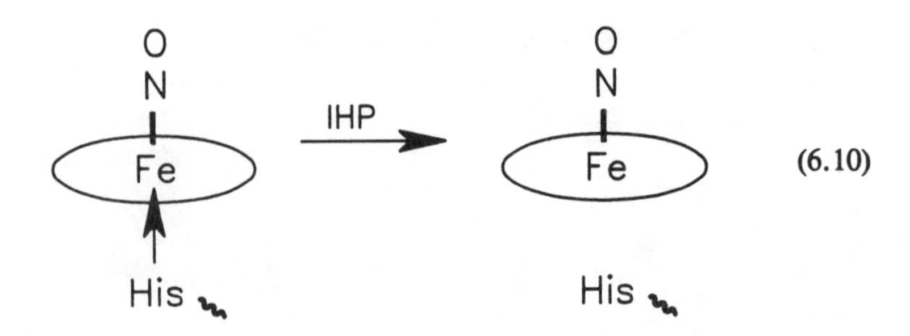

Thus, in general, the usual nine-line ESR pattern is replaced (at least to some degree) by a three-line pattern for the five-coordinate species. A similar effect is observed when the pH of solutions of HbNO and MbNO are lowered: protonation of the axial histidine results in the cleavage of the proximal Fe-N bond *(130-134)*.

The photolysis of HbNO (or MbNO) results in the expulsion of the NO ligand, although recombination rates are very fast *(135-142)*.

$$MbNO + h\nu \rightleftharpoons Mb + NO \qquad (6.11)$$

In picosecond absorption studies on the photodissociation of α- and β-nitrosylhemoglobin monomers *(143)*, the formation of five-coordinate HbNO (with the proximal imidazole detached) in the α- and β-monomers was found, which was not the case involving the tetramer. Furthermore, in a femtosecond absorption study of HbNO, the low yield of photodissociation of HbNO was attributed to a fast (2.5 ps) recombination of NO with the photogenerated (and unligated) hemoglobin *(144)*.

Nitrosylhemoglobin is somewhat air-sensitive, and reacts with oxygen to yield metHb and NO_3^- *(145)*. Interestingly, when fully-ligated nitrosylhemoglobin (i.e. $Hb(NO)_4$) is mixed with fully-ligated oxyhemoglobin (i.e. $Hb(O_2)_4$), an asymmetrical hybrid is produced, namely $(\alpha^{NO}\beta^{NO})(\alpha^{O_2}\beta^{O_2})$ *(146)*.

6.2.1.2: Metal Substitution in Hemoglobin and Myoglobin

The replacement of iron in Hb and Mb by other metal ions is usually done in order to study the stabilities and chemical reactivities of the metal-substituted biomolecules *(147)*. The only reported NO adducts of metal-substituted Mb are for cobalt *(148,149)* and manganese *(150)*. [Note that Co-NO is isoelectronic with Fe-O_2.] The MnNO derivatives of monomeric Hb *(151)* and insect hemoglobin CTT IV *(45)* have also been synthesized.

6.2.1.3: Cytochromes

The terminus of the respiratory chain is located at cytochrome oxidase, whose cytochrome a_3 center catalyzes the reduction of O_2 to H_2O *(152,153)*. Studies on the NO adduct of cytochrome *c* oxidase reveal the presence of the fully reduced (a^{2+}, a_3^{2+}NO) and the mixed valence (a^{3+}, a_3^{2+}NO) enzyme (154). Spectroscopic methods employed for the studies of these NO adducts include Resonance Raman *(154)*, ESR *(155-157)*, FT-ESR *(158)* and ENDOR *(156)*.

A number of nitrosyl adducts of other cytochromes have also been prepared, and these include the adducts of the cytochromes *c* *(159-161)*, cytochrome cd_1 *(162)* and cytochrome *c* peroxidase *(163)*. The crystal structure of the latter complex confirms the bent NO geometry of 143° *(163)*.

6.2.2: Biosynthesis of Nitric Oxide. Endothelium-Derived Relaxing Factor

Endothelial cells line the inner surface of the vessel in contact with the blood as it is pumped through the aorta *(164)*. These cells play a vital role in the relaxation and contraction of arteries and veins by producing a potent vasodilator *(165,166)*. It is known that a radical species is responsible for triggering the relaxation of smooth muscle (muscle that forms layers in the walls of the digestive tract, bladder, uterus, arteries, veins, and various ducts). This radical species is called the endothelium-derived relaxing factor (EDRF), and has been identified as the simple NO molecule *(164,167,168)*. The major pharmacological properties of EDRF (the physiological half-life of which is *ca.* 10 sec *(169)*) are relaxation of smooth muscle and inhibition of platelet aggregation *(167)*. In the latter case, endothelial cells are stimulated by aggregating platelets to produce EDRF which, in turn, limits the aggregation of more platelets and also may inhibit their adhesion to the vascular wall *(171)*. The NO synthesized in mammalian cells is derived from the unusual oxidation of the amino-acid L-arginine *(171)* as shown in Figure 6.4. Thus, as confirmed by labelling studies, oxidation occurs at the guanido nitrogen to produce NO, which may then (i) decompose to NO_2^- and NO_3^- *(167,170,173,174)* or (ii) activate the enzyme guanylate cyclase, an enzyme that catalyzes the formation of cyclic GMP from GTP *(175)*. Interestingly, increased levels of cyclic GMP have been associated with vascular smooth muscle relaxation *(176)*. This activation of guanylate

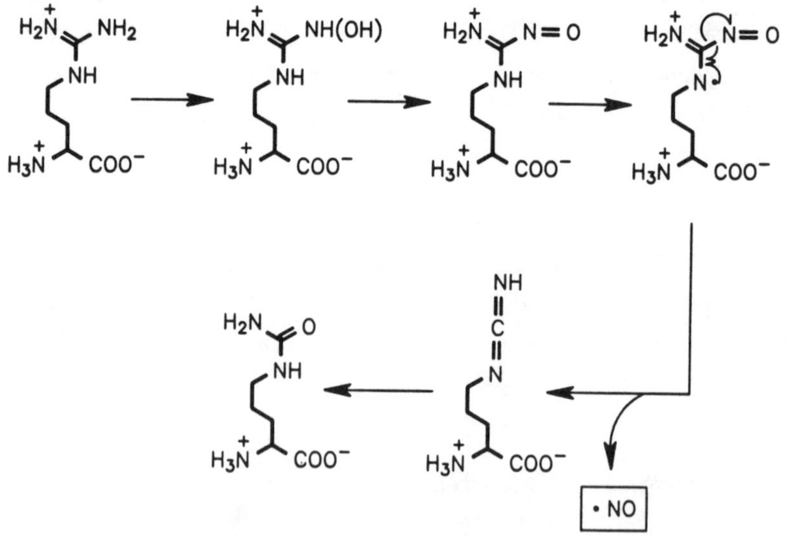

Figure 6.4. Biosynthesis of nitric oxide. [Adapted from reference 167.]

cyclase is believed to occur via the binding of NO to its heme iron *(177)*. Indeed, the binding of NO to the heme is necessary for the activation of this enzyme: Deactivation occurs when the enzyme is stripped of heme *(178)*, but the addition of heme, NO-heme, or even HbNO markedly reactivates the enzyme via the irreversible transfer of the heme or NO-heme to the enzyme *(179-185)*. The overall process of muscle relaxation may thus be schematically represented by

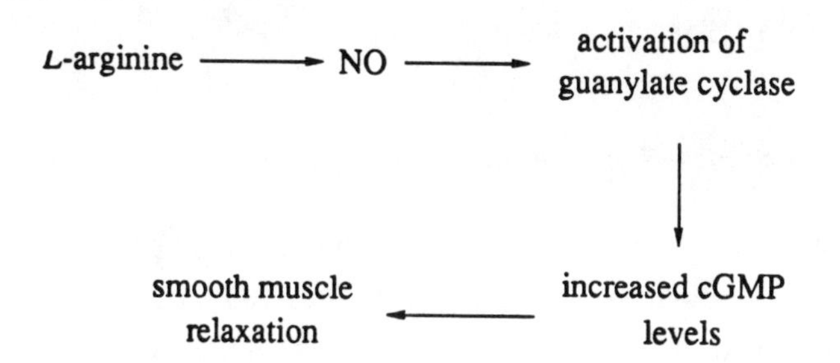

The oxidation of *L*-arginine may be circumvented by the administration of nitrovasodilator drugs such as nitroprusside, nitroglycerin, or amyl nitrite, since these compounds are capable of releasing NO in the presence of other species *(186)*. Other metal nitrosyl compounds employed for this purpose include $[Ru(NO)(NH_3)_5]Cl_2$ and $K_2[Ru(NO)Cl_5]$, but these are not as effective as is nitroprusside in decreasing arterial blood pressure and relaxing aortic strips from cats *(187,188)*. A number of pathways for the release of NO from nitroprusside may be envisaged *(188,189)*:

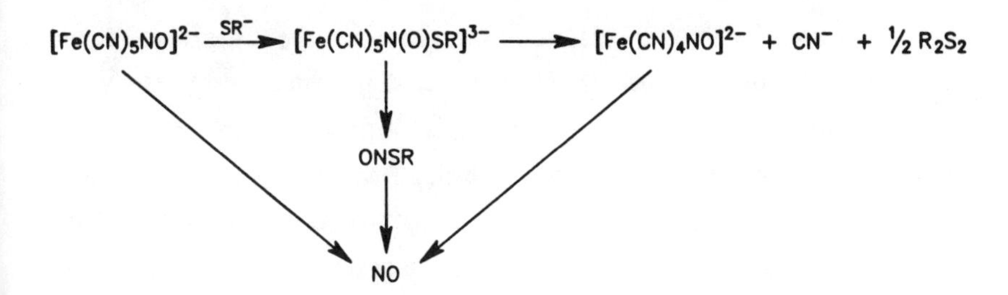

The toxic effect of free cyanide (which could lead to cyanide poisoning) can be overcome by the simultaneous administration of cyanide scavengers such

as thiosulfate. Free cyanide may also reverse the vasodilator effects of nitroprusside *(190)*.

The NO produced either from *L*-arginine or nitrovasodilators may also lead to the formation of HbNO *in vivo (191-194)* or may form nitrosylated iron-sulfur complexes in proteins or peptides *(174,195)*. Lastly, it has recently been shown that the endogenous release of NO is required for long-term synaptic depression in the cerebellum and that this depression is blocked by Hb *(196)*.

6.2.3: Biotransformation of Nitric Oxide

Although NO is not readily absorbed in mammalian air passages due to its somewhat low solubility in water (7.34 mL/100 mL cold water), its inhalation results in the formation of HbNO *(197,198)*. In fact, at concentrations of *ca.* 10 ppm, 80 - 90% of the NO is absorbed in normal or deep breathing, and since it is known that NO binds to Hb about 3 x 10^5 times stronger than does O_2, it is estimated that the oxygen transport ability of Hb exposed to 0.4 ppm of NO will decrease to less than the lethal level of 25% *(198)*. Under normal physiological conditions, however, the NO in HbNO is rapidly oxidized to NO_2^- and NO_3^-, and the Hb is converted to metHb *(198)*. Interestingly, metHb is quickly reduced to Hb by *metHb reductase* in erythrocytes. Thus, the oxidative conversion of HbNO to metHb and the subsequent reduction of metHb by *metHb reductase* provides a protective role for NO intoxication *(198)*.

In a study of ^{15}NO inhalation in rats (145 ppm for 2 h), it was found that about 55% of the inhaled ^{15}N is excreted in urine mostly as $^{15}NO_3^-$ (together with some ^{15}N-urea). Interestingly, a portion of the NO_3^- in the blood is transferred to the oral cavity through saliva and then reduced to NO_2^- by oral bacteria. Reaction of nitrite (under the acidic conditions of the stomach) with amines and proteins from ingested foods results in its conversion to N_2 gas, i.e.,

$$H_2N\text{-}CH(R)COOH + HNO_2 \longrightarrow HO\text{-}CH(R)COOH + N_2 + H_2O \quad (6.12)$$

It has been observed that $^{15}NO_3^-$ in the intestines is converted to $^{15}NH_3$ (via $^{15}NO_2^-$) by intestinal bacteria, the ammonia then being absorbed through the intestinal wall and metabolized to urea (Figure 6.5).

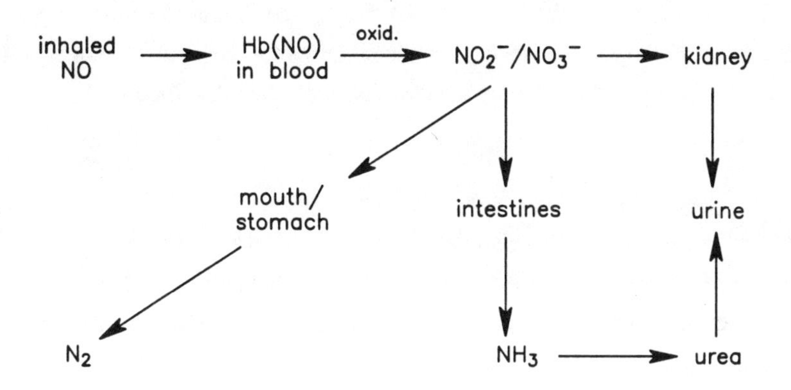

Figure 6.5. Proposed metabolic pathway of inhaled NO (from reference 197).

Enzymes capable of reducing nitrite (nitrite reductases *(199,200)*) apparently react with nitrite to form initially nitrosyl complexes *(200-205)*. Extracellular Hb is also able to trap NO released during denitrification as HbNO, a feature observed for six species of denitrifying bacteria *(206)*. Thus, the role that metal nitrosyls play in enzymatic processes cannot be underestimated, and studies aimed at determining the biodistribution of iron-nitrosyl complexes in the human body are therefore very important. Interestingly, an examination of the biodistribution of ^{99m}TcNO radiopharmaceuticals in rats (2 min after their injection) revealed that these metal-nitrosyl complexes were present in the liver, gastrointestinal tract, heart, muscle, lung, kidney and blood *(207)*.

Returning to the denitrification process (eq.6.10), the nitrite-nitrous oxide conversion is known to be catalyzed by two classes of nitrite reductase, containing either a heme cd_1 unit or copper *(210)*.

$$NO_3^- \longrightarrow NO_2^- \longrightarrow NO \longrightarrow N_2O \longrightarrow N_2 \qquad (6.13)$$

While NO-heme adducts have been the subject of much study (vide supra), not much is known about the role of copper nitrosyls in bioinorganic chemistry *(208-210)*. The first report of the role of copper nitrosyls was in the identification of a Cu-NO intermediate in the pathway for denitrification in *Achromobacter cycloclastes*, a species that utilizes a Cu-containing nitrite reductase *(209,210)*. However, its characterization depended mainly on ESR spectroscopy and by the gaseous NO/N_2O product distribution.

The reaction of nitrite (at pH = 5.7) with deoxyhemocyanin of *Astacus leptodactylus* yields methemocyanin and NO *(211)*, whereas the anaerobic treatment of deoxyhemocyanin with NO results in the formation of nitrosylhemocyanin *(212)*. In general, though, the reports of the reactions of NO with copper proteins are scattered, and the products are invariably not well characterized *(213)*.

6.2.4: NO and Food Chemistry

The "curing" of meat normally refers to a salting process that discourages microbial growth *(214)*. By the end of the nineteenth century, it was recognized that meat-curing brines contained nitrite as an impurity, and it was also recognized that it was the nitrite impurity that was responsible for giving the meat its bright-red color normally associated with its "freshness" *(215)*. It is now known that the species responsible for this bright-red color is MbNO. Skeletal muscle mitochondria which survive curing have been implicated in the reduction of nitrite to Mb-bound NO *(216)*. Interestingly, MbNO has a visible absorption spectrum similar to that of MbO_2: the α and β absorption bands of MbO_2 and MbNO are at 580 nm and 542 nm, and at 575 nm and 543 nm, respectively *(217)*. Indeed, MbNO and the low-spin ferrous complexes MbO_2 and MbCO all cause the bright-red color of "fresh" meat *(85,218-221)* (Table 6.1.).

As with many other nitrosyl complexes, MbNO is unstable under aerobic conditions. However, the stability of its color is enhanced if it is "denatured", usually by the addition of salts or by heating *(138,222,223)*. After the denaturing process, all that is left of MbNO is the (heme)Fe(NO) portion *(220,224-227)*, which is identical to the simple (P)Fe(NO) complexes discussed earlier in this Chapter. This (heme)Fe(NO) group is

Table 6.1: Pigments Found in Fresh, Cured or Cooked Meat

pigment	species	color
Mb	Fe(II)	purplish-red
MbO_2	Fe(II)	bright-red
metMb	Fe(III)	brown
MbNO	$\{FeNO\}^7$	bright-red
metMbNO	$\{FeNO\}^6$	crimson

Data taken in part from reference 214.

not entirely stable by itself, since it is paramagnetic and can easily be oxidized. However, the dinitrosyl species, (heme)$Fe(NO)_2$, is reportedly stable to oxidation and to ligand exchange *(228)*, and it is this dinitrosyl complex that is believed to constitute the stable pigment of cured meat, although its precise molecular structure is still uncertain *(229)*. This pigment is photosensitive in the presence of oxygen, hence appropriate packaging of the meat is necessary to exclude oxygen (e.g. vacuum-sealing of meats) *(228,230)*. Not surprisingly, a large variety of meats including beef *(231)*, sausages *(218,225,227,232)*, pork *(216,219)*, ham *(222,224)*, frankfurters *(219)* and turkey breast *(233)* have been "colored" by the curing process.

It must also be noted that the green color of some meats frequently results from (i) the addition of excess nitrite to cured meats to form nitrihemin, and (ii) the reaction of Mb with H_2S and O_2 to form sulfymyoglobin. The formation of the latter complex is more likely for meats with pH $>$ 6 since at lower pHs, the bacteria that produce H_2S are unable to do so *(214)*.

An advantage in the use of nitrite is that in addition to its reactions with Mb to form MbNO *(85,216,220,221,232-237)*, nitrite helps to prevent bacterial spoilage and food poisoning by inhibiting the growth of the bacterium *Clostridium botulinum* and other deleterious microorganisms which thrive under anaerobic conditions such as those that exist in sealed meat packages *(226,238-240)*. It has been noted that the toxins produced as *Clostridium botulinum* multiplies in food are highly poisonous: 1 mg of *botulinum toxin A* can kill three million mice *(226)*!

A major disadvantage in the use of nitrite has been its implication in *N*-nitrosation reactions of organic compounds at stomach pH levels of *ca.* 3.0 *(226,228)*. It is for this reason that alternatives to the use of nitrite have been considered. It is known in the food industry that in meat products the partial substitution of meat by vegetable proteins results in the color fading. Interestingly, the addition of authentic HbNO to the product restores its meat-like color *(230)*. Indeed, depending on the Hb concentration, the color can be set anywhere on a Frankfurter-to-red-meat color scale *(230)*. The iron-free heme derivative protoporphyrin-IX has been used as a "safe" coloring substitute due to the fact that it is stable and contains no FeNO group *(231)*. This porphyrin has been used as a colorant for lean beef and chicken breast. However, it readily loses its color with time and develops a purple hue *(231)*.

6.2.5: Biological Chemistry of Non-Heme Iron

6.2.5.1: g = 2.03 complexes

Paramagnetic iron nitrosyl complexes (g = 2.03) of the form $[Fe(NO)(SR)_2]^-$ are readily formed *in vivo* and *in vitro* and are commonly referred to as "2.03 complexes" (Section 3.7.2.2) *(241-244)*. In these complexes, the SR group is either cysteine or a derivative thereof, e.g.,

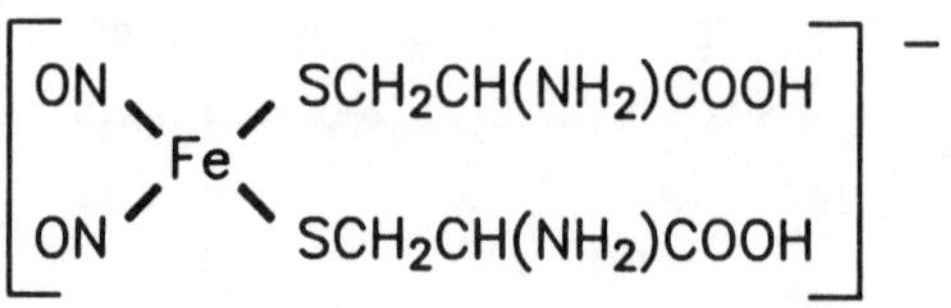

$$\left[\begin{array}{ccc} ON & & SCH_2CH(NH_2)COOH \\ & Fe & \\ ON & & SCH_2CH(NH_2)COOH \end{array}\right]^-$$

Addition of $FeSO_4$ and nitrite to the drinking water of randomly-bred rats results in the formation of such "2.03 complexes" in the liver, kidneys and small intestines of the test animals *(244-257)*. Such iron nitrosyl complexes are commonly associated with cancerous states (and corresponding mortality) in experimental animals *(241,259)*. For example, iron nitrosyl complexes have been found to form in the terminal period of *hepatocarcinogenisis* induced in rats and mice *(260)*. Furthermore, the related Roussin red methyl ester $[Fe_2(NO)_4(SMe)_2]$ has been implicated in cancer studies. For instance, in the Linxian valley of Henan province in northern China, the adult incidence of esophageal cancer is very high (one in four), and this has been attributed to dietary factors. In particular, some dietary staples such as cornbread, millet, and vegetables are stored in water for several weeks. Interestingly, the local water supply is high in nitrite/nitrate, and indeed, the $[Fe_2(NO)_4(SMe)_2]$ has been isolated (0.1 - 0.45 mg/kg of plant matter) from pickled vegetables preserved in such a manner *(226,241,261)*. This nitrosyl compound has also been found to be a natural mouse skin tumor promoter *(261)*.

It has already been mentioned that nitrite helps to prevent bacterial spoilage and food poisoning. In this context, it may be noted that $[Fe_4(NO)_7S_3]^-$ is formed when nitrite is heated with meat products and it inhibits *Clostridium* and *Salmonella* species *(226)*. Indeed, the antiseptic action of $K[Fe_4(NO)_7S_3]$ provides a good disinfectant for drinking water *(241)*.

In related studies, it has been shown that iron dinitrosyl complexes inhibit completely the aggregation of human thrombocytes suspended in an artificial protein-free medium, presumably via the action of NO-activated thrombocyte guanylate cyclase *(262)*. Furthermore, these iron dinitrosyl complexes (in doses of 1 - 10 mmol/kg) dose-dependently decrease blood pressure for 2 h in spontaneously hypertensive rats *(263,264)*, and their physiological action has also been linked to the activation of guanylate cyclase *(265)*. The ability of such iron dinitrosyl complexes to counteract platelet aggregation has been claimed to be superior to that of nitroprusside *(266)*.

The formation of nitrosyl complexes of non-heme iron has also been used to probe (a) the structure of *putidamonooxin* (a terminal oxygenase) *(267,268)*, (b) the redox components of the *photosystem II* reaction center *(269,270)*, and (c) the function of *protocatechuate 4,5-dioxygenase (271)*. However, these studies are very biologically oriented and will not be discussed any further.

6.2.6: NO as a Probe for Enzyme Function

At the beginning of this chapter, it was noted that many of the biological studies involving the use of NO were performed with the assumption that the binding of NO to metal centers in biologically important molecules should provide some insight as to the nature of O_2 binding to these biomolecules. To end this chapter, a few such studies that have resulted in some insight as to the nature of O_2 binding will be briefly illustrated.

Isopenicillin N synthase (IPNS) is a non-heme Fe-enzyme that catalyzes the formation of isopenicillin N from the tripeptide *d*-(*L*-α-aminoadipoyl)-*L*-cysteinyl-*D*-valine (ACV) *(272)*.

O H SH
HO N
NH2 O O N
H CO2H
O2
I PNS
O H S
HO N
NH2 O O N
H CO2H
2H2O

The intimate details of the mechanism of this oxidative bicyclization remain unknown. However, using NO as a probe (i.e., its substitution for O_2), evidence has accrued for the direct involvement of the Fe center in this unique O_2-dependent reaction *(273,274)*. Indeed, the active site of IPNS is currently viewed as shown:

His
His OH
Fe
His ACV
O_2(NO)

and again illustrates how NO has been applied to probe both heme and non-heme function.

The *bleomycins* are a family of antibiotics whose therapeutic activity is generally believed to correlate with their ability to bind to and degrade DNA *(275)*. However, this DNA strand-scission requires O_2 and a metal ion, with Fe(II) being the most active. Since the 1:1 bleomycin-Fe(II) complex and its oxygen adduct are ESR inactive, their nitrosyl analogues have been synthesized and characterized by ESR spectroscopy *(276-278)*. In general, their properties are similar to those of the nitrosyl adducts of hemoproteins, and will not be discussed further. Nitrosyl adducts of other important biomolecules such as *peroxidase (279-284)*, *prostaglandin H synthase (285)* and *cytochrome* P450*cam (286)* have also been prepared and studied with a view to gaining some insight as to their function.

6.3: References

1. Dickerson, R. E.; Geis, I. *Hemoglobin*; Benjamin/Cummings: California, 1983.
2. Kelly, S.; Lancon, D.; Kadish, K. M. *Inorg. Chem.* **1984**, *23*, 679.
3. Diebold, T.; Schappacher, M.; Chevrier, B.; Weiss, R. *J. Chem. Soc., Chem. Commun.* **1979**, 693.
4. Scheidt, W. R.; Hatano, D.; Rupprecht, G. A.; Piciulo, P. L. *Inorg. Chem.* **1979**, *18*, 292.
5. Scheidt, W. R.; Frisse, M. E. *J. Am. Chem. Soc.* **1975**, *97*, 17.
6. Scheidt, W. R.; Hoard, J. L. *J. Am. Chem. Soc.* **1973**, *95*, 8281.

7. Hoshino, M.; Kogure, M. *J. Phys. Chem.* **1989**, *93*, 5478.
8. Choi, I. K.; Ryan, M. D. *Inorg. Chim. Acta* **1988**, *153*, 25.
9. Mu, X. H.; Kadish, K. M. *Inorg. Chem.* **1988**, *27*, 4720.
10. Lancon, D.; Kadish, K. M. *J. Am. Chem. Soc.* **1983**, *105*, 5610.
11. Olson, L. W.; Schaeper, D.; Lancon, D.; Kadish, K. M. *J. Am. Chem. Soc.* **1982**, *104*, 2042.
12. Kon, H.; Chikira, M.; Smith, K. M. *J. Chem. Soc., Dalton Trans.* **1981**, 1726.
13. Yoshimura, T. *Nippon Kagaku Kaishi* **1988**, *(4)*, 537. CA109(6):45426d.
14. Kadish, K. M.; Cornillon, J. L.; Yao, C. L.; Malinski, T.; Gritzner, G. *J. Electroanal. Chem. Interfacial Electrochem.* **1987**, *235*, 189.
15. Feng, D. W.; Ryan, M. D. *Inorg. Chem.* **1987**, *26*, 2480.
16. Yoshimura, T. *Bull. Chem. Soc. Jpn.* **1987**, *60*, 1375.
17. Cao, X.; Mu. X. *Gaodeng Xuexiao Huaxue Xuebao* **1986**, *7(5)*, 387. CA106(20):167698c.
18. Yoshimura, T. *Inorg. Chim. Acta* **1986**, *125*, L27.
19. Scheidt, W. R.; Lee, Y. J.; Hatano, K. *J. Am. Chem. Soc.* **1984**, *106*, 3191.
20. Ohba, Y.; Yokoyama, Y.; Iwaizumi, M. *Inorg. Chim. Acta* **1983**, *78*, L57.
21. Mu, X. H.; Kadish, K. M. *Inorg. Chem.* **1990**, *29*, 1031.
22. Finnegan, M. G.; Lappin, A. G.; Scheidt, W. R. *Inorg. Chem.* **1990**, *29*, 181.
23. Settin, M. F.; Fanning, J. C. *Inorg. Chem.* **1988**, *27*, 1431.
24. Magliozzo, R. S.; McCracken, J.; Peisach, J. *Biochemistry* **1987**, *26*, 7923.
25. Fujita, E.; Chang, C. K.; Fajer, J. *J. Am. Chem. Soc.* **1985**, *107*, 7665.
26. Fujita, E.; Fajer, J. *J. Am. Chem. Soc.* **1983**, *105*, 6743.
27. Kelly, S.; Lancon, D.; Kadish, K. M. *Inorg. Chem.* **1984**, *23*, 1451.
28. Mochida, I.; Suetsugu, K.; Fujitsu, H.; Takeshita, K. *J. Chem. Soc., Chem. Commun.* **1982**, 166.
29. Kadish, K. M.; Mu, X. H.; Lin, X. Q. *Inorg. Chem.* **1988**, *27*, 1489.
30. Tsuji, K.; Imaizumi, M.; Oyoshi, A.; Mochida, I.; Fujitsu, H.; Takeshita, K. *Inorg. Chem.* **1982**, *21*, 721.
31. Rauchfuss, T. B.; Weatherill, T. D. *Inorg. Chem.* **1982**, *21*, 727.
32. Wayland, B. B.; Newman, A. R. *Inorg. Chem.* **1981**, *20*, 3093.

33. Modi, S.; Behere, D. V.; Shedbalkewr, V. P. *J. Chem. Res., Synop.* **1988**, *(8)*, 244.
34. Christahl, M.; Twilfer, H.; Gersonde, K. *Biophys. Struct. Mech.* **1982**, *9*, 61.
35. Kobayashi, K.; Tamura, M.; Hayashi, K. *Biochim. Biophys. Acta* **1982**, *702*, 23.
36. Yoshimura, T. *Inorg. Chem.* **1986**, *25*, 688.
37. Yoshimura, T.; Ozaki, T. *Arch. Biochem. Biophys.* **1984**, *229*, 126.
38. Yoshimura, T. *Bull. Chem. Soc. Jpn.* **1983**, *56*, 2527.
39. Yoshimura, T. *Arch. Biochem. Biophys.* **1983**, *220*, 167.
40. Yoshimura, T. *Arch. Biochem. Biophys.* **1982**, *216*, 625.
41. Yoshimura, T.; Ozaki, T.; Shintani, Y.; Watanabe, H. *Arch. Biochem. Biophys.* **1979**, *193(2)*, 301.
42. Dobler, W.; Federer, W.; Howorka, F.; Lindinger, W.; Durup-Ferguson, M.; Ferguson, E. *J. Chem Phys.* **1983**, 1543.
43. Yoshimura, T. *Inorg. Chim. Acta* **1982**, *57*, 99.
44. Yoshimura, T. *J. Inorg. Biochem.* **1983**, *18*, 263.
45. Yu, N. T.; Lin, S. H.; Chang, C. K.; Gersonde, K. *Biophys. J.* **1989**, *55*, 1137.
46. Guilard, R.; Lagrange, G.; Tabard, A.; Lancon, D.; Kadish, K. M. *Inorg. Chem.* **1985**, *24*, 3649.
47. Kadish, K. M.; Lancon, D.; Cocolios, P. *Inorg. Chem.* **1984**, *23*, 2372.
48. Lagrange, G.; Cocolios, P.; Guilard, R. *J. Organomet. Chem.* **1984**, *260*, C16.
49. Buchler, J. W.; Smith, P. D. *Chem. Ber.* **1976**, *109*, 1465.
50. Massoudipour, M.; Pandey, K. K. *Inorg. Chim. Acta* **1989**, *160*, 115.
51. Wayland, B. B.; Olson, L. W. *J. Am. Chem. Soc.* **1974**, *96*, 6037.
52. Srivastava, T. S.; Hoffman, L.; Tsutsui, M. *J. Am. Chem. Soc.* **1972**, *94*, 1385.
53. Yoshimura, T. *Inorg. Chim. Acta* **1984**, *83*, 17.
54. Bonnett, R.; Charalambides, A. A.; Martin, R. A.; Sales, K. D.; Fitzsimmons, W. *J. Chem. Soc., Chem. Commun.* **1977**, 884.
55. Kadish, K. M. In *Iron Porphyrins*; Lever, A. B. P.; Gray, H. B., Eds.; Addison-Wesley: Massachusetts, 1983, Part 2, Chapter 4.
56. Guillard, R.; Kadish, K. M. *Chem. Rev.* **1988**, *88*, 1121.

57. Choi, I.-K.; Liu, Y.; Feng, D. W.; Paeng, K.-J.; Ryan, M. D. *Inorg. Chem.* **1991**, *30*, 1832.
58. Hoshino, M.; Konishi, S. *Chem. Phys. Lett.* **1985**, *115*, 511.
59. Addison, A. W.; Stephanos, J. J. *Biochemistry* **1986**, *25*, 4104.
60. Palmer, G. In *The Porphyrins*; Dolphin, D. D., Ed.; Academic: New York, 1979; Vol. 4B, p 313.
61. Palmer, G. In *Iron Porphyrins*; Lever, A. B. P.; Gray, H. B., Eds.; Addison-Wesley: Massachusetts, 1983, Part 2, p 77.
62. Morse, R. H.; Chan, S. I. *J. Biol. Chem.* **1980**, *255*, 7876.
63. Waleh, A.; Ho, N.; Chantranupong, L.; Loew, G. H. *J. Am. Chem. Soc.* **1989**, *111*, 2767.
64. Keeton, W. T.; Gould, J. L. *Biological Science*, 4th. ed.; Norton: New York, 1986; p 347.
65. Desideri, A.; Chiancone, E.; Ascoli, F. *J. Inorg. Biochem.* **1985**, *25*, 225.
66. Darnell, J.; Lodish, H.; Baltimore, D. *Molecular Cell Biology*; Freeman: New York, 1986; pp 884-889.
67. *Cytochrome Oxidase: Structure, Function and Physiopathology*; Brunori, M.; Chance, B., Eds.; Annals of the New York Academy of Sciences: New York, 1988; Vol. 550.
68. Di Iorio, E. E. *Methods Enzmol.* **1981**, *76*, 57.
69. Sanches, R. *Biochim. Biophys. Acta* **1988**, *955*, 310.
70. Yoshimura, T.; Watanabe, H. *Taiki Osen Gakkaishi* **1984**, *19*, 359. CA103(9):66136a.
71. Maida, N.; Kon, K.; Imaizumi, K.; Shiga, T. *Taiki Osen Gakkaishi* **1984**, *19*, 239. CA101(23):205582s.
72. Sharma, V. S.; Isaacson, R. A.; John, M. E.; Waterman, M. R.; Chevion, M. *Biochemistry* **1983**, *22*, 3897.
73. John, M. E.; Waterman, M. R. *J. Biol. Chem.* **1979**, *254*, 11953.
74. Morris, R. J.; Gibson, Q. H. *J. Bio. Chem.* **1980**, *255*, 8050.
75. Chiodi, H.; Mohler, J. G. *Environ. Res.* **1985**, *37*, 355.
76. DiFeo, T. J.; Addison, A. W. *Biochem. J.* **1989**, *260*, 863.
77. Gibson, Q. H. In *The Porphyrins*; Dolphin, D. D., Ed.; Academic: New York, 1978; Vol. 5C, p 187.
78. Neto, L. M.; Nascimento, O. R.; Tabak, M.; Caracelli, I. *Biochim. Biophys. Acta* **1988**, *956*, 189.

79. Doyle, M. P.; Mahapatro, S. N.; Broene, R. D.; Guy, J. K. *J. Am. Chem. Soc.* **1988**, *110*, 593.
80. Bazylinski, D. A.; Hollocher, T. C. *J. Am. Chem. Soc.* **1985**, *107*, 7982.
81. Bazylinski, D. A.; Hollocher, T. C. *Inorg. Chem.* **1985**, *24*, 4285.
82. Doyle, M. P.; Mahapatro, S. N. *J. Am. Chem. Soc.* **1984**, *106*, 3678.
83. Stolze, K.; Nohl, H. *Biochem. Pharmacol.* **1989**, *38*, 3055.
84. Young, L. J.; Siegel, L. M. *Biochemistry* **1988**, *27*, 2790.
85. Kamarei, A. R.; Karel, M. *J. Food. Sci.* **1982**, *47*, 682.
86. Yamamoto, T.; Nozawa, T.; Kaito, A.; Hatano, M. *Bull. Chem. Soc. Jpn.* **1982**, *55*, 2021.
87. Doyle, M. P.; Van Doornik, F. J.; Funckes, C. L. *Inorg. Chim. Acta* **1980**, *46*, L111.
88. Doetschman, D. C.; Rizos, A. K.; Szumowski, J. *J. Chem Phys.* **1984**, *81*, 1185.
89. Utterback, S. G.; Doetschman, D. C.; Szumowski, J.; Rizos, A. K. *J. Chem Phys.* **1983**, *78*, 5874.
90. Friedman, J. M.; Stepnoski, R. A.; Stavola, M.; Ondrias, M. R.; Cone, R. L. *Biochemistry* **1982**, *21*, 2022.
91. Chevion, M.; Peisach, J.; Blumberg, W. E. *Int. J. Biol. Macromol.* **1979**, *1(5)*, 208.
92. John, M. E.; Waterman, M. R. *FEBS Lett.* **1979**, *106*, 219.
93. Tsuneshige, A.; Imai, K.; Hori, H.; Tyuma, I.; Gotoh, T. *J. Biochem. (Tokyo)* **1989**, *106*, 406.
94. Bhuyan, A. K.; Lemtur, A.; Subramanian, J.; Lalthantluanga, R. *Biochim. Biophys. Acta* **1989**, *997*, 36.
95. Bartnick, D. E.; Mizukami, H.; Romero-Herrera, A. E. *J. Biol. Chem.* **1983**, *258*, 1599.
96. Mishra, K. C.; Mishra, S. K.; Roy, J. N.; Ahmad, S.; Das, T. P. *J. Am. Chem. Soc.* **1985**, *107*, 7898.
97. Desideri, A.; Meier, U. T.; Winterhalter, K. H.; Di Iorio, E. E. *FEBS Lett.* **1984**, *166*, 378.
98. Blum, H.; Bowyer, J. R.; Cusanovich, M. A.; Waring, A. J.; Ohnishi, T. *Biochim. Biophys. Acta* **1983**, *748*, 418.
99. Christahl, M.; Gersonde, K. *Biophys. Struct. Mech.* **1982**, *8*, 271.
100. Louro, S. R. W.; Bemski, G. *FEBS Lett.* **1982**, *142*, 293.

101. Louro, S. R. W.; Ribeiro, P. C.; Bemski, G. *Biochim. Biophys. Acta* **1981**, *670*, 56.
102. John, M. E.; Lalthantluanga, R.; Liljeqvist, G.; Paleus, S.; Braunitzer, G. *Z. Naturforsch., B: Anorg. Chem., Org. Chem.* **1982**, *37B*, 744.
103. Muench, P. J.; Stapleton, H. J. *J. Chem Phys.* **1985**, *82*, 2828.
104. Caracelli, I.; Meirelles, N. C.; Tabak, M.; Filho, O. B.; Nascimento, O. R. *Biochim. Biophys. Acta* **1988**, *955*, 315.
105. Hori, H.; Ikeda-Saito, M.; Yonetani, T. *J. Biol. Chem.* **1981**, *256*, 7849.
106. Scholler, D. M.; Wang, M.-Y. R.; Hoffman, B. M. *J. Biol. Chem.* **1979**, *254*, 4072.
107. Dickinson, L. C.; Symons, M. C. R. *Chem. Soc. Rev.* **1983**, *12(4)*, 387.
108. Mackin, H. C.; Benko, B.; Yu, N. T.; Gersonde, K. *FEBS Lett.* **1983**, *158*, 199.
109. Morishima, I.; Hara, M. *Biochem. Biophys. Res. Commun.* **1984**, *121*, 229.
110. Knowles, F. C. *Arch. Biochem. Biophys.* **1984**, *230*, 327.
111. Yoshimura, T. *Kagaku Zokan (Kyoto)* **1982**, *(95)*, 39. CA97(13):105640q.
112. Huang, T.-H. *J. Biol. Chem.* **1979**, *254*, 11467.
113. Hoehn, M.; Huettermann, J.; Chien, J. C. W.; Dickinson, L. C. *J. Am. Chem. Soc.* **1983**, *105*, 109.
114. Kappl, R.; Huettermann, J. *Isr. J. Chem.* **1989**, *29*, 73.
115. Deatherage, J. F.; Moffat, K. *J. Mol. Biol.* **1979**, *134*, 401.
116. Ascenzi, P.; Coletta, M.; Desideri, A.; Polizio, F.; Condo, S. G.; Giardina, B. *J. Inorg. Biochem.* **1990**, *40*, 157.
117. Stephanos, J. J.; Addison, A. W. *Eur. J. Biochem.* **1990**, *189*, 185.
118. Ascenzi, P.; Desideri, A.; Amiconi, G.; Bertollini, A.; Bolognesi, M.; Castagnola, M.; Coletta, M.; Brunori, M. *J. Inorg. Biochem.* **1988**, *34*, 19.
119. Ascenzi, P.; Santucci, R.; Desideri, A.; Amiconi, G. *J. Inorg. Biochem.* **1988**, *32 (4)*, 225.
120. Krishna, Y. V. S. R.; Aruna, B.; Narayana, P. A. *Biochim. Biophys. Acta* **1987**, *916*, 48.
121. Desideri, A.; Ascenzi, P.; Chiancone, E.; Amiconi, G. *J. Inorg. Biochem.* **1987**, *29*, 131.
122. Morishma, I.; Hara, M.; Ishimori, K. *Biochemistry* **1986**, *25*, 7243.

123. Friedman, J. M.; Rousseau, D. L.; Ondrias, M. R.; Stepnoski, R. A. *Science* **1982**, *218 (4578)*, 1244.

124. Ristau, O.; Greschner, S.; Stolley, P.; Oddoy, A.; Rein, H. *Stud. Biophys.* **1984**, *103*, 149.

125. Hille, R.; Olson, J. S.; Palmer, G. *J. Biol. Chem.* **1979**, *254*, 12110.

126. Friedman, J. M.; Scott, T. W.; Stepnoski, R. A.; Ikeda-Saito, M.; Yonetani, T. *J. Biol. Chem.* **1983**, *258*, 10564.

127. Walters, M. A.; Spiro, T. G.; Scholler, D. M.; Hoffman, B. H. *J. Raman Spectrosc.* **1983**, *14*, 162.

128. Stong, J. D.; Burke, J. M.; Daly, P.; Wright, P.; Spiro, T. G. *J. Am. Chem. Soc.* **1980**, *102*, 5815.

129. John, M. E.; Waterman, M. R. *J. Biol. Chem.* **1980**, *255*, 4501.

130. Ascenzi, P.; Coletta, M.; Desideri, A.; Brunori, M. *Biochim. Biophys. Acta* **1985**, *829*, 299.

131. Walters, M. A.; Spiro, T. G. *Biochemistry* **1982**, *21*, 6989.

132. John, M. E. *Eur. J. Biochem.* **1982**, *124*, 305.

133. John, M. E. *Biochim. Biophys. Acta* **1981**, *669*, 113.

134. Coletta, M.; Boffi, A.; Ascenzi, P.; Brunori, M.; Chiancone, E. *J. Biol. Chem.* **1990**, *265*, 4828.

135. Houde, D.; Petrich, J. W.; Rojas, O. L.; Poyart, C.; Antonetti, A.; Martin, J. L. *Springer Ser. Chem. Phys.* **1986**, *46*, 419.

136. Jongeward, K. A.; Marsters, J. C.; Magde, D. *Springer Ser. Chem. Phys.* **1986**, *46*, 427. CA106(9):63276x.

137. Martin, J. L.; Migus, A.; Poyart, C.; Lecartentier, Y.; Astier, A.; Antonetti, A. *Springer Ser. Chem. Phys.* **1984**, *38*, 447. CA102(7):58024b.

138. Kamerei, A. R.; Karel, M. *Int. J. Radiat. Biol. Relat. Stud. Phys., Chem. Med.* **1983**, *44*, 135.

139. Cornelius, P. A.; Hochstrasser, R. M.; Steele, A. W. *J. Mol. Biol.* **1983**, *163*, 119.

140. Cornelius, P. A.; Hochstrasser, R. M. *Springer Ser. Chem. Phys.* **1982**, *23*, 288. CA98(5):29922m.

141. Linhares, M. P.; El-Jaick, L. J.; Bemski, G.; Wajnberg, E. *Int. J. Biol. Macromol.* **1990**, *12*, 59.

142. Chantranupong, L.; Loew, G. H.; Waleh, A. In *Porphyrins: Excited States and Dynamics*; Gouterman, M.; Rentzepis, P. M.; Straub, K. D., Eds; ACS Symposium Series 321; Washington D. C., 1986; pp 2-19.
143. Guest, C. R.; Noe, L. J. *Biophys. J.* **1988**, *54*, 731.
144. Petrich, J. W.; Poyart, C.; Martin, J. L. *Biochemistry* **1988**, *27*, 4049.
145. Kon, K.; Maeda, N.; Suda, T.; Shiga, T. *Taiki Osen Gakkaishi* **1980**, *15(10)*, 401. CA95(1):2179a.
146. Miura, S.; Morimoto, H. *J. Mol. Biol.* **1980**, *143*, 213.
147. Hoffman, B. M. In *The Porphyrins*; Dolphin, D. D., Ed.; Academic: New York, 1979; Vol. 7B, p 403.
148. Yu, N. T.; Thompson, H. M.; Mizukami, H.; Gersonde, K. *Eur. J. Biochem.* **1986**, *159*, 129.
149. Hori, H.; Ikeda-Saito, M.; Leigh, J. S., Jr.; Yonetani, T. *Biochemistry* **1982**, *21*, 1431.
150. Parthasarathi, N.; Spiro, T. G. *Inorg. Chem.* **1987**, *26*, 2280.
151. Lin, S. H.; Yu, N. T.; Gersonde, K. *FEBS Lett.* **1988**, *229*, 367.
152. Malmström, B. G. *Chem. Rev.* **1990**, *90*, 1247.
153. Brunori, M.; Antonini, G.; Malatesta, F.; Sarti, P.; Wilson, M. T. *Eur. J. Biochem.* **1987**, *169*, 1.
154. Rousseau, D. L.; Singh, S.; Ching, Y. C.; Sassaroli, M. *J. Biol. Chem.* **1988**, *263*, 5681.
155. Scholes, C. P.; Janakiraman, R.; Taylor, H.; King, T. E. *Biophys. J.* **1984**, *45*, 1027.
156. LoBrutto, R.; Wei, Y. H.; Mascarenhas, R.; Scholes, C. P.; King, T. E. *J. Biol. Chem.* **1983**, *258*, 7437.
157. Barlow, C.; Erecinska, M. *FEBS Lett.* **1979**, *98*, 9.
158. Twilfer, H.; Gersonde, K.; Christahl, M. *J. Magn. Reson.* **1981**, *44*, 470.
159. Yoshimura, H.; Suzuki, S.; Nakahara, A.; Iwasaki, H.; Matsubara, T.; Iwaizumi, M. *Inorg. Chim. Acta* **1987**, *136*, L17.
160. Suzuki, S.; Yoshimura, T.; Nakahara, A.; Iwasaki, H.; Shidara, S.; Matsubara, T. *Inorg. Chem.* **1987**, *26*, 1006.
161. Izumi, K.; Cassens, R. G.; Greaser, M. L. *J. Food Sci.* **1982**, *47*, 1419.
162. Liu, M. C.; Huynh, B. H.; Payne, W. J.; Peck, H. D., Jr.; Dervartanian, D. V.; Legall, J. *Eur. J. Biochem.* **1987**, *169*, 253.
163. Edwards, S. L.; Kraut, J.; Poulos, T. L. *Biochemistry* **1988**, *27*, 8074.
164. Butler, A. R. *Chem. Britain* **1990**, 419.

165. Olah, M. E.; Rahwan, R. G. *Pharmacology* **1988**, *37*, 305. CA110:18326k.

166. Bossaller, C.; Hehlert-Friedrich, C.; Jost, S.; Rafflenbeul, W.; Lichtlen, P. *Eur. Heart J.* **1989**, *10 (Suppl. F)*, 44.

167. Marletta, M. A. *TIBS* **1989**, *14*, 488.

168. Moncada, S.; Herman, A. G.; Vanhoutte, P. *Trends in Pharmacological Sciences* **1987**, *8*, 365.

169. Halliwell, B.; Gutteridge, J. M. C. *Free Radicals in Biology and Medicine*, 2nd. ed.; Clarendon: Oxford, 1989; p 408.

170. Tayeh, M. A.; Marletta, M. A. *J. Biol. Chem.* **1989**, *264*, 19654.

171. Förstermann, U.; Mugge, A.; Alheid, U.; Bode, S. M.; Frölich, J. C. *Eur. Heart J.* **1989**, *10 (Suppl. F)*, 36.

172. Marletta, M. A.; Yoon, P. S.; Iyengar, R.; Leaf, C. D.; Wishnok, J. S. *Biochemistry* **1988**, *27*, 8706.

173. Yu, W. C.; Goff, E. U. *Anal. Chem.* **1983**, *55*, 29.

174. Lancaster, J. R., Jr.; Hibbs, J. B., Jr. *Proc. Natl. Acad. Sci. U.S.A.* **1990**, *87*, 1223.

175. Walter, U.; Waldmann, R.; Nieberding, M. *Eur. Heart J.* **1988**, *9 (Suppl. H)*, 1.

176. Ignarro, L. J.; Kadowitz, P. J. *Ann. Rev. Pharmacol. Toxicol.* **1985**, *25*, 171.

177. Boehme, E.; Grossman, G.; Herz, J.; Muelsch, A.; Spies, C.; Schultz, G. *Adv. Cyclic Nucleotide Protein Phosphorylation Res.* **1984**, *17*, 259.

178. Craven, P. A.; DeRubertis, F. R. *Biochim. Biophys. Acta* **1983**, *745*, 310.

179. El Dieb, M. M. R.; Parker, C. D.; White, A. *Biochim. Biophys. Acta* **1987**, *928*, 83.

180. Ignarro, L. J.; Adams, J. B.; Horwitz, P. M.; Wood, K. S. *J. Biol. Chem.* **1986**, *261*, 4997.

181. Tsai, S. C.; Adamik, R.; Manganiello, V. C.; Vaughan, M. *Biochem. J.* **1983**, *215*, 447.

182. Wolin, M. S.; Wood, K. S.; Ignarro, L. J. *J. Biol. Chem.* **1982**, *257*, 13312.

183. Edwards, J. C.; Barry, B. K.; Gruetter, D. Y.; Ohlstein, E. H.; Baricos, W. H.; Ignarro, L. J. *Biochem. Pharmacol.* **1981**, *30*, 2531.

184. Craven, P. A.; DeRubertis, F. R.; Pratt, D. W. *J. Biol. Chem.* **1979**, *254*, 8213.

185. Craven, P. A.; DeRubertis, F. R. *J. Biol. Chem.* **1978**, *253*, 8433.
186. Feelisch, M.; Noack, E.; Schroder, H. *Eur. Heart J.* **1988**, *(Suppl. A)*, 57.
187. Kruszyna, H.; Kruszyna, R.; Hurst, J.; Smith, R. P. *J. Toxicol. Environ. Health* **1980**, *6*, 757.
188. Butler, A. R.; Glidewell, C. *Chem. Soc. Rev.* **1987**, *16*, 361.
189. Vanin, A. F.; Mordvintsev, P. I.; Shabarchina, M. M.; Kubrina, L. N.; Aliyev, D. I.; Moshkovskii, Y. S. *Biophysics (Moscow)* **1983**, *28*, 1052.
190. Kruszyna, H.; Kruszyna, R.; Smith, R. P. *Anesthesiology* **1982**, *57*, 303.
191. Kruszyna, R.; Kruszyna, H.; Smith, R. P.; Wilcox, D. E. *Toxicol. Appl. Pharmacol.* **1988**, *94*, 458.
192. Kruszyna, H.; Kruszyna, R.; Smith, R. P.; Wilcox, D. E. *Toxicol. Appl. Pharmacol.* **1987**, *91*, 429.
193. Kuropteva, Z. V.; Pastushenko, O. N. *Dokl. Akad. Nauk SSSR* **1985**, *281*, 189. CA103(5):31937k.
194. Avdeeva, O. S.; Pulatova, M. K.; Vanin, A. F.; Emanuel, N. M. *Dokl. Akad. Nauk SSSR* **1979**, *249*, 224. CA92(5):34400v.
195. Pellat, C.; Henry, Y.; Drapier, J. C. *Biochem. Biophys. Res. Commun.* **1990**, *166*, 119.
196. Shibuki, K.; Okada, D. *Nature* **1991**, *349*, 326.
197. Yoshida, K. Kasama, K. *Environ. Health Perspect.* **1987**, *73*, 201.
198. Maeda, N.; Imaizumi, K.; Kon, K.; Shiga, T. *Environ. Health Perspect.* **1987**, *73*, 171.
199. Sadana, J. C.; Khan, B. M.; Fry, I. V.; Cammack, R. *Biochem. Cell Biol.* **1986**, *64*, 394.
200. Fry, I. V.; Cammack, R.; Hucklesby, D. P.; Hewitt, E. J. *FEBS Lett.* **1980**, *111*, 377.
201. Morgenstern, M. A.; Schowen, R. L. *Z. Naturforsch., A: Phys. Sci.* **1989**, *44*, 450.
202. Kim, C. H.; Hollocher, T. C. *J. Biol. Chem.* **1984**, *259*, 2092.
203. Cammack, R.; Fry, I. V. *Biochem. Soc. Trans.* **1980**, *8*, 642.
204. Johnson, M. K.; Thomson, A. J.; Walsh, T. A.; Barber, D.; Greenwood, C. *Biochem. J.* **1980**, *189*, 285.
205. Kanayama, Y.; Yamamoto, Y. *Plant Cell Physiol.* **1990**, *31*, 207.
206. Goretski, J.; Hollocher, T. C. *J. Biol. Chem.* **1988**, *263*, 2316.
207. Latham, I. A.; Thornback, J. R.; Newman, J. L. *Eur. Patent 291281 A1*, 1988. CA111(11):93017j.

208. (a) Tyeklár, Z.; Karlin, K. D. *Acc. Chem. Res.* **1989**, *22*, 241. (b) Partha, P. P.; Karlin, K. D. *J. Am. Chem. Soc.* **1991**, *113*, 6331.

209. Hulse, C. L.; Tiedje, J. M.; Averill, B. A. *J. Am. Chem. Soc.* **1989**, *111*, 2322.

210. Suzuki, S.; Yoshimura, T.; Kohzuma, T.; Shidara, S.; Masuko, M.; Sakurai, T.; Iwasaki, H. *Biochem. Biophys. Res. Commun.* **1989**, *164*, 1366.

211. Tahon, J. P.; Van Hoof, D.; Vinckier, C.; Witters, R.; De Ley, M.; Lontie, R. *Biochem. J.* **1988**, *249*, 891.

212. Deleersnijder, W.; Witters, R.; Lontie, R. *Inorg. Chim. Acta* **1983**, *80*, L45.

213. Gorren, A. C. F.; de Boer, E.; Wever, R. *Biochim. Biophys. Acta* **1987**, *916*, 38.

214. Lawrie, R. A. *Meat Science*, 3rd. ed.; Pergamon: Oxford, 1979; pp 263-272 and 300-311.

215. Walters, C. L. In *Meat*; Cole, D. J. A.; Lawrie, R. A., Eds.; Butterworths: U.K., 1974; Chapter 20.

216. Lougovois, V.; Houston, T. W. *Food Chem.* **1987**, *24*, 241.

217. Giddings, G. G. *J. Food Sci.* **1977**, *42*, 288.

218. Bello, J.; Gorospe, O.; Sanchez-Monge, J. M. *An. Bromatol.* **1988**, *40*, 121. CA110(9):74042b.

219. Cantoni, C.; Comi, G.; D'Aubert, S.; Perlasca, M. *Ind. Aliment* **1978**, *17(156)*, 940. CA90(19):150384a.

220. Bonnett, R.; Charalambides, A. A.; Martin, R. A. *J. Chem. Soc., Perkin Trans. 1* **1978**, 974.

221. Livingston, D. J.; Brown, W. D. *Food Technol.* **1981**, *35*, 244.

222. Vada-Kovacs, M.; Czibula, E.; Rekasi, E.; Incze, Z. *Fleischwirtschaft* **1988**, *68*, 1358 and 1365. CA110(7):56235n

223. Kamerai, A. R.; Karel, M. *Int. J. Radiat. Biol. Relat. Stud. Phys., Chem. Med.* **1983**, *44*, 123.

224. Sakata, R.; Nagata, Y. *Nippon Chikusan Gakkaiho* **1986**, *57*, 103. CA105(5):41443b.

225. Sakata, R.; Nagata, Y. *Nippon Chikusan Gakkaiho* **1983**, *54*, 667. CA100(9):66693t.

226. Glidewell, C. *Chem. Britain* **1990**, 137.

227. Slinde, E. *J. Food Sci.* **1987**, *52*, 1152.

228. Rubin, L. J.; Diosady, L. L.; O'Boyle, A. R. *Can. Chem. News* **1990**, *42(7)*, 18.

229. Shahidi, F.; Pegg, R. B. *Can. Chem. News* **1991**, *43(2)*, 12.

230. Noel, P.; Culioli, J.; Melcion, J. P.; Goutefongea, R.; Coquillet, R. *Lebensm. -Wiss. Technol.* **1984**, *17(6)*, 305. CA102(11):94472e.

231. Smith, J. S.; Burge, D. L., Jr. *J. Food Sci.* **1987**, *52*, 1728.

232. Cantoni, C.; Redaelli, A.; Perlasca, M.; Apostolo, M. *Ind. Aliment* **1978**, *17*, 942. CA90:150385b.

233. Ahn, D. U.; Maurer, A. J. *Poult. Sci.* **1989**, *68*, 100.

234. Lougovois, V.; Houston, T. W. *Food Chem.* **1989**, *32*, 47.

235. Sakata, R.; Nagata, Y. *Fleischwirtschaft* **1988**, *68*, 1146. CA110(1):6575c.

236. Culioli, J.; Noel, P.; Goutefongea, R. *Sci. Aliments* **1981**, *1(2)*, 169. CA95(13):113656y.

237. Woolford, G.; Cassens, R. G. *J. Food Sci.* **1977**, *42*, 586.

238. Ashworth, J.; Hargreaves, L. L.; Jarvis, B. *J. Food Technol.* **1973**, *8*, 477.

239. Roberts, T. A.; Ingram, M. *J. Food. Technol.* **1973**, *8*, 467.

240. Roberts, T. A.; Garcia, C. E. *J. Food Technol.* **1973**, *8*, 463.

241. Butler, A. R.; Glidewell, C.; Li, M. *Adv. Inorg. Chem.* **1988**, *32*, 335.

242. Michalski, W. P.; Nicholas, D. J. D. *Arch. Microbiol.* **1987**, *147(3)*, 304.

243. Vanin, A. F. *Biophysics (Moscow)* **1987**, *32*, 136.

244. Vanin, A. F.; Varich, V. Y. *Stud. Biophys.* **1981**, *86*, 177.

245. Mordvintsev, P. I.; Vanin, A. F. *Izv. Akad. Nauk SSSR, Ser. Biol.* **1988**, 942. CA110(9):73019n.

246. Kurbanov, I. S. *Stud. Biophys.* **1987**, *120(2)*, 145.

247. Aliev, D. I.; Vanin, A. F. *Izv. Akad. Nauk As. SSR, Ser Biol. Nauk* **1986**, *(5)*, 123. CA107(9):74864j.

248. Mordvitsev, P. I.; Kubrina, L. N.; Kleshchev, A. L.; Vanin, A. F. *Stud. Biophys.* **1984**, *103*, 63. CA102(19):162416s.

249. Vanin, A. F.; Mordvintsev, P. I.; Kleshchev, A. L. *Stud. Biophys.* **1984**, *102*, 135. CA102(7):57409u.

250. Vanin, A. F.; Aliev, D. I. *Stud. Biophys.* **1983**, *97*, 223.

251. Varich, V. Y.; Vanin, A. F. *Biophysics (Moscow)* **1983**, *28*, 1125.

252. Vanin, A. F.; Aliev, D. I. *Stud. Biophys.* **1983**, *93*, 63.

253. Varich, V. Y. *Biophysics (Moscow)* **1981**, *26*, 906.

254. Vanin, A. F.; Kubrina. L. N.; Aliev, D. I. *Stud. Biophys.* **1980**, *80*, 221.

255. Varich, V. Y. *Biophysics (Moscow)* **1979**, *24*, 1146.
256. Vanin, A. F.; Varich, V. Y. *Biophysics (Moscow)* **1979**, *24*, 686.
257. Varich, V. Y. *Biophysics (Moscow)* **1979**, *24*, 359.
258. Vanin, A. F.; Kiladze, S. V.; Kubrina, L. N. *Biophysics (Moscow)* **1978**, *23*, 479.
259. Varfolomeyer, V. N.; Bogdanov, G. N.; Sidorenko, L. I.; Boikov, P. Y.; Todorov, I. N. *Biophysics (Moscow)* **1983**, *28*, 654.
260. Sidorik, E. P.; Korchevaya, L. M.; Burlaka, A. P. *Eksp. Onkol.* **1980**, *2(5)*, 37. CA94(7):42408k.
261. Liu, J.; Li, M. *Carcinogenesis (London)* **1989**, *10*, 617.
262. Kuznetsov, V. A.; Mordvintsev, P. I.; Dank, E. K.; Yurkiv, V. A.; Vanin, A. F. *Vopr. Med. Khim.* **1988**, *34 (5)*, 43. CA110(3):18331h.
263. Galagan, M. E.; Oranovskaya, E. V.; Mordvintsev, P. I.; Medvedev, O. S.; Vanin, A. F. *Byull. Vses. Kardiol. Nauchn. Tsentra AMN SSSR* **1988**, *12*, 75. CA110(15):128338y.
264. Kleshchev, A. L.; Mordvintsev, P. I.; Vanin, A. F. *Stud. Biophys.* **1985**, *105*, 93.
265. Vanin, A. F.; Kleshchev, A. L.; Mordvintsev, P. I.; Sedov, K. R. *Dokl. Akad. Nauk SSSR* **1985**, *281*, 742. CA103(11):81468d.
266. Mordvintsev, P. I.; Rudneva, V. G.; Vanin, A. F.; Shimkevich, L. L.; Khodorov, B. I. *Biochemistry (Moscow)* **1987**, *51*, 1584.
267. Bill, E.; Bernhardt, F. H.; Trautwein, A. X.; Winkler, H. *Eur. J. Biochem.* **1985**, *147*, 177.
268. Twilfer, H.; Bernhardt, F. H.; Gersonde, K. *Eur. J. Biochem.* **1985**, *147*, 171.
269. Diner, B. A.; Petrouleas, V. *Biochim. Biophys. Acta* **1990**, *1015*, 141.
270. Petrouleas, V.; Diner, B. A. *Biochim. Biophys. Acta* **1990**, *1015*, 131.
271. Arciero, D. M.; Lipscomb, J. D. *J. Biol. Chem.* **1986**, *261*, 2170.
272. Baldwin, J. E.; Bradley, M. *Chem. Rev.* **1990**, *90*, 1079.
273. Chen, V. J.; Orville, A. M.; Harpel, M. R.; Frolik, C. A.; Surerus, K. K.; Munck, E.; Lipscomb, J. D. *J. Biol. Chem.* **1989**, *264*, 21677.
274. Ming, L.-J.; Que, L. Jr.; Kriauciunas, A.; Frolik, C. A.; Chen, V. J. *Inorg. Chem.* **1990**, *29*, 1111.
275. Stubbe, J.; Kozarich, J. W. *Chem. Rev.* **1987**, *87*, 1107.
276. Sugiura, Y.; Takita, T.; Umezawa, H. *J. Antibiot.* **1981**, *34*, 249.

277. Antholine, W. E.; Petering, D. H. *Biochem. Biophys. Res. Commun.* **1979**, *91*, 528.
278. Sugiura, Y.; Ishizu, K. *J. Inorg. Biochem.* **1979**, *11*, 171.
279. Luka, G. S.; Rodgers, K. R.; Goff, H. M. *Biochemistry* **1987**, *26*, 6927.
280. Lukat, G. S.; Jabro, M. N.; Rodgers, K. R.; Goff, H. M. *Biochim. Biophys. Acta* **1988**, *954*, 265.
281. Smulevich, G.; Spiro, T. G. *Biochim. Biophys. Acta* **1985**, *830*, 80.
282. Bolscher, B. G. J. M.; Wever, R. *Biochim. Biophys. Acta* **1984**, *791*, 75.
283. Sievers, G.; Peterson, J.; Gadsby, P. M. A.; Thomson, A. J. *Biochim. Biophys. Acta* **1984**, *785*, 7.
284. Hayashi, K.; Kobayashi, K. *Mem. Inst. Sci. Ind. Res., Osaka Univ.* **1985**, *42*, 91. CA103(3):18758h.
285. Karthein, R.; Nastainczyk, W.; Ruf, H. H. *Eur. J. Biochem.* **1987**, *166*, 173.
286. Hu, S.; Kincaid, J. R. *J. Am. Chem. Soc.* **1991**, *113*, 2843.

7

Reactions of Metal Nitrosyl Complexes

The reactions of metal nitrosyl complexes may be divided into two distinct classes: (i) reactions that chemically transform one or more NO ligands, and (ii) reactions that do not transform any NO ligands. The latter reactions may occur at the metal center or at ancillary ligands, although such reactions are often facilitated by the bound nitrosyl group(s).

Earlier descriptions of the former category treat the linear NO ligand as an electrophile (formally NO^+) which essentially undergoes nucleophilic attack at the nitrosyl nitrogen or, conversely, consider the bent NO ligand as a nucleophile (formally NO^-) which undergoes electrophilic attack at the nitrogen atom. These simplistic views have often been very useful in understanding the reactivity patterns of the MNO group in most coordination complexes. However, the reactions of the MNO group with nucleophiles and electrophiles are necessarily *bimolecular*. Therefore, in each case, the relative energies and orbital constitutions of the $M(NO)_x$ group and of the reagent must be taken into account in order to assign electrophilicity or nucleophilicity to either species.

In this chapter, the wide range of reactivity of the MNO moiety, ranging from simple conformation changes to reactions at the N and O atoms is presented. Furthermore, the characteristic patterns of reactivity of metal nitrosyl complexes in which the NO groups in these complexes serve mainly as "spectator" ligands, but still provide unique chemical reactivity to the metal centers, are discussed. Most of the stoichiometric reactions of other organic groups bound to metal nitrosyls have been considered in Chapter 4, and will not be repeated here. Other treatments and earlier reviews on the reactivity of metal nitrosyl complexes may be found in the literature *(1-8)*.

7.1: Conformation Changes in MNO Bonds

7.1.1: Linear-bent Nitrosyl Interchange

The ability of the NO ligand to adopt linear or bent coordination geometries allows it to function as a formal three-electron or one-electron donor, respectively.

M—N≡O

three-electron donor

M—N̈=O

one-electron donor

A linear-to-bent structural conversion should, therefore, allow a metal center to achieve coordinative unsaturation. Alternatively, an electron-rich metal center may be able to release some of its electronic strain by backdonating increased amounts of electron density to the π^* orbitals of the NO ligand, thereby "forcing" the MNO group to bend. Indeed, most bent MNO linkages occur in electron-rich metal complexes. The linear-to-bent conversion may also occur as a result of an intramolecular fluxional process or may be achieved by photolytic, electrochemical, or chemical means. For example, the $Cp_2V(NO)X$ molecules (X = Br, I) are fluxional due to linear-bent conformation changes of the bound NO groups *(9,10)*. The five-coordinate $[RuCl(NO)_2(PPh_3)_2]BF_4$ complex possesses both linear and bent NO ligands in the solid state but is fluxional in solution due to a rapid, intramolecular interconversion of the linear and bent NO ligands *(11)*. The Os congener also undergoes a similar linear-bent interconversion in solution *(12,13)*. The behavior of $[Ir(NO)(\eta^3\text{-}C_3H_5)(PPh_3)_2]^+$ in solution also reveals a facile, well-defined linear-bent NO interchange for this compound *(14)*. Surprisingly, this linear-bent interchange occurs without an accompanying η^3-η^1 allyl fluxionality (eq 7.1).

$$\left[(\text{C}_8\text{H}_{12})(\text{Ph}_3\text{P})_2\text{Ir}{-}\text{N}{\equiv}\text{O} \right]^+ \rightleftharpoons \left[(\text{C}_8\text{H}_{12})(\text{Ph}_3\text{P})_2\text{Ir}{-}\ddot{\text{N}}{=}\text{O} \right]^+ \quad (7.1)$$

Interestingly, both linear and bent nitrosyl forms of the complex may be crystallized as their PF_6^- and BF_4^- salts, respectively.

Occasionally, photolysis of nitrosyl complexes results in the bending of the NO ligand(s) in the excited state complexes. For example, the photolysis of $Co(NO)(CO)_3$ *(15,16)*, CpNi(NO) *(17)*, or $CpV(NO)_2CO$ *(18)* results in the bending of NO ligands. Furthermore, the photolysis of $Cr(NO)_4$, $Mn(NO)_3CO$ and $Mn(NO)(CO)_4$ in inert matrices also results in the bending of nitrosyl ligands in the excited state complexes *(20-22)*. Finally, the electrochemical one-electron reductions of $CpM(NO)(CO)_2$ (M = Cr, Mo) also result in the extra electrons being localized in the MNO groups, generating bent MNO linkages *(19)*.

On occasion, linear-to-bent NO conformation changes may be induced chemically by the addition of neutral or anionic nucleophiles (L) to the metal center (eq 7.2), i.e.,

$$\text{M}{-}\text{N}{\equiv}\text{O} + \text{L} \longrightarrow (\text{L}){-}\text{M}{-}\ddot{\text{N}}{=}\text{O} \quad (7.2)$$

Indeed, it is probably as a result of such processes that metal carbonyl nitrosyls undergo associative rather than dissociative substitution reactions. Thus, whereas the "18-electron" binary metal carbonyls such as $Ni(CO)_4$ react by dissociative mechanisms, $Co(NO)(CO)_3$ and $Fe(NO)_2(CO)_2$ (which are isoelectronic and isostructural with $Ni(CO)_4$) react by associative mechanisms *(23)* (eq 7.3).

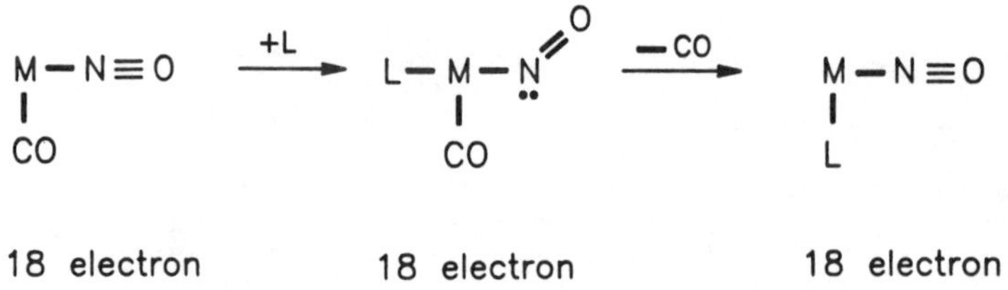

$$\underset{\substack{|\\ CO}}{M}-N\equiv O \xrightarrow{+L} L-\underset{\substack{|\\ CO}}{M}-\ddot{N}{=}O \xrightarrow{-CO} \underset{\substack{|\\ L}}{M}-N\equiv O \qquad (7.3)$$

18 electron   18 electron   18 electron

A classic example of how the addition of a nucleophile can affect the conformation of an NO ligand in a nitrosyl complex is provided in the following example of stereochemical control of valence. The five-coordinate $[Co(NO)(das)_2]^{2+}$ complex (das = *o*-phenylene-bis(dimethyl)arsine) possesses a linear nitrosyl group (179°), whereas the $[Co(NO)(das)_2(NCS)]^+$ complex adopts bent NO geometry (132°) *(24)*. As expected, nucleophiles react with the nitrosyl nitrogen of the dicationic complex, whereas electrophiles react with the nitrosyl nitrogen of the monocationic complex.

7.1.2: Terminal-Bridging Nitrosyl Interchange

Another kind of conformation change of a bound NO group is the terminal-bridging transformation. For example, it has long been known that $[CpCr(NO)_2]_2$ undergoes a *cis-trans* isomerization via an intermediate species that possesses only terminal nitrosyl ligands (eq 7.4).

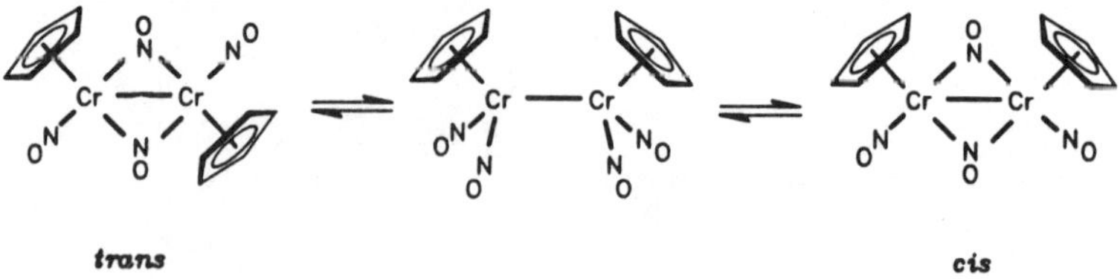

(7.4)

trans   *cis*

In polar solvents the equilibrium is displaced in favor of the *cis*-isomer, whereas in non-polar solvents the complex exists almost exclusively as the centrosymmetric *trans*-isomer. Similar dynamic terminal-bridging conversions occur in isoelectronic $[CpFe(CO)_2]_2$ *(25)*, and in $[CpMn(CO)(NO)]_2$ *(25-27)*. Chemically-induced terminal-bridging NO conversions such as the reactions described in eqs 7.5 and 7.6 have also been noted *(28,29)*.

$$[CpFe]_2(\mu\text{-}NO)_2 + CH_2N_2 \longrightarrow [CpFe(NO)]_2(\mu\text{-}CH_2) + N_2 \quad (7.5)$$

$$[(ON)Ir(COD)(\mu\text{-}pz)_2Ir(COD)]^+ \xrightarrow{HCl} [\{Ir(COD)Cl\}_2(\mu\text{-}NO)(\mu\text{-}pz)_2]^+ \quad (7.6)$$

7.2: Reactivity at the Nitrosyl Oxygen

7.2.1: Solvent Interactions

Solvents interact with nitrosyl complexes via (i) the metal center, especially in the case of coordinatively unsaturated complexes, (ii) the nitrosyl group, or (iii) any ancillary ligands present, and it is known that the NO stretching-frequency (ν_{NO}) of a given metal-nitrosyl complex is dependent on the solvent used to record its IR spectrum *(30-34)*.

The two most common solvent parameters used to rationalize solvent-ν_{NO} trends are the *Gutmann donor number* (DN) and the *Gutmann acceptor number* (AN) of the solvent employed *(35)*. These parameters may be used to rank organic solvents in order of their donor and acceptor abilities, respectively. For instance, solvent molecules coordinate *trans* to the NO group in the $[(P)Fe(NO)(S)]^+$ complexes (P = porphyrin; S = solvent), and the ν_{NO}s of these complexes have been shown to correlate linearly with the DN of the solvents as shown in Figure 7.1 *(36)*.

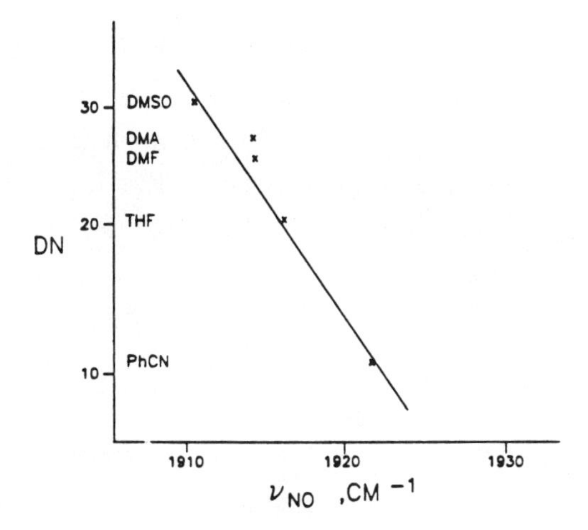

Figure 7.1. Effect of solvent donor number on ν_{NO} of $[(P)Fe(NO)(solv)]^+$ complexes (adapted from reference 36).

The *greater* the DN of the solvent, the lower the ν_{NO} observed, a fact that is consistent with the *increased* backdonation of electron density from the metal to the nitrosyl ligand due to the enhanced electron richness of the metal center. However, of more relevance to this section is the linear correlation of AN with ν_{NO}. In these instances, the solvent acceptor orbitals interact with the "basic" nitrosyl oxygen. Fig 7.2 illustrates such a

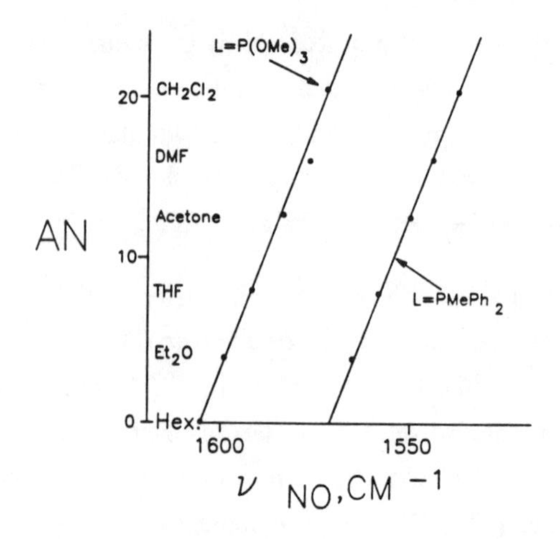

Figure 7.2. Effect of solvent acceptor number on ν_{NO} of $CpMo(NO)L_2$ ($L = P(OMe)_3$, $PMePh_2$) (redrawn from reference 37).

correlation of ν_{NO}s of the $CpMo(NO)L_2$ complexes with the AN of some common organic solvents *(37)*. This variation in ν_{NO} reflects the ability of the NO ligands in these complexes to function as Lewis bases (via lone pairs on the oxygen atom) towards *acceptor* organic solvents. The DN and AN solvent parameters are also known to affect the redox potentials of several metalloporphyrin complexes *(38)*.

7.2.2: Lewis Acid Adducts

A number of metal nitrosyls react with Lewis acids to form adducts via (i) the nitrosyl ligand, (ii) the metal, or (iii) the ancillary ligands such as CO or halide. Representative examples of metal nitrosyl/Lewis acid adducts are listed in Table 7.1. In general, when the Lewis acid binds via the nitrosyl oxygen, electron density is drained from the NO group,

Table 7.1: Lewis Acid Adducts of Metal Nitrosyls

Complex	Lewis Acid	Site of reactivity	Ref
CpNi(NO)	$GaCl_3$, $SbCl_3$	NO	*39*
	$TiCl_4$	Ni	*39*
$CpCr(NO)(CO)_2$	Cp_3Sm, $AlCl_3$	NO	*40,41*
	$AlBr_3$	CO or NO	*42,43*
$CpMo(NO)(CO)_2$	$AlBr_3$, Cp'_3Sm	NO	*42,44*
$CpMo(NO)(CO)PPh_3$	$AlCl_3$, Cp'_3Er	NO	*43*
	$SnCl_4$	Mo	*43*
$CpW(NO)(CO)_2$	Cp'_3Er	NO	*44*
$CpCr(NO)_2Cl$	Cp_3Yb	NO	*44*
	$AlCl_3$,$TiCl_4$	Cl	*43*
$[CpCr(NO)_2]_2$	Cp_3Sm	NO	*40*
$CpW(NO)R_2$	$AlMe_3$	NO	*45*
$[CpFe(NO)]_2$	Cp_3Sm	μ-NO	*40*
$[Cp'Mn(NO)(CO)]_2$	Cp_3Sm	NO, μ-CO	*40*

and the metal compensates by donating more electron density to this group via backbonding to the π^* orbitals. Thus, the ν_{NO} is effectively *lowered.* This lowering of ν_{NO} may range from 35 to 264 cm^{-1}, depending on the strength of the interaction between the NO ligand and the Lewis acid. In contrast, when the Lewis acid binds via the metal, electron density is drained from the π^* orbitals of the nitrosyl ligand onto the metal, resulting in a *higher* ν_{NO}. A similar argument is used when the Lewis acid binds via the ancillary ligands in a metal nitrosyl complex. The data presented in Table 7.2 illustrate these points *(46)*.

Table 7.2. Comparison of Nitrosyl Stretching Frequencies for $CpRe(NO)(PPh_3)Me$ (Re-NO in the Table) and its Boron Trichloride Adduct.

Compound	ν_{NO} (cm^{-1})
Re-NO	1656
$Re\text{-}NO\text{-}BCl_3$	1392
$ON\text{-}Re\text{-}BCl_3$	1752

As may be seen from the Table, adduct formation via the nitrosyl linkage significantly *decreases* the ν_{NO} by 264 cm^{-1}, whereas adduct formation via the metal *increases* the ν_{NO} by 96 cm^{-1}. The crystal structure of $CpRe(NO\text{-}BCl_3)(PPh_3)(SiMe_2Cl)$ unambiguously reveals the presence of the isonitrosyl linkage in this compound *(47)*.

7.2.3: Protonation and Other Electrophilic Additions

Protonic reagents may also add to the nitrosyl oxygen of a nitrosyl complex. In some instances, only hydrogen-bonding is observed. For example, the nitrosyl oxygen in $CpM(NO)(CO)_2$ (M = Cr, Mo) hydrogen-bonds to the proton of $(CF_3)_3COH$ or HCl in liquid xenon *(48)*. However, actual complexation of H^+ occurs in other cases to yield hydroxyimido compounds *(49,50)*.

$$[Ru_3(CO)_{10}(\mu_2\text{-}NO)]^- + H^+ \longrightarrow Ru_3(CO)_{10}(\mu_3\text{-}NOH) \qquad (7.7)$$

$$[Cp'Mn(NO)]_3(\mu_3\text{-}NO) + H^+ \longrightarrow [Cp'Mn(NO)]_3(\mu_3\text{-}NOH)^+ \qquad (7.8)$$

Reaction 7.7 occurs with the concomitant transformation of a μ_2-NO ligand to a μ_3-NOH ligand, whereas in reaction 7.8, it is the μ_3-NO ligand that is attacked by the proton. Similarly, the protonation of the asymmetric μ_2-NO group in $[CpWFe_2(CR)(CO)_7(\mu_2\text{-}NO)]^-$ yields a μ_3-NOH species *(51)*, and the methylation of the μ_2-NO group in $[Ru_3(CO)_{10}(NO)]^-$ also occurs to yield the corresponding methoxyimido (μ_3-NOMe) derivative *(49,52)*.

7.2.4: Nitrosyl Bond Activation Resulting In The Loss of Oxygen

A common feature in the reactivity of metal nitrosyls is the ability of the coordinated nitrosyl to lose its oxygen to oxygen-acceptors such as phosphines, forming intermediate nitride complexes which are sometimes isolable. For example, both the N and O atoms of the 'reacting' NO group are maintained in the metal's coordination sphere in reaction 7.9 *(53)*, whereas only the N of the 'reacting' NO group is retained in reaction 7.10 *(54,55)*.

$$Re(NO)_2Cl_3 + 2PPh_3 \longrightarrow Re(NO)Cl_3(NPPh_3)(OPPh_3) \qquad (7.9)$$

$$Mo(NO)_2(dttd) + 2PPh_3 \longrightarrow Mo(NO)(NPPh_3)(dttd) + OPPh_3 \qquad (7.10)$$

(dttd = 2,3,8,9-dibenzo-1,4,7,10-tetrathiadecan dianion)

Migrations of coordinated CO ligands to intermediate nitrides to form isocyanate species have also been documented *(56,57)*, e.g.,

$$Ir(CO)(NO)(PPh_3)_2 + 2PPh_3 + h\nu \longrightarrow Ir(NCO)(PPh_3)_3 + OPPh_3 \qquad (7.11)$$

$$CpMo(NO)(CO)_2 + 2PPh_3 + h\nu \longrightarrow CpMo(NCO)(CO)(PPh_3)_2 \qquad (7.12)$$

Photolysis of $CpW(NO)(CO)_2$ in an argon matrix results in the *intramolecular* formation of an isocyanate ligand *(58)*. Although the mechanism of formation of the NCO ligand is not clear, it has been conclusively determined by isotopic labelling (using ^{13}CO, $C^{18}O$, $N^{18}O$) that the oxygen of the NCO ligand *does not* derive from the nitrosyl ligand, but rather derives from a coordinated CO group. However, the fate of the nitrosyl oxygen is uncertain, since it does *not* react with photoejected CO to form CO_2.

- CO
W
OC
CO
N
O
N:

Many stable nitrides have been isolated as a result of the fragmentation of a coordinated NO ligand and, indeed, the primary sources of nitrogen atoms in nitrido clusters are coordinated NO groups *(59-62)*. For example, the deoxygenation reaction described in eq 7.13 is first-order in the nitrosyl cluster, and leads to the production of a nitride complex *(63)*.

$$[FeRu_3(NO)(CO)_{12}]^- + CO \longrightarrow [FeRu_3(N)(CO)_{12}]^- + CO_2 \quad (7.13)$$

The $Cp^*_3Mo_3Co_2(CO)_8(\mu_3\text{-}NH)(\mu_4\text{-}N)$ nitrido cluster is obtained as a minor product from the photolysis of a mixture of $Co(NO)(CO)_3$ and $[Cp^*Mo(CO)_2]_2$ *(64)*. Similarly, the thermolysis (200°C) of a mixture of $CpM(NO)(CO)_2$ and $[CpM(CO)_3]_2$ (M = Mo, W) results in the production of the nitrido-oxo clusters, $Cp_3M_3(CO)_4(\mu_3\text{-}N)(O)$ (M_3 = Mo_3, Mo_2W, MoW_2, W_3). However, isotopic labelling in the latter reactions reveal that the oxo ligands in these trimetallic clusters do not derive from the NO groups *(65)*. Hydride reagents can also "reduce" bound NO groups to nitrides in monometallic *(66)* or cluster compounds *(59)*. The reduction of bound NO is covered in Section 7.3.2.

The unprecedented nitrosyl-aryl to oxo-imido conversion occurs in a mononitrosyl-diaryl complex as shown *(67)*:

Furthermore, when $Re(O)_2(mes)_2$ is treated with NO gas, an oxo-amido complex, $Re(O)_3[N(mes)_2]$, forms *(68)*. As mentioned in Chapter 5, such NO-splitting reactions have been invoked in the disproportionation and catalytic reduction of NO over metal surfaces, but rarely have such reactions been observed with monomeric complexes (Section 7.4.1.3).

7.3: Reactivity at the Nitrosyl Nitrogen

7.3.1: Reactions with Electrophiles

In Section 7.2.3, it was mentioned that protons may bind to the oxygen atoms of bound NO groups, forming NOH species. However, when the N atom of an MNO group is sufficiently electron rich, it may also undergo electrophilic attack by protons. Such reactions normally occur with bent nitrosyl groups, since the nitrogen is formally sp^2 hybridized and possesses a lone pair of electrons *(2,69-73)*. Dinitrosyl complexes may also undergo electrophilic attack at the N atom(s) if the $M(NO)_2$ group is sufficiently electron rich *(74)*.

$$Mo(NO)F(dppe) + H^+ \longrightarrow Mo(HNO)F(dppe) \quad (7.15)$$

$$Os(NO)Cl(CO)(PPh_3)_2 + HCl \longrightarrow Os(HNO)Cl_2(CO)(PPh_3)_2 \quad (7.16)$$

$$Re(NO)_2(PPh_3)_2(I) + HCl \longrightarrow Re(NO)(HNO)(PPh_3)_2Cl(I) \quad (7.17)$$

The coordinated HNO ligand in eq 7.16 has been established by X-ray methods *(75)*. The intermolecular conversions of bound NO to the η^2-bound HNOH, H_2NOH, and H_2NO ligands have also been observed *(73,76)*:

$$Os(NO)_2(PPh_3)_2 + 2\ HCl \longrightarrow Os(NO)(HNOH)(PPh_3)_2Cl_2 \quad (7.18)$$

$$Ir(NO)(PPh_3)_3 + 3\ HCl \longrightarrow Ir(H_2NOH)(PPh_3)_2Cl_3 + PPh_3 \quad (7.19)$$

$$(s\text{-}s)Mo(NO)_2 + 1/2\ N_2H_4 \longrightarrow (s\text{-}s)Mo(NO)(H_2NO) + 1/2\ N_2 \quad (7.20)$$

(s-s = 2,2'-(ethylenedithio)dibenzenethiolate)

The cationic $[Co(NO)(das)_2Br]^+$ compound also reacts with H^+/MeOH to form $[Co(HNO)(das)_2Br]^{2+}$ *(24)*.

7.3.2: Reactions with Nucleophiles

Nucleophilic attack on the coordinated NO in metal nitrosyls commonly proceeds via initial attack of the nucleophile on the nitrosyl nitrogen in the essentially linear MNO group. The nitroprusside ion, $[Fe(NO)(CN)_5]^{2-}$, is the most studied nitrosyl complex in this regard *(1)*. Although this ion has a *double negatively charge,* it has a relatively high ν_{NO} of 1939 cm^{-1} (*c.f.* NO at 1878 cm^{-1} and NO^+ at 2220 cm^{-1}). Interestingly, an analogy has been drawn between the reactivity of the free nitrosonium ion, NO^+, and that of the coordinated NO group in nitroprusside *(77)*:

$$NO^+ + 2\,OH^- \rightleftharpoons NO_2^- + H_2O \qquad (7.21)$$

$$[Fe(NO)(CN)_5]^{2-} + 2\,OH^- \rightleftharpoons [Fe(CN)_5NO_2]^{4-} + H_2O \qquad (7.22)$$

Consequently, the ability of the coordinated NO group in a metal nitrosyl complex to behave as an electrophile (i.e., as NO^+) has been correlated with the observed ν_{NO} of the complex, and it has been proposed that metal nitrosyl complexes having $\nu_{NO} \geq$ 1886 cm^{-1} or F(N-O) > 13.8 mdyn $Å^{-1}$ will behave as electrophiles and will undergo nucleophilic attack at the coordinated nitrosyl group *(71,72)*.

Calculations (SINDO method) on some first-series transition metal pentacyanonitrosyls, $[M(NO)(CN)_5]^{2-}$ (M = Fe, Mn, Cr), have been performed to determine the net atomic charges of the N and O atoms of the coordinated nitrosyl group in these compounds *(78)*. The results are tabulated in Table 7.3.

Table 7.3. Net Charges on the N and O Atoms of the NO Group in Some $[M(NO)(CN)_5]^{n-}$ Complexes Calculated by the SINDO Method.[a]

Complex ion	qN (e)	qO (e)	qNO (e)	ν_{NO} (cm^{-1})
$[Fe(NO)(CN)_5]^{2-}$	+0.503	-0.757	-0.254	1939
$[Mn(NO)(CN)_5]^{2-}$	+0.334	-0.806	-0.472	1885
$[Mn(NO)(CN)_5]^{3-}$	+0.192	-0.883	-0.691	1725
$[Cr(NO)(CN)_5]^{3-}$	+0.179	-0.887	-0.708	1645
$[Cr(NO)(CN)_5]^{4-}$	+0.122	-0.941	-0.819	1515

[a] Data taken in part from reference 78.

Reading down Table 7.3, one notes the large decrease in ν_{NO} which may be correlated with increasing metal-to-π^*(NO) intramolecular transfer of electron density *(79)*, which in turn is consistent with the significant decrease in the net charges on both the N and O atoms of the NO group. This result is consistent with the experimental observation that of these compounds, only $[Fe(NO)(CN)_5]^{2-}$ and $[Mn(NO)(CN)_5]^{2-}$ appear to be susceptible to nucleophilic attack. It is not surprising, therefore, why the reactions of nitroprusside *(77)* with nucleophiles (L) such as amines *(80-85)*, mono- and diaminoacids *(86,87)*, carbanions *(88)*, thiols *(89,90)*, and acetone *(91)* all proceed via initial attack of the nucleophile at the nitrosyl nitrogen, i.e.,

$$[Fe(CN)_5NO]^{2-} + L \longrightarrow [Fe(CN)_5\{N(=O)L\}]^{2-} \qquad (7.23)$$

The N(=O)L group may then be removed from the metal's coordination sphere, a process which forms the basis of various nitrosation-diazotization reactions well known for the nitroprusside ion. This kind of reactivity has also been linked to the biological activity of nitroprusside (Section 6.2).

The Ru and Os analogues of the nitroprusside ion appear to exhibit chemistry similar to that of nitroprusside. For example, the reactions of $[Ru(NO)(CN)_5]^{2-}$ with OH^- and SH^- parallel those of nitroprusside, i.e., via the initial attack of the nucleophile at the nitrosyl nitrogen of the coordinated NO group *(92)*. However, no reactions of this type are observed with other nucleophiles such as piperidine, $[CH(CN)_2]^-$, $[CH(COMe)_2]^-$, MeS^- or $[S_2O_4]^{2-}$, all of which are known to react readily with nitroprusside. Indeed, these observations are consistent with the view that the relative energies and orbital constitutions of the MNO complex and the reagents have to be considered before predicting reactivity patterns for the coordinated NO group. Nevertheless, alkoxides attack the coordinated nitrosyl in $[Ru(NO)(bpy)_2(py)]^{3+}$ to yield the $[Ru(bpy)_2(py)\{N(O)OR\}]^{2+}$ alkyl nitrite complexes (R = Me, Et, *n*-Bu, *i*-Pr) *(93)*. Also, the $[M(NO)XL_4]^{2+}$ complexes (M = Ru, Os; X = halide; L = py, bpy) react with nucleophiles such as ketones *(94,95)*, alkylidenearylhydrazones RCH=NNHR *(96)*, amines *(97)*, and sulfite *(98)* via initial attack of the nucleophile on the nitrosyl nitrogen.

On the basis of molecular orbital calculations, it has been proposed that it is the nitrosyl ligand, and *not* the carbonyl ligand, in

$[CpRe(NO)(CO)(PPh_3)]^+$ that is initially attacked by the hydride anion to form the kinetic HNO intermediate compound as shown *(99)*:

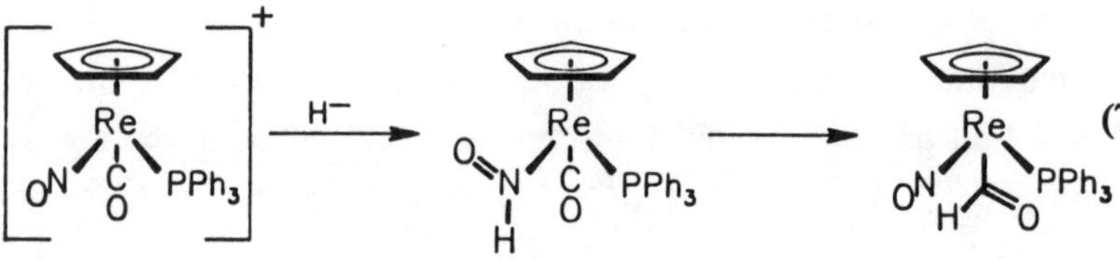

$$[CpRe(NO)(CO)(PPh_3)]^+ \xrightarrow{H^-} CpRe(HNO)(CO)(PPh_3) \longrightarrow CpRe(NO)(CHO)(PPh_3) \qquad (7.24)$$

This kinetic HNO compound then isomerizes to the thermodynamic formyl product. However, only the thermodynamically more stable formyl complex has been observed under normal experimental conditions.

The reaction of H-atom sources with coordinated NO groups formally leads to a *reduction* of the NO ligand. For example, hydrogenation of $[CpCo(NO)]_2$ generates $[CpCo]_2(NO)(NH_2)$, which reacts further with hydrogen to liberate ammonia *(100)*. The sequential transformation of a μ_3-NO ligand to a μ_3-NH group has been carried out as shown in the following sequence of reactions *(50)*:

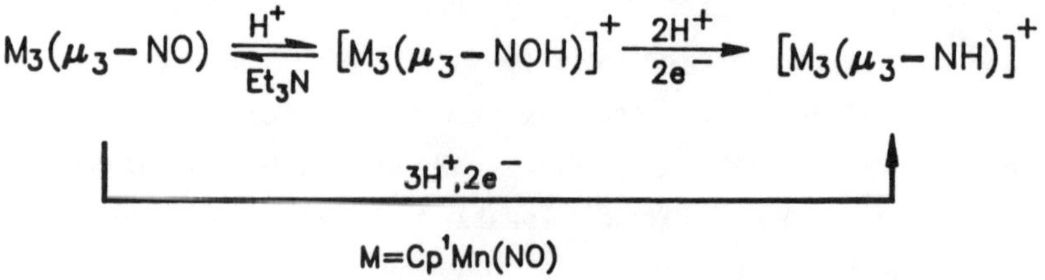

$$M_3(\mu_3\text{-}NO) \underset{Et_3N}{\overset{H^+}{\rightleftharpoons}} [M_3(\mu_3\text{-}NOH)]^+ \xrightarrow[2e^-]{2H^+} [M_3(\mu_3\text{-}NH)]^+$$

$$M_3(\mu_3\text{-}NO) \xrightarrow{3H^+,\,2e^-} [M_3(\mu_3\text{-}NH)]^+$$

$$M = Cp'Mn(NO)$$

Scheme 7.1

Furthermore, the reaction of a cobalt dinitrosyl complex with borohydride (eq 7.25) leads to the formation of ammonia *(101)*,

$$[Co(NO)_2(PPh_3)_2]^+ \xrightarrow{BH_4^-} Co(NO)(PPh_3)_3 + NH_3 \qquad (7.25)$$

Reductions of coordinated NO groups in some nitrosyl complexes of Fe, Ru, and Os have served as useful models for the studies of the mechanism of reduction of bound nitrite to ammonia *(102,103)*. For example, electrochemical studies of the redox behavior of $[(tpy)(bpy)M(NH_3)]^{2+}$ and $[(tpy)(bpy)M(NO_2)]^+$ (M = Ru, Os), provide direct experimental evidence for the formation of M(NH), M(N), $M(NH_2O)$ and M(NHO) intermediates in the overall conversion of bound nitrite to ammonia *(103)*. A critical feature also noted is the acid-base interconversion of the starting reagents to their nitrosyl derivatives, $[(tpy)(bpy)M(NO)]^{3+}$. The related [(Hedta)Ru(NO)] and $[(Hedta)Fe(NO)]^-$ complexes are effective catalysts for the reduction of NO_2^- to N_2O, N_2, $[NH_3OH]^+$ or NH_4^+ *(102)*.

7.3.3: Carbon-Nitrogen Bond Formation

In the previous section, the reactions of external nucleophiles with the nitrogen of the coordinated NO group were detailed. When carbon nucleophiles are employed in such reactions, new carbon-nitrogen bonds are formed within the coordination spheres of the metals. In this section, the *inter-* and *intramolecular* formations of such carbon-nitrogen bonds are discussed.

7.3.3.1: Intermolecular C-N Bond Formation via Bound NO Groups

It was stated earlier that carbanions react with the nitroprusside ion via attack at the coordinated nitrosyl group *(88)*. Such *intermolecular* "couplings" of carbon nucleophiles with *bound* nitrosyls are fairly common. However, many such reactions are not as straightforward as they may seem. Depending on the nature of the reaction or the reaction conditions, many of the reactions occur with simultaneous rearrangement of the newly-formed "RNO" group. For instance, the reduction of the coordinated NO sometimes occurs when metal-nitrosyls are reacted with anionic nucleophiles *(100,104)*:

$$[CpCo(NO)]_2 \xrightarrow{RLi} [CpCo]_2(NO)(NHR) \qquad (7.26)$$

$$CpNi(NO) \xrightarrow{RLi} (CpNi)_3(\mu_3\text{-}NR) \qquad (7.27)$$

(R = Ph, *t*-Bu)

The osmium complex (**A**) shown below is obtained from a novel nitrosyl-diazoalkane coupling reaction between $Os(NO)Cl(PPh_3)_3$ and $N_2C(I)CO_2Et$ in the presence of hydrochloric acid *(105)*. Also, the osmium complex (**B**) is obtained by reacting $Os(NO)_2(PPh_3)_2$ with perfluorocarboxylic acids, RCO_2H (R = CF_3, C_2F_5) *(106,107)*.

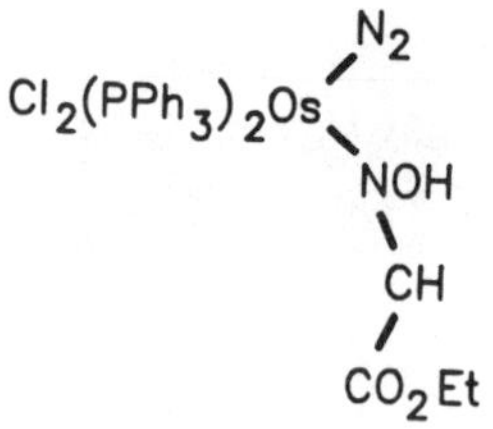

$Cl_2(PPh_3)_2Os(N_2)(NOH{-}CH{-}CO_2Et)$

(A)

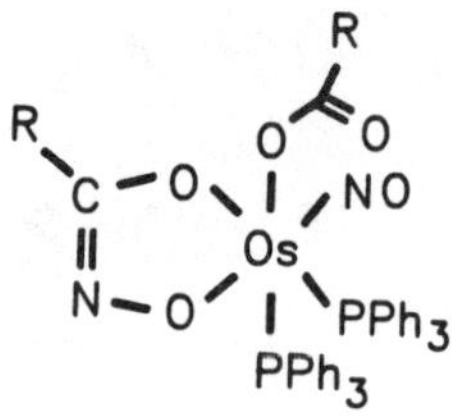

(B)

An interesting reaction leading to the formation of *two* carbon-nitrogen bonds involves treating $[CpCo(NO)]_2$ with NO in the presence of alkenes *(108-111)*.

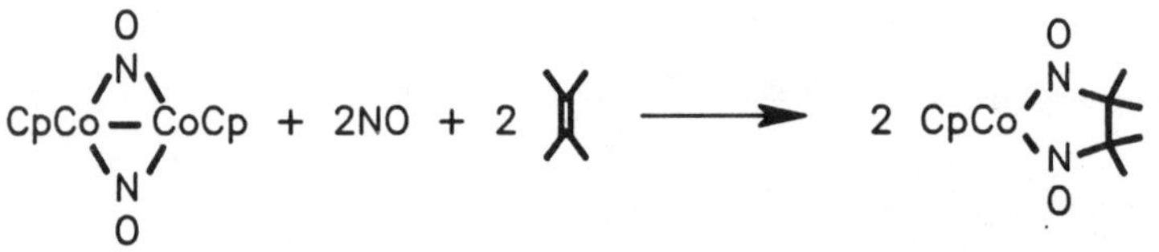

$$CpCo(\mu\text{-}NO)_2CoCp + 2NO + 2\ \text{alkene} \longrightarrow 2\ CpCo(NO)_2C_2R_4 \qquad (7.28)$$

It has been proposed that the $[CpCo(NO)]_2$ dimer reacts initially with NO to generate $CpCo(NO)_2$ and, indeed, spectroscopic evidence for the thermally stable dinitrosyl intermediate has been obtained. [Although the $C_5H_4(CO_2Me)$ and C_5Me_5 analogues of the dinitrosyl complex have been synthesized and characterized spectroscopically, they are too reactive to be isolated *(108)*.] Nevertheless, the $CpCo(NO)_2$ complex then reacts with one molecule of the olefin to yield the dinitrosoalkane complex as shown in eq 7.28. Although the $CpCo(NO)_2$ complex has not yet been characterized structurally, it has been proposed that one or both NO ligands may be *bent* in the "reactive" species. It has been calculated that the orbitals which change most in energy during such bending are precisely those that might be expected to interact with the π and π^* orbitals of the alkene to form the dinitrosoalkane ligands *(112)*.

A wide range of alkenes have been employed in reaction eq 7.28, and the sequence shown below describes the order of *increasing* stability of the dinitrosoalkane complexes that they form:

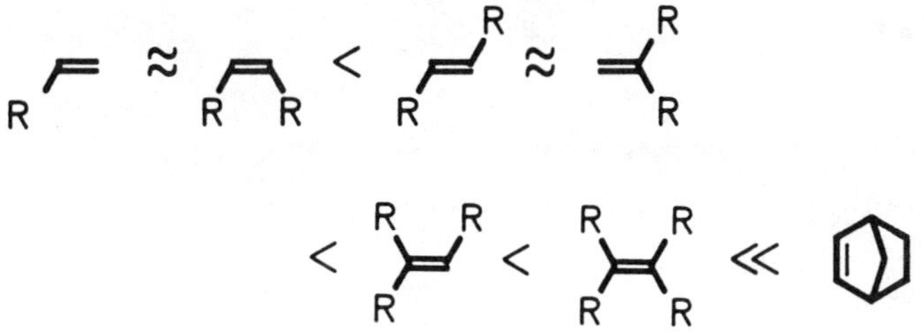

Reaction 7.28 is also completely stereospecific. For example, when the trisubstituted olefins (*E*)- and (*Z*)-3-methyl-2-pentene are each reacted with $[CpCo(NO)]_2$ and NO, they yield their respective stable, isomeric, cobalt dinitrosoalkane complexes:

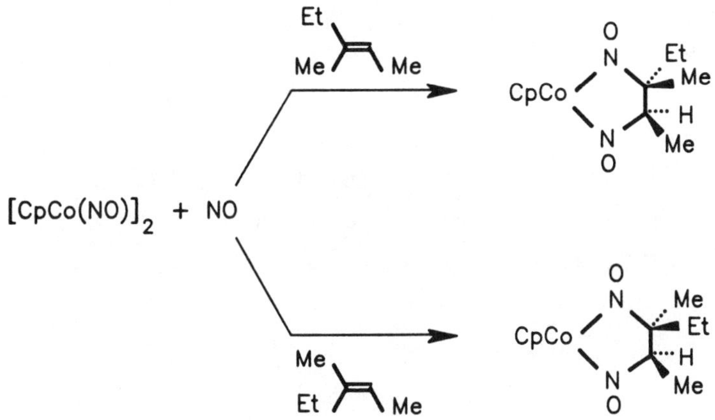

(7.29)

Although the dinitrosoalkane ligands are not displaced by CO or phosphines even under forced conditions, the alkane portions of these ligands undergo remarkable, stereospecific, alkene-exchange reactions (eq 7.30) *(108)*.

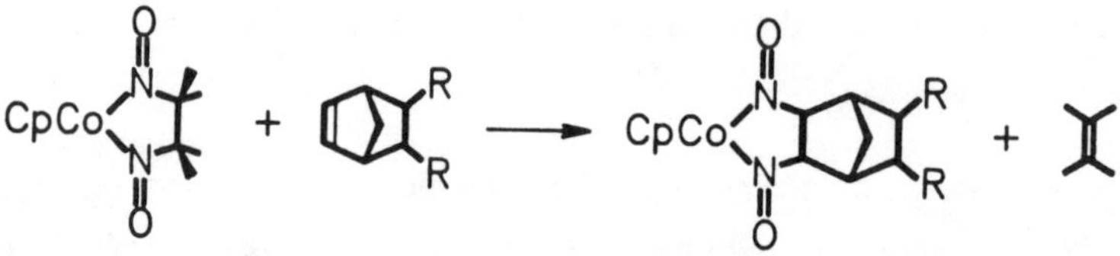

(7.30)

Results of kinetic experiments are consistent with a mechanism involving reversible dissociation of the Co-dinitrosoalkane complex into $CpCo(NO)_2$ and the alkene, followed by reaction of $CpCo(NO)_2$ with another alkene. Reduction of the Co-dinitrosoalkanes with $LiAlH_4$ results in the production of primary vicinal diamines (eq 7.31).

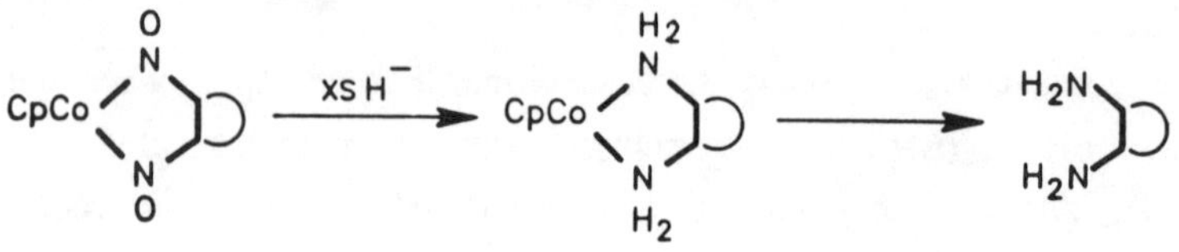

(7.31)

A variety of organic compounds resulting from C-N bond formation may also be generated during the reactions of nitrosyl complexes with alkylating agents. For example, the alkylation of $Ru(NO)_2(PPh_3)_2$ with benzyl bromide under an atmosphere of CO yields, in addition to other organic products, PhCH=NOH, PhCN, and $PhCONH_2$. Such reactions of other metal nitrosyl complexes with alkylating agents resulting in carbon-nitrogen bond formation have been reviewed *(113)*.

7.3.3.2: Intramolecular C-N Bond Formation via Bound NO Ligands

One of the earlier documented examples of the *intramolecular* insertion of a bound NO group into a metal-carbon bond was the report that $CoMe_2(PMe_3)_3$ reacts with NO to produce $Co(NO)Me_2(PMe_3)_2$, which then isomerizes via NO insertion to $[CoMe(\mu\text{-}MeNO)(PMe_3)_2]_2$ *(114)*. It is now also known that the related $CoMe_3(PMe_3)_3$ compound reacts with NO to form $[Co(NO)Me(MeNO)(OPMe_3)]_2$ and $Co(NO)Me(OPMe_3)_3$ *(115)*. Indeed, many such reactions of metal alkyl complexes with NO gas result in carbon-nitrogen bond formation, presumably via the initial coordination of NO *(1)*. In a detailed study of the interaction of NO with a number of diamagnetic and paramagnetic metal-alkyl complexes, it has been found that generalizations concerning the nature of the products obtained from these reactions can be made *(116)*: (i) diamagnetic metal-alkyl(aryl) complexes react with NO to produce *N*-alkyl-*N*-nitrosohydroxylaminato, ON(R)NO, complexes *(45,117)*, e.g.,

$$Cp_2ZrPh_2 + 2NO \longrightarrow Cp_2Zr(Ph)[\kappa^2\text{-}O,O'\text{-}ON(Ph)N{=}O] \quad (7.32)$$

(ii) paramagnetic alkyls (with one unpaired electron) react with NO to initially produce M{RNO} complexes (η^2 or bridged), which may then decompose to oxo compounds *(116)*.

The first direct observation of the conversion of a well-defined alkyl-nitrosyl complex into a simple nitrosoalkane complex was provided by the report that the CpCo(NO)R (R = Me, Et, *i*-Pr, *p*-$CH_2C_6H_4Me$) complexes react with phosphines (L) in an overall intramolecular NO insertion (R migration?) reaction as shown in eq 7.33.

$$CpCo(NO)(R) \longrightarrow [CpCo\text{–}N(O)R] \xrightarrow{L} CpCo(L)\text{–}N(O)R \quad (7.33)$$

Kinetic measurements of this reaction (R = Me) in the presence of PPh_3 indicate that the reaction follows pseudo-first-order kinetics. Thus, a 40-fold increase in $[PPh_3]$ produces only a very weak response in the rate of the reaction. This implies the rate-determining formation of an intermediate such as CpCo{N(=O)R}, or its solvate, which is then trapped by the phosphine to yield the products. However, when the phosphine employed is PEt_3, a second reaction pathway leading to the production of the $CpCo(NO)Me(PEt_3)$ adduct (with a *bent* NO ligand) is also observed. Nevertheless, phosphine dissociation precedes any NO insertion reaction, and the insertion then follows a pathway similar to that of eq 7.33. *(118)*.

Another class of compounds that undergo intramolecular insertions of NO are the $Cp'Fe(NO)Me_2$ complexes (Cp' = Cp, Cp*) *(119,120)*.

$$Cp'Fe(NO)Me_2 + PMe_3 \longrightarrow Cp'Fe\{N(=O)Me\}(Me)(PMe_3) \quad (7.34)$$

As with the CpCo(NO)R complexes described earlier, kinetic measurements (Cp' = Cp) are consistent with a rate-determining migratory insertion reaction followed by rapid trapping of the intermediate nitrosoalkane complex by phosphine *(119)*. The ruthenium congeners, $Cp^*Ru(NO)R_2$ (R = Me, Et, Ph) also undergo intramolecular insertion reactions in the presence of phosphines:

$$Cp^*Ru(NO)(Et)(Ph) + PPhMe_2 \longrightarrow Cp^*Ru\{N(=O)Et\}(Ph)(PPhMe_2) \quad (7.35)$$

Results of kinetic studies also imply first-order processes with no observed dependence on phosphine ($PPhMe_2$) concentration. Thermolysis of the nitrosoethane complex in the presence of added $PPhMe_2$ results in the elimination of benzene and the formation of an oximate complex (eq 7.36), a reaction presumably initiated by base abstraction of the acidic methylene proton of the nitrosoethyl group.

$$Cp^*Ru(N(O)Et)(Ph)(PPhMe_2) + PPhMe_2 \xrightarrow{-PhH} Cp^*Ru(N(\rightarrow O){=}CHMe)(PPhMe_2)_2 \quad (7.36)$$

The thermal reactions of the products of reactions 7.35 and 7.36 lead to the generation of oximate, carboxamide and cyano compounds of ruthenium *(121)*.

The intramolecular insertion of NO into a methinic C-H bond is also observed when an iridium complex containing a 2-pyridinecarbaldehyde azine ligand is reacted with RCN (R = Me, Ph) in the presence of ethanol *(122,123)*:

(7.37)

L = Me(EtO)C=NH (acetoimidate); [Ir] = $Ir(PPh_3)_2$

The $[Co(NO)(CO)_2X]^-$ anions (X = halide) are also known to react with benzyl halides to form benzaldoximes, possibly via intramolecular migration of the bound NO group into the Co-benzyl bond *(124)*.

Finally, a remarkable and unprecedented *alkyne insertion into a metal-nitrosyl bond* resulting in the formation of a C-N bond has been documented. This reaction results when $[\{CpRh(CO)\}_2(\mu\text{-}NO)]^+$ is reacts with PhC_2CR as shown *(125)*:

$$[CpRh(CO)(\mu\text{-}NO)RhCp(CO)]^+ \xrightarrow[-CO]{Ph-\equiv-Ph} [CpRh\{\mu\text{-}N(O)C(Ph)=C(Ph)\}RhCp(CO)]^+$$

The coordination chemistry of *C*-nitroso compounds has been reviewed *(126)*.

7.3.4: The M(NO) ⟶ M(NO$_x$) Conversion and the Oxidation of Substrates

Several metal nitrosyls are oxygenated at the MNO group to produce the corresponding $M(NO_x)$ (x = 2, 3) species. Generally, these oxygenations result in the production of nitro (NO_2) or nitrito (ONO) compounds *(127-129)*. In a few instances, nitrato (x = 3) species may be formed instead *(130)*. Such oxygenations of nitrosyl groups may be effected *intramolecularly*, e.g. *(131)*,

$$Ru(NO)(O_2)(PPh_3)_2Cl + 2CO \longrightarrow Ru(NO_3)(PPh_3)_2(CO)_2Cl \quad (7.38)$$

or *intermolecularly* by employing stoichiometric amounts of oxygen-transfer agents such as iodosylbenzene or dioxygen (L = dtc, benacen; B = Im, py) *(132-135)*, e.g.,

$$Co(L)(B)NO + PhIO \longrightarrow Co(L)(B)(NO_2) + PhI \quad (7.39)$$

$$RhBr_2(NO)(Pcy_2H)_2 + O_2 \longrightarrow RhBr_2(NO_3)(Pcy_2H)_2 \quad (7.40)$$

$$Ni(NO)Cl(PPh_3)_2 + 1/2\ O_2 \longrightarrow Ni(NO_2)Cl(PPh_3)_2 \quad (7.41)$$

The mechanism of the reaction of the MNO group with dioxygen remains unknown, although peroxy-nitrato species such as $M\{N(=O)OO\}$ are possible intermediates in these reactions.

The $M(NO_x)$ groups may be deoxygenated by oxidizable substrates to regenerate the MNO precursors,

$$\mathrm{S} + \mathrm{M{-}(NO_x)} \rightarrow \mathrm{SO} + \mathrm{M{-}NO}; \quad \mathrm{M{-}NO} + \tfrac{x}{2}\,\mathrm{O_2} \rightarrow \mathrm{M{-}(NO_x)}$$

S = oxidizable substrate

a feature that permits the construction of a stoichiometric or catalytic cycle for the oxygenation of substrates. By far the majority of research in this area involves complexes of the Groups 8, 9 and 10 metals *(136)*.

(i) Iron, Ruthenium and Osmium

Oxidation of $[Fe(NO)_2X]_2$ (X = Cl, I) by 1 atm of dioxygen in the presence of $OPPh_3$ yields $Fe(NO_3)_2(OPPh_3)_2X$, which contains bidentate nitrato (not nitro) ligands *(137)*:

$Fe(\kappa^2\text{-}O_2NO)_2Cl\,L_2$ (L = $OPPh_3$)

The nitrato ligands in the chloro complex transfer oxygen atoms to phosphines to yield the corresponding phosphine oxides, with the concomitant regeneration of the "$Fe(NO)_2$" dinitrosyl fragment in $Fe(NO)_2Cl(OPPh_3)$. This O-transfer reaction is also used in oxidizing cyclohexene to the corresponding epoxide; a reaction that constituted the first documented example of O-transfer from a nitrato ligand to an olefin *(137)*.

The related $Fe(NO_3)_2Cl(hmpa)_2$ and $Fe(NO_3)Cl_2(hmpa)_2$ compounds (hmpa = hexamethylphosphoric triamide, a non-oxidizable labile ligand) also transfer oxygen atoms to phosphines and catalyze the autoxidation of cyclohexene, producing cyclohex-1-en-3-one, cyclohex-1-en-3-ol and cyclohexene oxide *(130)*. The $[Ru(NO_2)(PR_3)_2(trpy)]^{2+}$ complexes (R = alkyl, aryl) are unstable, and decompose to their nitrosyl analogues. Furthermore, $[Ru(NO_2)(PMe_3)_2(trpy)]^{2+}$ (trpy = 2,2',2''-terpyridine), formed by the reaction of its nitrosyl precursor with oxygen, oxidizes norbornene to *exo*-2,3-epoxynorbornane with the concomitant formation of the RuNO group *(138)*.

Finally, whereas *uv* photolysis of $CpRu(CO)_2(NO_2)$ at relatively long wavelengths ($\lambda > 350$ nm) results in the formation of a Ru-ONO species, photolysis at shorter wavelengths results in an O-transfer reaction to produce CpRu(NO)(CO) and CO_2 *(139)*.

(ii) Cobalt and Rhodium

Many cobalt complexes of the form $(L)Co(NO_2)$ (L = a tetradentate ligand) have been employed in the stoichiometric and catalytic oxidations of phosphines, alcohols and alkenes *(140-146)*. For example, $(py)Co(saloph)(NO_2)$ (saloph = *N,N'*-bis(alicylidene-*o*-phenylene)diamino) oxidizes phosphines to phosphine oxides. It has been noted that the oxidation occurs only in hydrogen-bonding solvents, whereas weakly basic solvents such as acetonitrile prevent the oxidation. Amplification of the solvent effect by employing strong Lewis acids (A), such as BF_3 and Li^+, allows the use of these cobalt-nitro complexes as oxidants for dialkyl sulfides, alcohols and cyclohexadiene *(146)*. In the case of the sulfides, the corresponding oxides are formed, whereas in the case of primary and secondary alcohols, the carbonyl compound is formed (eq 7.42). Similarly, in the case of diene oxidation, benzene is the ultimate product of the reaction (eq 7.43).

$$Co-NO_2 \cdot A + R_2CHOH \longrightarrow Co-NO \cdot A + R_2C=O + H_2O \quad (7.42)$$

$$Co-NO_2\cdot A + \text{(cyclohexadiene)} \longrightarrow \text{(benzene)} + Co-NO\cdot A + H_2O \quad (7.43)$$

The Lewis acid-activated Co-NO_2 complexes fail to oxidize monoolefins. However, success is achieved by activating the olefins towards nucleophilic attack by coordination to a second metal complex *(144,145)*. The Pd complex, $(MeCN)_2PdCl_2$, or thallium benzoate are commonly used for this purpose and are usually added to the reaction mixture containing the olefin and the $Co(NO_2)$ compound *(141,143,148)*. For example, employing such methods thus allows the ethylene in $[(C_2H_4)PdCl_2]_2$ to be oxidized to acetaldehyde, and propylene to acetone.

Due to the fact that $Pd(MeCN)_2Cl(NO_2)$, by itself, is known to oxidize alkenes *(147)*, a mechanistic study of this latter O-transfer to olefins was undertaken to elucidate the reaction pathways for the mixed Co/Pd system *(142)*. This was done indirectly, by employing $(py)Co(TPP)(NO_2)$ as the cobalt catalyst and by accurately measuring the product distributions in these oxidation reactions. The reactions shown in Scheme 7.2 account for the range of reactivity observed for the mixed Co/Pd system. Thus, transfer of a nitro group from Co to Pd generates a small concentration of the $Pd(MeCN)_2Cl(NO_2)$ catalyst (step *a*). Subsequently, metallacyclic oxidation of the olefin occurs (step *b*). [When norbornene is the olefin, this metallacycle is observed spectroscopically.] Release of the oxidized olefin (step *c*) then generates a Pd-nitrosyl that would rapidly transfer the NO ligand back to Co (step *d*) to generate the observed intermediate Co-nitrosyl. This Co complex then reacts with dioxygen to regenerate the starting Co-nitro complex (step *e*). Even though other complex mechanisms cannot be ruled out, experimental evidence for the individual steps in the reaction scheme have been obtained.

The corresponding $(TPP)Rh(NO_2)$ and $[(bipy)(MeCN)_3Rh(NO_2)]^{2+}$ complexes also oxidize olefins in the presence of Pd(II) *(145)*.

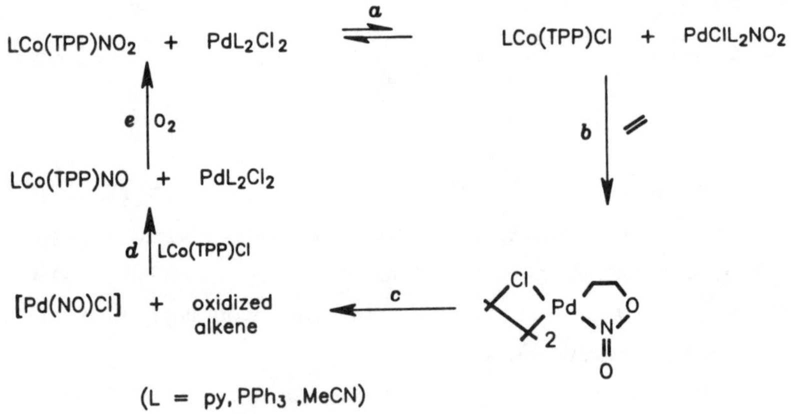

$LCo(TPP)NO_2$ + PdL_2Cl_2 ⇌ (a) $LCo(TPP)Cl$ + $PdClL_2NO_2$

(b) ∥ ↓

(c) [Pd(NO)Cl] + oxidized alkene ← [Cl–Pd–N(=O)–O metallacycle]$_2$

(d) $LCo(TPP)Cl$ ↑ $LCo(TPP)NO$ + PdL_2Cl_2

(e) O_2 ↑

(L = py, PPh_3, MeCN)

Scheme 7.2

(iii) Nickel, Palladium and Platinum.

The reactions of $M(NO_2)XL_2$ (M = Ni, Pt; X = halide; L = phosphine) with CO produce $M(NO)XL_2$ and CO_2 by O-transfer reactions *(135,149)*. In the preceding section, it was mentioned that the $(MeCN)_2PdCl(NO_2)$ complex oxidizes olefins. Definitive proof that oxygen-transfer from the nitro group to the olefin occurs is found in quantitative IR measurements, where the ^{18}O label from the $Pd(NO_2)$ complex ends up in the oxidized olefin, with no detectable dilution of the isotopic label *(150)*. When the oxidation of terminal olefins is carried out by the Pd complex in the presence of acetic acid in air, approximately equal amounts of 2-acetoxy-1-alkanol and 1-acetoxy-2-alkanol are the major products *(151)* (eq 7.44).

$$\text{RHC=CHD} \longrightarrow \text{HO, OAc product (R, H; H, D)} + \text{AcO, OH product (R, H; H, D)} \qquad (7.44)$$

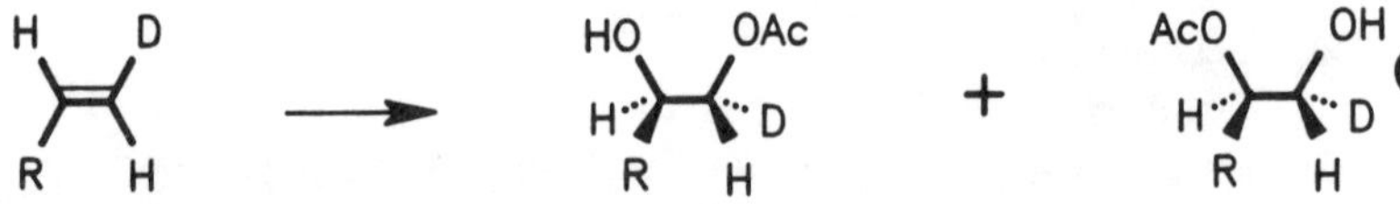

R = n-C_8H_{17}

Quantum-chemical methods have been used to investigate the detailed mechanism of the oxidation of ethylene by $(MeCN)_2PdCl(NO_2)$ *(152)*, and the following mechanism has been proposed which is consistent with experimental observations *(153)*. The first step consists

$$Cl(L)_2Pd(ONO) + C_2H_4 \longrightarrow Cl(L)Pd(C_2H_4)(ONO) + L$$

$$\downarrow$$

$$Cl(L)_2Pd{-}CH_2CH_2ONO$$

$$\downarrow$$

$$Cl(L)Pd(CH_2CH_2ON{=}O) + L \longrightarrow (L)_2Pd(CH_2CHO)(NO) + HCl$$

$$\uparrow$$

$$(L)_2Pd(NO)Cl + CH_3CHO$$

of displacement of a solvent molecule (L) by ethylene, followed by subsequent insertion of the ethylene into a Pd-O bond, producing a nitro-ethane complex of Pd. Ring closure then occurs to form a metallacycle which undergoes a ring-opening reaction with the concomitant release of HCl. The HCl released is then immediately consumed by the intermediate Pd complex to produce acetaldehyde and regenerate the Pd-nitrosyl. The $(MeCN)_2PdCl(NO_3)$, $(MeCN)_2Pd(NO_2)_2$ and $Na_2[Pd(NO_2)Cl_3]$ complexes also oxidize ethylene and propylene via similar mechanisms *(154,155)*. However, it must also be pointed out that some metal-nitrosyl complexes react with oxygen to ultimately transform their M(NO) groups into M(O) oxo species *(156-158)*.

7.4: Reactivity at the Metal Center

7.4.1: Involving Bound Nitrosyl

7.4.1.1: Nitrosyl Exchange

The gas phase nitrosyl exchange reactions of $Fe(NO)_2(CO)_2$ and $Co(NO)(CO)_3$ have been studied *(159)*. The rate laws are first order in both the metal nitrosyl and NO. Also, the activation energies are in the usual range found for associative (rather than dissociative) reactions. Consequently, a simple associative mechanism such as

$$^{15}NO + M(NO) \longrightarrow "(^{15}NO)M(NO)" \longrightarrow M(^{15}NO) + NO \qquad (7.45)$$

is consistent with this mode of substitution, where "$(^{15}NO)M(NO)$" may be either a transition state or an intermediate of increased coordination number.

The kinetics of ^{15}NO exchange with the $M(NO)_2IL$ complexes (M = Fe, Co; L = PPh_3, $AsPPh_3$) establishes that these exchange processes also occur by simple associative mechanisms *(160)*. A notable exception is $Co(NO)_2I(PPh_3)$, a case in which the solvent actively participates in the exchange process. Fast nitrosyl exchange occurs between ^{14}NO and either $Fe(^{15}NO)(S_2CNMe_2)_2$ (< 2 min) or *cis*-$Fe(^{15}NO)(S_2CNMe_2)_2I$ (1.5 h in the dark) *(161)*. The exchange process of the latter complex is speeded up (0.5 h) when carried out in common laboratory lighting. This enhancement is probably due to the photolysis of the Fe-I bond in this six-coordinate complex. A fast nitrosyl exchange process also occurs when $Co(NO)(qn)_2$ (qn = quinolinolate) reacts with NO gas *(162)*.

The nitrosyl exchange reactions of pentahalo- or pentacyanonitrosyls of the form $[M(NO)X_5]^{n-}$ in aqueous solution have been studied. However, these reactions are very complex and are pH dependent, indicating that they may proceed via nitrite intermediates. Indeed, such nitrosyl exchanges also occur when $^{15}NO_2^-$, in aqueous alkaline solution, is employed *(163)*. Earlier claims for a related nitrosyl (M-NO) to thionitrosyl (M-NS) interchange by using $(NSCl)_3$ *(164,165)* have been discounted *(166)*.

7.4.1.2: Expulsion (Displacement) of Bound NO

Some metal complexes reversibly bind nitric oxide (Section 2.1). A consequence of this reversible binding is that some metal nitrosyl complexes undergo transformations that result in the dissociation of their NO ligands. Such dissociations may be induced thermally, photolytically, or chemically. Thus, even though some thermal reactions of MNO complexes result in the formal disproportionation of NO (see next section), some of the thermal reactions result in a simple expulsion of the bound NO group. For example, NO dissociates from the solid phase of $(L_4)Co(NO)$ (L_4 = tetradentate Schiff base) in a first order reaction, independent of the nature of L_4 *(167)*. Also, NO dissociates from the $[M(NO)(CN)_5]^{n-}$ complexes (M = Fe, Cr, Mn, Co) photolytically *(168,169)* or thermally at temperatures between 240°C and 340°C *(170,171)*. For the series of complexes $[Ru(NH_3)_5L]Cl_n$, *(130,172)*, the cleavage temperatures of the Ru-L bonds are, in increasing order, MeCN (191°C) < N_2 (255°C) < CO (286°C) < NO^+ (324°C). This trend parallels the order of π-backbonding and strength of the Ru-L bond *(172)*.

The photodissociation of $Co(NO)(CO)_3$ in matrices occurs via initial loss of CO and then NO *(173)*. In contrast, the photolysis of $Rh(NO)(CO)(PPh_3)_2$ in dichloromethane, and in the presence of PPh_3, results in the expulsion of NO rather than CO to form $Rh(CO)Cl(PPh_3)_2$ *(174)*. Similarly, the thermal- and photoreactions of $[Ir(NO)(CO)Cl(PPh_3)_2]^+$ in the presence of PPh_3 lead to NO dissociation via the homolytic cleavage of the possibly weak, bent, Ir-NO bond *(174)*.

Coordinating solvents such as MeCN and THF displace NO ligands in $[V(NO)_3Cl_2]_2$ to produce $V(NO)_2Cl(solvent)_2$ *(175)*. Also, a range of strong electron donors L (L = py, phosphines, arsines, isonitriles) react with $CpCr(NO)_2Cl$ or $W(NO)_4Cl_3$ resulting in NO expulsion *(176,177)*.

$$CpCr(NO)_2Cl + L \longrightarrow CpCr(NO)(L)Cl + NO \qquad (7.46)$$

$$WCl_3(NO)_4 + L \longrightarrow WCl_3(NO)L_2 + WCl_2(NO)_2L_2 \qquad (7.47)$$

Denitrosylation occurs when $Ru(NO)_2(PPh_3)_2$ or $[Ru(NO)_2(PPh_3)_2Cl]^+$ are reacted with the dttd dianion, resulting in the formation of $Ru(dttd)(PPh_3)_2$

(66). Similar denitrosylations occur for $[Cr(NO)_2(MeCN)_4]^{2+}$ *(178,179)* and $[Ru(NO)(SR)_4]^-$ (R = aryl) *(180)* when they are reacted with thiolate-type reagents. Whereas the use of the BF_4^- anion (from $AgBF_4$) does not disrupt the "$Co(NO)_2$" core in the $Co(NO)_2I(PR_3)$ complex, the use of $NaBPh_4$ induces the denitrosylation of the complex to its dimeric $[(NO)Co(\mu\text{-}NO)(\mu\text{-}PR_3)_2Co(NO)]^+$ derivative *(181)*. In a similar manner, lithium halides (or halogens) react with some Pt-nitrosyl complexes resulting in denitrosylation and the formation of the corresponding Pt-X_2 compounds *(182)*.

In a series of interesting reactions, perfluorocarboxylic acids R_fCO_2H react with the $M(NO)_2(PPh_3)_2$ complexes (M = Ru, Os) resulting in denitrosylation *(106)*:

$$Ru(NO)_2(PPh_3)_2 + R_fCO_2H \longrightarrow Ru(O_2CR_f)_3(NO)(PPh_3)_2 \quad (7.48)$$

$$Os(NO)_2(PPh_3)_2 + R_fCO_2H \longrightarrow OsH(O_2CR_f)_2(NO)(PPh_3)_2 \quad (7.49)$$

In these reactions, the NO ligands are lost presumably as HNO, resulting from their protonations by the acid.

7.4.1.3: Decomposition of Metal Nitrosyls

NO is thermodynamically unstable and, at elevated temperatures (about 1100 - 1200°C), decomposes to N_2 and O_2. However, at high pressure and moderate temperatures (about 50°C) NO disproportionation occurs to give mixtures of N_2O and NO_2 *(183)*. One of the reasons for studying the reactivity of the coordinated NO group is the need to develop efficient metal catalysts that transform the environmentally undesirable NO molecule (Chapter 8) to less harmful products such as N_2 and O_2. However, as yet, many of the catalysts developed to effect such transformations normally require the presence of added fuels such as CO, H_2, and CH_4. New means are constantly being sought to construct the "perfect" catalyst that would effect the decomposition or disproportionation of NO without requiring the addition of expensive secondary fuels.

In previous sections, some aspects of the chemistry of the MNO group such as (i) the reactions of metal nitrosyls resulting in the expulsion of

NO, (ii) the reactions of metal nitrosyls with O-acceptors resulting in the formal expulsion of "O", leading to the formation of metal nitrides, and (iii) the reactions of metal nitrosyls with O-donors resulting in the "acquisition" of O_x (x = 1, 2) and the production of $M(NO_x)$ species. For the most part, attention was focused on the fate of the metal-containing fragments. Here, the discussion concerns how such reactions may be utilized to transform the NO molecule into other gaseous nitrogen oxides, and is centered on the decompositions of metal nitrosyl complexes (i) in the absence of added NO gas, (ii) in the presence of added NO gas, and (iii) in the presence of reducing ligands.

(i) In the Absence of Added NO Gas

Many metal nitrosyl complexes decompose upon heating to their metal-oxo derivatives, and the fate of the N and O atoms of the once-coordinated NO group depends largely on the presence, or absence, of a coordinated "reducing" ligand, e.g.,

$$[Mo(NO)_2Cl_2]_n \longrightarrow [Mo(O)Cl_2] + N_2O \qquad (7.50$$

$$Mo(NO)_2Cl_2L_2 \longrightarrow MOIL_2(OL)_2 + N_2 \qquad (7.51)$$

(L = reducing ligand, e.g., PPh_3)

$$Mo(NO)_2Cl_2L_2 \longrightarrow [Mo(O)Cl_2] + 2L + N_2O \qquad (7.52)$$

(L = non-reducing ligand, e.g., $OPPh_3$)

Thus, dinitrogen is produced from the thermal decomposition of the Mo-complex (eq 7.51) in the presence of a reducing ligand such as triphenylphosphine *(184,185)*. In contrast, N_2O is produced when no reducing ligand is present (eqs 7.50 and 7.52) *(184-186)*. Thus, the reducing ligands scavenge the oxygen atoms in such decomposition reactions.

Such decomposition reactions of metal nitrosyl complexes are extremely complex, and it is important that control reactions be carried out in order to obtain meaningful data for scientific evaluation. For example, the thermal decomposition of $Co(NO)(CO)_3$ results in the fragmentation of the cobalt-ligand bonds *(187)* (eq 7.53). However, metallic cobalt also activates the reactions described by eqs 7.54 and 7.55. Furthermore, the NO_2 and O_2 produced convert the cobalt into its oxide (eqs 7.56 and 7.57).

$$Co(NO)(CO)_3\ (vap) \longrightarrow Co(s) + 3CO(g) + NO(g) \qquad (7.53)$$

$$CO(g) + NO(g) \longrightarrow C(s) + NO_2(g) \qquad (7.54)$$

$$2NO(g) \longrightarrow N_2O(g) + 1/2\ O_2(g) \qquad (7.55)$$

$$Co(s) + 1/2\ O_2(g) \longrightarrow CoO(s) \qquad (7.56)$$

$$Co(s) + NO_2(g) \longrightarrow CoO(s) + NO(g) \qquad (7.57)$$

Thus, any measurements of the ratios of gaseous products evolved from thermal decompositions of metal nitrosyls must take into account possible side reactions such as those shown above. In addition, it has also been shown that some metallic oxides are themselves efficient catalysts for NO disproportionation *(188)*. Nevertheless, the thermal decomposition of $Co(NO)(CO)_3$ has been utilized to generate CoO films prepared by chemical deposition from the gas phase. Thus, pyrolysis (at 450 K) of $Co(NO)(CO)_3$ on Si(111) or (100) substrates *(189)* produces thin films of CoO. However, in the presence of organic hydroxyl groups, Co_3O_4 films are produced instead *(190)*.

A mechanism for the decomposition of two molecules of NO to N_2 and O_2 has been proposed *(191)*. In experiments on the thermal degradation of a number of metal dinitrosyl complexes, the nature of the metal and the ancillary ligands were varied systematically in order to determine which factors affected the ratios of the gaseous decomposition products shown in eq 7.58.

$$M(NO)_2 \longrightarrow N_2 + NO + N_2O \qquad (7.58)$$

$$[M = Co, Rh, Ru, Pd, Fe, W, Cr, Mo]$$

From the results of these studies, it appears that the systems that lead to N_2 and O_2 formation possess a *cis*-coplanar, *gem*-dinitrosyl microstructure, containing one bent ("activated") and one linear nitrosyl in the reactive species, i.e.,

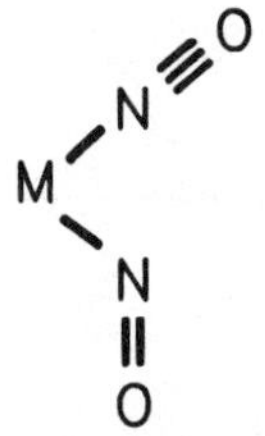

For instance, the thermal decomposition of $[RuCl(NO)_2(PPh_3)_2]PF_6$ results in rapid, quantitative yields of N_2 (and presumably O_2). It is also known that the atoms in the $Ru(NO)_2$ moiety of this complex are coplanar, with one nitrosyl bonded in a linear fashion (178°) and the other in a bent (138°) fashion. Furthermore, it has been suggested that if the $M(NO)_2$ unit can be forced into the *cis*-coplanar structure, it should undergo decomposition to N_2 and O_2 *(191)*. The results of the thermal decompositions of some metal dinitrosyls are shown in Table 7.4. As shown by the first six entries in the Table, the Pd and Co dinitrosyl complexes have the greater propensity to release the bound NO in simple dissociation reactions. It is also interesting to note that the $[Co(NO)_2X]_2$ complexes (X = Cl, Br) are employed as nitrosylating agents for the syntheses of other metal nitrosyl complexes (Section 2.1). It is tempting, therefore, to speculate that the $Pd(NO)_2Cl_2$ complexes should be better nitrosylating agents in this regard, since they lose their coordinated NO groups more efficiently.

The last four entries in Table 7.4 show the effect of replacing the diamino ligands (tmed, capable of donating 4 electrons to the metal) with the triamino-chelating ligands (pmdt, capable of donating 6 electrons). The increased electron count in the pmdt complexes serve to sufficiently "activate" the $Co(NO)_2$ group towards N_2O production rather than a simple expulsion of NO. One may speculate that the dinitrosyl intermediates involved in the pmdt cases may involve some NO orbital rehybridization (such as the bending of one NO ligand to result in the formation of 18-electron complexes). Nevertheless, these latter results are entirely consistent with a *general* observation that first-row transition metal dinitrosyls decompose to N_2O rather than N_2 *(191)*.

Table 7.4 Ratios of Nitrogen Oxides Produced in the Thermolysis (150 °C) of Some Metal Dinitrosyl Complexes.

compound	yield (%)		
	N_2	NO	N_2O
$Pd(NO)_2Cl_2$	0	100	0
$[Co(NO)_2Br]_2$	2	98	0
$[Co(NO)_2Cl]_2$	8	83	9
$[Rh(NO)_2Cl]_n$	30	20	50
$[(tmed)Co(NO)_2]PF_6$	7	93	0
$[(tmed)Co(NO)_2]I$	10	52	38
$[(pmdt)Co(NO)_2]Cl$	7	16	77
$[(pmdt)Co(NO)_2]I$	7	12	81

pmdt = pentamethyldiethylenetriamine (a σ-donor chelating ligand) [Data from reference 191].

(ii) In the Presence of Added NO Gas

In addition to its use in nitrosation reactions, NO may also oxygenate metal centers to produce metal-oxo complexes. Related to this fact is that NO gas has been shown to react with a number of cobalt dinitrosyl complexes, the net result being the *oxidation* of the bound nitrosyl ligands (L = PPh_3) *(192-194)*, e.g.,

$$Co(NO)(en)_2Cl_2 + 2\ NO \longrightarrow Co(NO_2)(en)_2Cl_2 + N_2O \quad (7.59)$$

$$Co(NO)L_3 + 7\ NO \longrightarrow Co(NO)_2L(NO_2) + 1/2\ N_2 + 2N_2O + 2OL \quad (7.60)$$

$$[Co(NO)(NH_3)_5]Cl_2 + 2\ NO \longrightarrow [Co(NH_3)_5NO_2]Cl_2 + N_2O \quad (7.61)$$

It appears that such reactions proceed by the electrophilic attack on a coordinated (and possibly bent) nitrosyl ligand by free NO. Presumably, a subsequent decomposition of the $M(N_2O_2)$ unit then results in the net disproportionation of NO. Whatever the mechanism is, it bears a striking resemblance to the thermal degradation of the $M(NO)_2$ unit described in the previous section.

(iii) In the Presence of Reducing Ligands

The decomposition of metal nitrosyls in the presence of (coordinated) reducing (e.g., PPh_3) or non-reducing (e.g., $OPPh_3$) ligands has already been commented on, i.e.,

$$Mo(NO)_2Cl_2(PPh_3)_2 \longrightarrow MOIL_2(OPPh_3)_2 + N_2 \qquad (7.51)$$

$$Mo(NO)_2Cl_2(OPPh_3)_2 \longrightarrow Mo(O)Cl_2(OPPh_3)_2 + N_2O \qquad (7.52)$$

The nature of the coligands drastically affects the pathway for NO decomposition. Studies on the nature of the decomposition of $[Rh(NO)_2Cl]_n$ have also indicated that the presence of different types of phosphines or amines may significantly affect the types of gaseous products produced *(191)*, and the results presented in Table 7.5 illustrate this point well.

Table 7.5 Effect of Added Bases on the Thermal Decomposition (150 °C) of $[Rh(NO)_2Cl]_n$ in Dichlorobenzene.

added base	yield (%)		
	N_2	NO	N_2O
no phosphine	13	54	33
$Ph_2PCH_2CH_2PPh_2$	25	0	75
Bu_3PO	15	1	84
pmdt	100	0	0
hmtt	100	0	0
Bu_3N	43	0	57

hmtt = hexamethyltriethylenetetraamine [Data from reference 191].

As may be noted, (i) the presence of added bases strongly disfavors the simple dissociation of NO in these thermal reactions, (ii) the presence of the non-chelating Bu_3N results in a relatively "equal" distribution of N_2 and N_2O, whereas (iii) the presence of the *chelating* amines during the decomposition of the Rh complex result in the exclusive formation of N_2.

7.4.1.4: The Catalytic Reduction of NO

The *catalytic* reduction of NO over a wide variety of metal complexes or surfaces has been studied extensively. Many of these studies have employed reducing molecules such as carbon monoxide, hydrogen, ammonia, or even hydrocarbons. This is not so surprising because the "reduction" reactions outlined below are generally more thermodynamically favorable than are those involving the direct decomposition of NO to N_2 and O_2.

$$NO + 5/2\ H_2 \longrightarrow NH_3 + H_2O \quad (7.62)$$

$$NO + 1/2\ CO \longrightarrow 1/2\ N_2O + 1/2\ CO_2 \quad (7.63)$$

Another practical reason for the use of reducing co-molecules is that, when NO is decomposed over a catalyst, the residual oxygen normally forms an oxide coating on the surface of the catalyst, thus deactivating it. When reductants are present, they act as scavengers for the residual oxygen by forming molecules such as H_2O and CO_2 *(188)*.

A number of homogeneous systems catalyze the reduction of NO by CO to produce N_2O and CO_2 (eq 7.63); such systems include nitrosyl complexes of Rh and Ir of the form $[M(NO)_2(PPh_3)_2]^+$ *(195)*. Furthermore, an aqueous $PdCl_2$-$CuCl_2$-HCl system also catalyzes this reduction at a rapid rate *(196)*. Indeed, the related $PtCl_4^{2-}$-$CuCl_2$-HCl (with added CuCl) system catalyzes this reaction with a greater turnover rate than for other catalysts previously employed *(197)*, and presumably does so by the following reactions:

$$CuCl_2^- + CO \longrightarrow CuCl(CO) + Cl^- \quad (7.64)$$

$$CuCl(CO) + PtCl_4^{2-} \longrightarrow PtCl_3(CO)^- + CuCl_2^- \quad (7.65)$$

$$PtCl_3(CO)^- + 2\ NO + Cl^- \longrightarrow PtCl_4^{2-} + 2\ N_2O + CO_2 \quad (7.66)$$

Although the mechanisms of such reactions are not known with certainty, labelling studies have shown that the oxygen in the CO_2 generated comes from water, possibly via the intermediate $[Pt(NO)Cl_3(COOH)]^-$ species formed during reaction 7.66. Presumably, the N_2O produced derives from a Pt-dinitrosyl species *(197)*.

The related homogeneous catalytic reduction of NO by olefins in the presence of an aqueous Pd-$CuCl_2$ system, i.e.,

$$2\ NO + RCH{=}CH_2 \longrightarrow N_2O + RCOCH_3 \qquad (7.67)$$

is also known to occur *(198)*. In this rather complex system, the oxidation of the olefin to the ketone is catalyzed by Pd(II), whereas Pd(0) and Cu(I) catalyze the reduction of NO to N_2O via eqs 7.68 - 7.71.

$$Pd^{2+} + RCH{=}CH_2 + H_2O \longrightarrow Pd^0 + RCOCH_3 + 2\ H^+ \qquad (7.68)$$

$$Pd^0 + 2\ NO + 2\ H^+ \longrightarrow Pd^{2+} + N_2O + H_2O \qquad (7.69)$$

$$Pd^0 + 2\ Cu^{2+} \longrightarrow Pd^{2+} + 2\ Cu^+ \qquad (7.70)$$

$$2\ Cu^+ + 2\ NO + 2\ H^+\ (Pd^{2+}\ cat) \longrightarrow 2\ Cu^{2+} + N_2O + H_2O \qquad (7.71)$$

The $Rh_2(CO)_4Cl_2$ complex in the presence of a base (e.g., KOH) catalyzes the reduction of N_2O (or NO) by CO to eventually produce N_2 *(199)*.

7.4.2: Not Involving Bound Nitrosyl

There are many important reactions in which metal nitrosyl complexes function as stoichiometric reactants or specific catalysts. In this section, the wide scope, utility and applications of metal nitrosyl complexes in reactions resulting in the production of new organic compounds will be considered.

7.4.2.1: Olefin Metathesis

The net result of the olefin metathesis reaction is the rupture and formation of carbon-carbon double bonds:

$$R_1CH{=}CHR_2 + R_3CH{=}CHR_4 \rightleftharpoons R_1CH{=}CHR_3 + R_2CH{=}CHR_4 \qquad (7.72)$$

This type of reaction can be applied to substituted alkenes, dienes, polyenes and even to alkynes *(200)*, and is especially important in industry. For example, cycloalkenes may be ring-enlarged to produce macrocyclic compounds and eventually polyalkenes, e.g.,

$$n\ \text{cyclo-}[CH{=}CH(CH_2)_x] \rightleftharpoons \left[{=}{-}(CH_2)_x\right]_n \qquad (7.73)$$

which may possess properties ranging from those of amorphous rubbers to those of highly crystalline polymers.

The initial report on the use of metal nitrosyl complexes as catalysts for the metathesis reaction centered on the use of the $M(NO)_2Cl_2L_2$ complexes (M = Mo, W; L = PPh_3, py, $OPPh_3$, etc) in combination with a variety of alkylaluminum halides in chlorobenzene *(201)*. It was also noted at the time that, in contrast to the other soluble catalyst systems derived from WCl_6 in which only terminal olefins were satisfactory substrates, the Mo and W nitrosyl catalysts effect the metathesis of α-olefins and even ethylene *(202)*, e.g.,

$$\text{2-pentene} \overset{[M]}{\rightleftharpoons} \text{3-hexene} + \text{2-butene} \qquad (7.74)$$

([M] = $Mo(NO)_2Cl_2(PPh_3)_2$ + $Me_3Al_2Cl_3$ in PhCl)

although when $EtAlCl_2$ is used as cocatalyst, simple *isomerization* of the olefin also occurs *(202)*. The results of kinetic studies of the metathesis reactions involving 2-pentene show that the rate of disappearance of the olefin is first order in catalyst and variable order in olefin *(203)*. Information as to the nature of the active catalyst for the olefin metathesis is obtained from the observation that mixing of the Mo-nitrosyl complex with an aluminum-containing cocatalyst in PhCl results in the formation of methane and ethylene (L = PPh_3) *(204)*, i.e.,

$$Mo(NO)_2L_2Cl_2 + Me_3Al_2Cl_3 \longrightarrow CH_4 + CH_2{=}CH_2 + \text{Mo-cocat.} \quad (7.75)$$

When deca-2,8-diene is mixed with the catalyst solution, the initial product is propene, followed by the production of 2-butene and cyclohexene (eq 7.76).

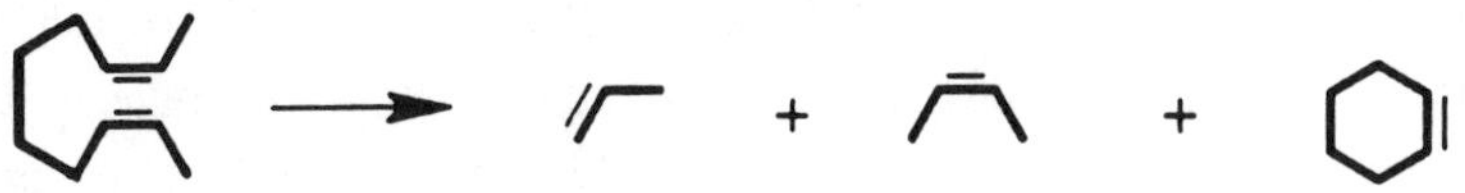

(7.76)

The results of other studies with labelled terminal groups indicate a possible mechanism for the initiation of the olefin metatheses, which involves the participation of a carbene species *(204)*.

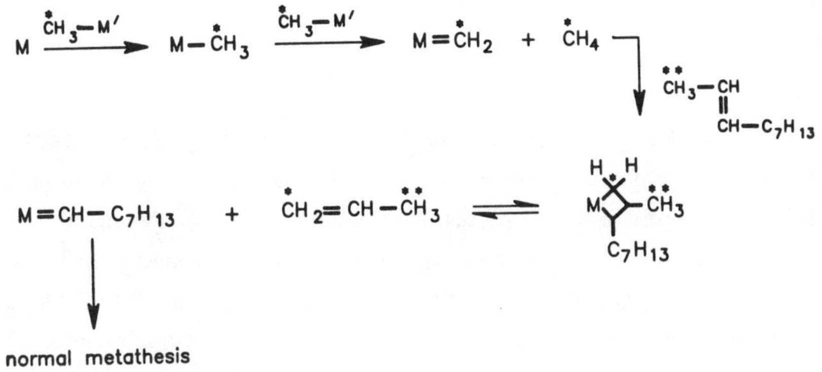

$$M \xrightarrow{\overset{*}{C}H_3-M'} M-\overset{*}{C}H_3 \xrightarrow{\overset{*}{C}H_3-M'} M{=}\overset{*}{C}H_2 + \overset{*}{C}H_4$$

$$\downarrow \; \overset{**}{C}H_3-CH{=}CH-C_7H_{13}$$

$$M{=}CH-C_7H_{13} + \overset{*}{C}H_2{=}CH-\overset{**}{C}H_3 \rightleftharpoons \text{(metallacyclobutane: } M,\ H,\ H,\ \overset{**}{C}H_3,\ C_7H_{13})$$

$$\downarrow$$

normal metathesis

The production of ethylene in the initial mixture probably results from the dimerization of the methylene groups (from $M{=}CH_2$) during the initiation step.

Interestingly, it has been observed that the nitridomolybdenum complex, $Mo(N)Cl_3(OPPh_3)_2$, in combination with $EtAlCl_2$, constitutes an excellent metathesis catalyst. This has raised the intriguing possibility that a nitridocarbene molybdenum compound is the active catalyst in these reactions *(205-209)* especially since the catalytic activity of the nitrido species has been found to be comparable to that of the dinitrosylmolybdenum species discussed above. Whatever the fate of the $Mo(NO)_2$ moiety in these reactions, it has been well established that the metathesis reactions described above do indeed proceed via carbene intermediates. Other related Mo and W complexes such as $W(CO)_4(NO)X$, $W(CO)_3(NO)(PPh_3)X$ and $W(CO)_2(NO)(PPh_3)_2X$ (X = halide) have also been used for these reactions *(177,210)*. The W complexes show substantially higher activities in the metathesis of 1,7-octadiene to cyclohexene and ethylene than does $Mo(NO)_2Cl_2(PPh_3)_2$ *(211)*. A number of molybdenum dinitrosyl alkoxide, acetate, and azide complexes of the general formula $\{Mo(NO)_2X_2\}$, in combination with $EtAlCl_2$ and/or $Et_4Sn/AlCl_3$, also catalyze the olefin metathesis reaction *(212-218)* via the possible involvement of carbene species *(218-220)*. Lastly, the attachment of the $Mo(NO)_2Cl_2(PPh_3)_2$ complex to a polystyrene support (and activated by $Me_3Al_2Cl_3$) has also been found to serve as a hybrid catalyst for olefin metathesis *(221)*.

7.4.2.2: Oligomerizations and Related Coupling Reactions

The nitrosyl ligand in a metal nitrosyl complex is a strong π-acid. Thus, it drains electron density away from the metal center, imparting electrophilic behavior to many of its complexes. This section deals with the use of metal nitrosyls in various isomerizations, dimerizations, oligomerizations, and polymerization reactions of olefins. In these reactions, the dominant interaction between the olefin and the metal nitrosyl complex probably involves a substantial transfer of electron density from the π-bond to the metal, resulting in the formation of incipient carbocations, i.e.,

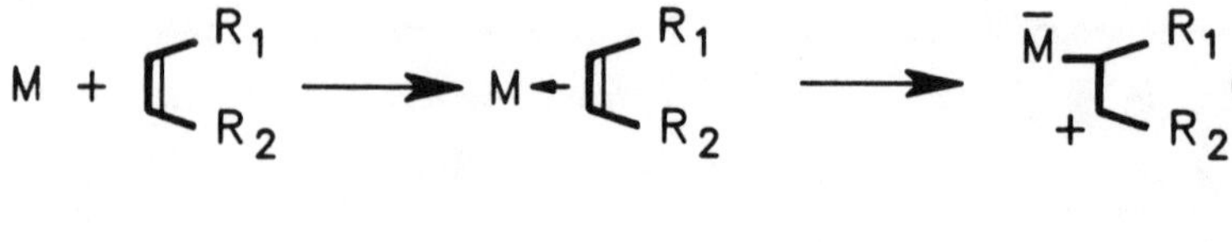

$$M + \left[\!\!\begin{array}{l} R_1 \\ R_2 \end{array}\right. \longrightarrow M \leftarrow \left[\!\!\begin{array}{l} R_1 \\ R_2 \end{array}\right. \longrightarrow \begin{array}{l} \bar{M}-C(R_1) \\ +C(R_2) \end{array} \quad (7.77)$$

Therefore, electrophilic metal nitrosyls would be expected to catalyze carbocationic rearrangements of alkenes *(222)*. Most of the work done in this area has been with cationic dinitrosyl complexes of Mo, W, and Fe. For example, the isomerization of olefins (eq 7.78), the catalytic dimerization of 1,1-diphenylethylene (eq.7.79), and even the skeletal rearrangement of *tert*-butylethylene (eq 7.80),

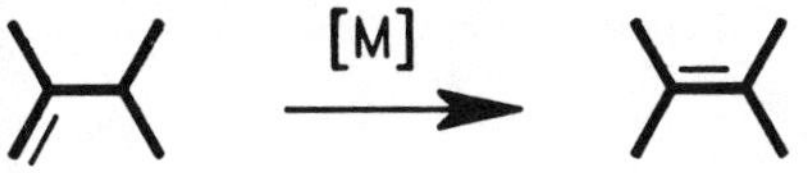

[M] (7.78)

[M] = $Mo(NO)_2(AN)_4^{2+}$, $CpW(NO)_2^+$, $W(NO)_2(AN)_4^{2+}$

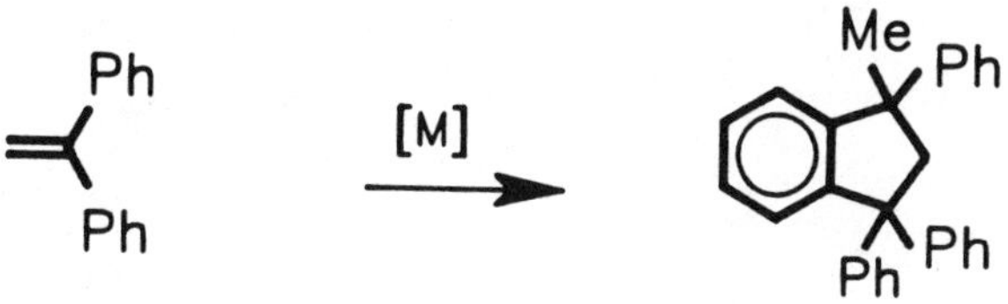

Ph Ph [M] Me Ph Ph Ph (7.79)

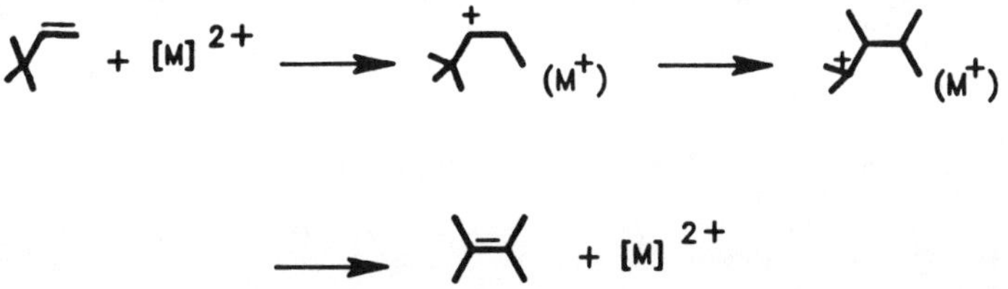

+ $[M]^{2+}$ → + (M^+) → (M^+)

(7.80)

→ + $[M]^{2+}$

[M] = $M'(NO)_2(AN)_4^{2+}$; M' = Mo, W

have all been carried out *(223,224)*. Furthermore, solvated dinitrosyl dications of Mo, W, Co and Fe effect the cationic polymerizations of olefins and dienes *(223,225-228)* and the $[CpM(NO)_2]^+$ cations of Mo and W also effect the [2+2] cycloaddition reaction of 2,3-dimethyl-2-butene and phenylacetylene to produce the cyclobutene *(229)*. The electrophilic behavior of the $[Cp'M(NO)_2]^+$ cations (Cp' =Cp, Cp*; M = Cr, Mo, W) has been exploited in the regiospecific condensation of methyl propiolate (a heterosubstituted alkyne) and 2,3-dimethyl-2-butene to produce organometallic lactone complexes *(229)*.

(7.81)

This latter reaction is believed to proceed via the initial coordination of the acetylenic ester to the metal center via the triple bond, followed by nucleophilic attack of the olefin on the coordinated acetylenic ester.

The $[(PMe_3)(CO)_3(NO)W]FSbF_5$ complex polymerizes cyclopentadiene and isoprene, and is a catalyst for the Diels-Alder reaction of cyclopentadiene or butadiene with α,β-unsaturated enones *(230)* (eq 7.82).

(7.82)

(R = H,Me,Et,OMe)

Similarly, the electrophilic $[HC(py)_3M(NO)_2]^{2+}$ cations of Mo and W catalyze Diels-Alder reactions *(231)*.

The "$Fe(NO)_2$" group has also been used extensively in reactions leading to the cyclodimerization of diolefins *(228,232-234)* (eq 7.83).

(7.83)

This "$Fe(NO)_2$" moiety may be generated by a number of methods, such as (i) the electrochemical reduction of $FeCl_3$ in the presence of NO, (ii) the electrochemical reduction of $[Fe(NO)_2X]_2$ (X = Cl, I), or (iii) the reaction of $[Fe(NO)_2Cl]_2$ with either $Ni(COD)_2$, zerovalent metal carbonyls, or even bare metal *(235-237)*. The catalytic species in THF is formulated as $Fe(NO)_2(THF)_n$, which results from chloride expulsion in the electrochemical reduction of the $Fe(NO)_2Cl(THF)$ monomer *(238,239)*. Consistent with this catalyst formulation is that addition of phosphine to the "catalyst solution" results in the formation of the known $Fe(NO)_2(PR_3)_2$ complexes.

Interestingly, the photochemical reaction of $Fe(CO)_2(NO)_2$ (itself a highly specific catalyst for reaction 7.83) with 1,3-butadiene in liquid xenon produces the $Fe(NO)_2(C_4H_6)_2$ complex, which is likely to be the catalytically active intermediate during the cyclodimerization reaction , since the three coordinated C=C bonds could then easily couple to form 4-vinylcyclohexene (eq 7.83) *(240)*.

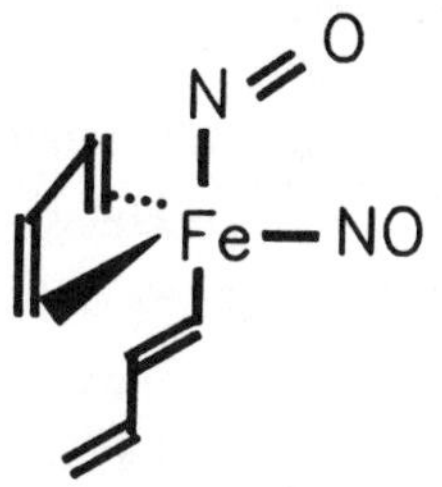

N≡O

Fe—NO

A well-documented example of the oligomerization of acetylenes is provided by the reaction of $Ru(H)(NO)(PPh_3)_3$ with diphenylacetylene to produce 1-benzylidene-2,3-diphenylindene,

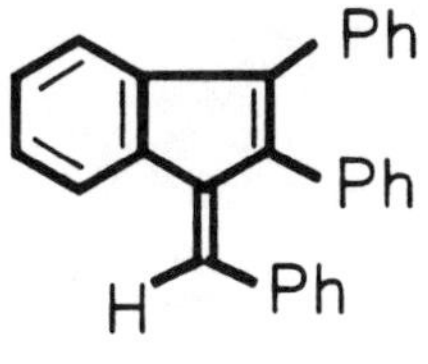

Ph

Ph

H Ph

Monosubstituted acetylenes and hexafluorobut-2-yne are polymerized by the Ru complex. Furthermore, phenylacetylene is trimerized *(241)*, presumably via a Ru-acetylide intermediate. The CpNi(NO) compound also catalyzes the cyclooligomerization of 1,3-butadiene *(242)*.

7.4.2.3: Epoxidation

Epoxidation of olefins by *t*-BuOOH (eq 7.84) is catalyzed by the $Mo(NO)_2Cl_2L_2$ complexes (L = hmpt, $OPPh_3$, DMF, etc) *(243-245)*. The most active catalyst for the epoxidation of 1-octene is the complex in which L is hmpt *(246)*.

$$\text{>C=C<} + \text{ROOH} \longrightarrow \text{>C(O)C<} + \text{ROH} \qquad (7.84)$$

7.4.2.4: Fischer-Tropsch Synthesis

The sorption of $Co(NO)(CO)_3$ (molecular diameter, 6.3 Å) into the pores of NaY zeolite (pore diameter, 7.4 Å) generates, after appropriate activation by H_2, a novel shape-selective Fischer-Tropsch catalyst for the direct conversion of hydrocarbon-carbon monoxide mixtures (synthesis gas) to alkenes with a cut-off at C_8 *(247)*. The high alkene:alkane product ratios observed are consistent with alkene formation in a hydrogen-poor environment within the zeolite pores.

7.4.2.5: Carbon-Nitrogen Bond Formation

The nitrosonium cation inserts into the Cr-R bonds of the $CpCr(NO)_2R$ complexes (R = Me, CH_2SiMe_3, Ph) to produce cationic RNO complexes *(248,249)*:

$$\text{Cr—R} + \text{NO}^+ \longrightarrow \left[\text{Cr}\leftarrow\text{N(=O)—R}\right]^+ \qquad (7.85)$$

When R = Me, a rearrangement of the RNO ligand occurs to generate the corresponding formaldoxime complex (eq 7.86).

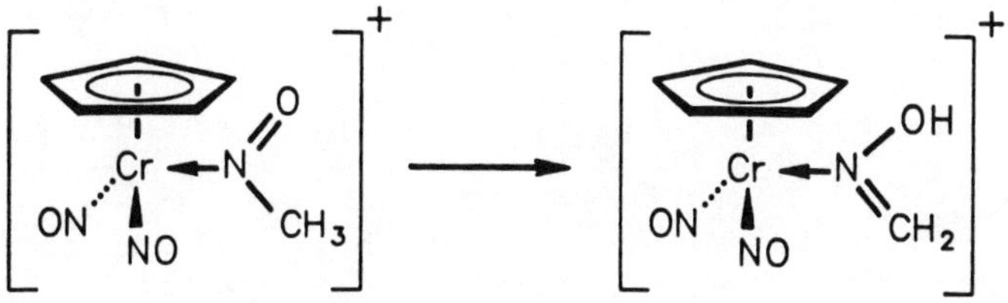

(7.86)

Studies with the perdeuterated Cr-CD_3 complex show that the hydroxyl proton does indeed originate from the methyl group. The aryldiazonium cation $[p\text{-}O_2NC_6H_4N_2]^+$ also undergoes an insertion into the Cr-R bonds of these complexes (R = Me, Ph) with the formation of the *cis*-diazene product (eq 7.87).

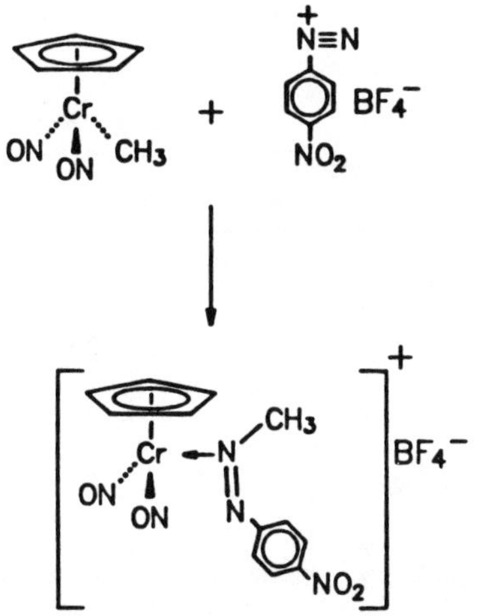

(7.87)

The resulting RNO or RN=NR' ligands formed in eqs 7.85 and 7.87 may then be displaced from the coordination spheres of the metals, resulting in an overall stoichiometric cycle for C-N bond formation, as shown below for the case of NO^+:

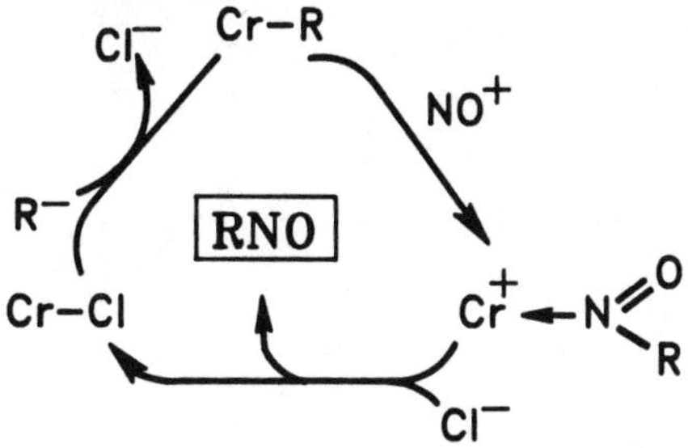

The insertion of the 'NO' moiety into a metal-carbon bond of a cluster compound has also been observed *(250)*.

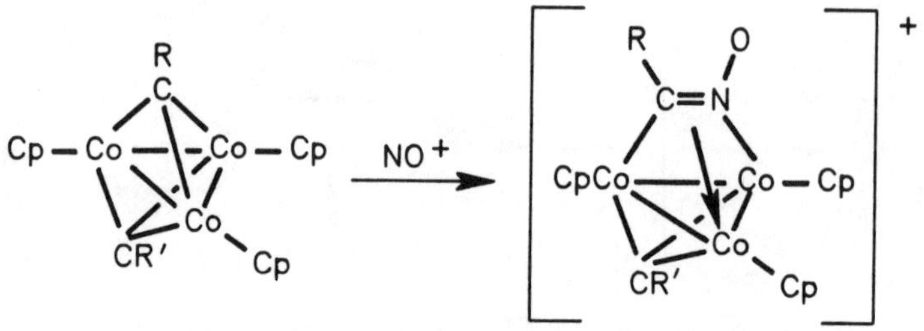

(7.88)

(R, R' = Me, Ph, H, Pr, Bu, CO_2Me)

Carbon-nitrogen bond formation also occurs when a Pt-nitrosobenzene complex is reacted with the nitrosonium cation *(251)*.

$$Pt(PhNO)(PPh_3)_2 + NO^+ \longrightarrow [Pt(ONN(Ph)O)(PPh_3)_2]^+ \quad (7.89)$$

This latter reaction represents a formal insertion of the nitrosonium ion into a Pt-N bond.

7.4.2.6: Transfer of Coordinated Ligands

Some Group 6 metal nitrosyl carbenes serve as unique carbene-transfer agents for the synthesis of otherwise unobtainable metal carbenes *(252,253)* (eq 7.90).

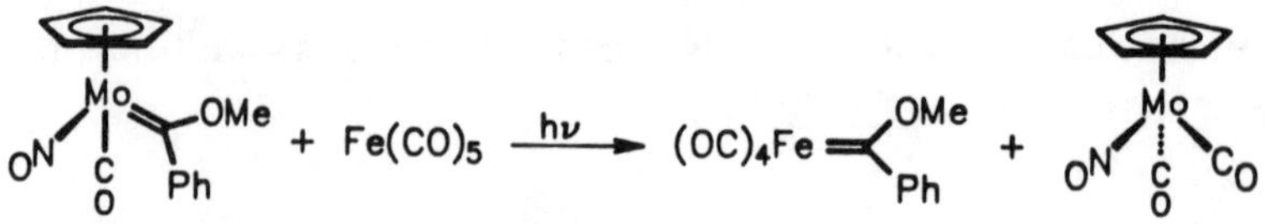

(7.90)

The carbene-transfer to $Ni(CO)_4$ also occurs, yielding the analogous Ni-carbene complex.

Acyl group transfer from a cobalt complex also occurs to achieve the acylation of allylic halides, conjugated enones, and quinones *(254)*, i.e.,

$Co(NO)(CO)_2(PPh_3)$ —R⁻ (−40°)→ [RC(O)–$Co(NO)(CO)(PPh_3)$]⁻ (R=Me,Bu)

R′ R′ O R″ R R′ O

R′ X

R′ R O + R′ R′

HO R O O (R=Me)

The tricarbonylnitrosylferrate anion, $[Fe(NO)(CO)_3]^-$, catalyzes the alkylation of allylic carbonate with malonate anion. Good regioselectivity is achieved, since the nucleophile attacks the carbon bearing the leaving group *(255)* i.e.,

R R R OCO_2R —$NaCH(CO_2Me)_2$→ R R R $CH(CO_2Me)_2$ (7.91)

It is believed that a σ-allyl complex of the type (σ-allyl)$Fe(NO)(CO)_2$ forms upon initial coordination of the allylic carbonate as shown *(255)*.

CO

[Fe − CO]⁻

OCO_2R

− CO

[(CHX_2)Fe]⁻

CHX_2

[(OCO_2R)Fe]⁻

[CHX_2]⁻

− [OCO_2R]⁻

(σ − allyl)Fe

Fe ≡ $Fe(CO)_2NO$

X = CO_2Me

It has also been shown that (allyl)$Fe(NO)(CO)_2$ complexes transfer their coordinated allyl groups to many organic functionalities such as allylic and acyl halides *(256)* to give 1,5-dienes and β,γ-unsaturated ketones, respectively *(257)*. The related $Fe(CO)_2(NO)_2$ and $Co(CO)_3NO$ complexes also catalyze the alkylation of crotyl and α-methyallyl acetates and chlorides by sodium dimethylmalonate *(258)*. Other reactions of coordinated allyls have already been discussed in Section 4.7.

The $[Fe(NO)(CO)_3]^-$ complex is also an efficient reagent or catalyst for the high yield carbonylation of benzyl, allyl, and alkyl halides *(259)* (eq 7.92), possibly via the initial alkylation of the Fe complex followed by CO insertion, and finally attack of methoxide to form the methyl ester (eq 7.93).

$$RX \xrightarrow{CO, K_2CO_3, MeOH} RCO_2Me \qquad (7.92)$$

$$[Fe]^- \xrightarrow{RX} Fe{-}R \xrightarrow{CO} Fe{-}C(=O)R \xrightarrow{OMe^-} RCO_2Me + [Fe]^- \qquad (7.93)$$

Lastly, $Co(NO)(CO)_3$ is an efficient catalyst for the phase transfer catalyzed carbonylation of alkyl halide to carboxylic acids *(260)*,

$$ArCH_2X + CO \xrightarrow[R_4NCl]{NaOH} ArCH_2CO_2H \qquad (7.94)$$

a process which probably occurs via the mechanism shown on the next page (in this mechanism, the $Co(NO)(CO)_3$ complex is simply represented as Co-CO).

The dinitrosyl complex, $Ru(NO)_2(PPh_3)_2$, is a catalyst for the transformation of benzyl bromides to oxime ethers under phase-transfer conditions and in the presence of NO *(261)*.

$$Co{-}CO \xrightarrow{R_4NOH} [Co{-}COOH]^- \xrightarrow{-CO_2} [Co{-}H]^-$$

$[Co{-}H]^-$ → (RX) → $RCo{-}H$ → (CO) → $RC(O){-}Co{-}H$ → (OH^-) → RCO_2H + $[Co{-}H]^-$

$Co \equiv Co(CO)_2NO$

7.4.2.7: Oxidative Additions Involving H-H and C-H Bonds

The electron-rich $M(NO)(PR_3)_3$ complexes of the Group 9 metals (M = Co, Rh, Ir; R = alkyl, aryl) undergo oxidative-addition reactions *(262,263)*. These complexes typically possess tetrahedral coordination geometries in solution, and are not prone to ligand dissociation reactions which would result in the formation of the coordinatively unsaturated "$M(NO)(PR_3)_2$" species. Nevertheless, the Rh complexes are moderately active homogeneous catalysts for the hydrogenation of olefins, allenes, and non-conjugated dienes. Presumably, the catalytically active Rh-dihydride species is formed reversibly from the oxidative addition of H_2 to Rh, with the concomitant loss of phosphine due to its slight degree of dissociation. Consistent with this proposed pathway is that addition of phosphine, or the use of the tripodal, chelating $MeC(CH_2PPh_2)_3$ phosphine, inhibits the reaction. The catalytic activity for $Rh(NO)(PR_3)_3$ decreases in the order $P(p\text{-tol})_3 > PPh_3 > P(p\text{-}C_6H_4F)_3 > PMePPh_2$.

An investigation of the catalytic hydrogenation of alkenes by the closely related $Rh(NO)(PPh_3)_2(L)$ (L = *p*-benzoquinone) compound *(264)* reveals that the reaction is very complex. Although the homogeneous catalytic activity of the Rh nitrosyl complex is extremely low, the *decomposition* of the "$Rh(NO)(PPh_3)_2$" intermediate produces a species which is catalytically very active. Thus, the initial homogeneous process is very slow but is contaminated after a certain period by a heterogeneous process of higher activity due to the separation of a finely divided solid, probably finely divided metallic Rh *(264)*.

A number of electron-rich alkyl hydride complexes of tungsten initiate the *intramolecular* and *intermolecular* activation of C-H bonds under mild conditions *(265)*.

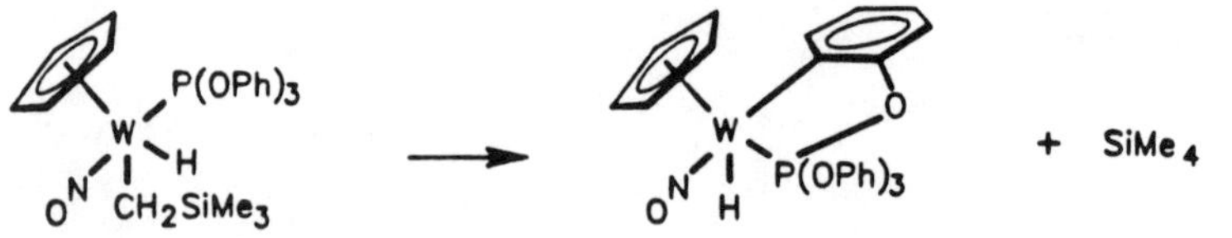

(7.95)

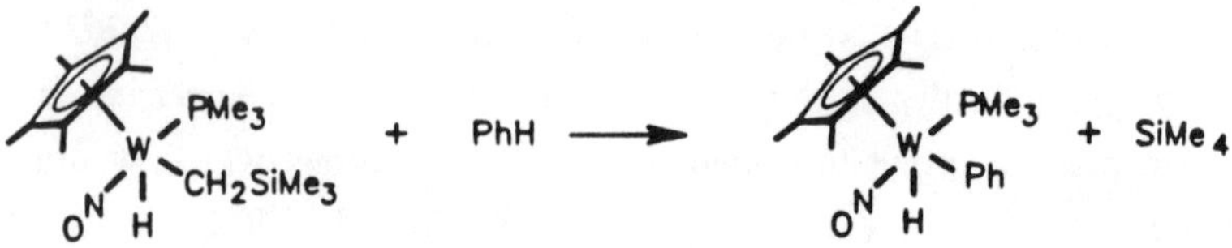

(7.96)

The bimetallic complex shown below results from the cyclometallation of two $[Ru(SR)_4NO]^-$ fragments. The reaction formally involves the C-H activation of ortho methyl groups of the SR ligands (R = Me_4C_6H or $Me_2C_6H_3$) *(180)*.

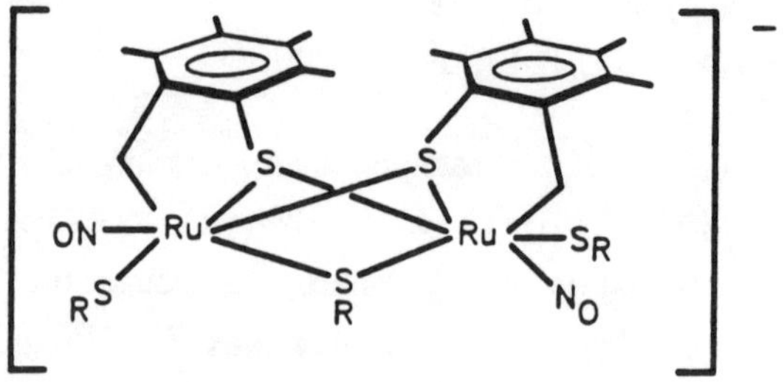

It has been proposed that oxidative addition of an ortho C-H bond at the vacant sixth coordination site of the anion induces the reductive elimination of one molecule of RSH. This is then followed by dimerization to produce the complex shown.

It has also been observed that the hydride, $HFe(NO)(CO)_2(PPh_3)$, is an active stoichiometric reagent for the hydrogenation of olefins *(266)*. Finally, a heterobimetallic Re-Rh complex (based on a chiral Re template) catalyzes the asymmetric hydrogenation of enamide precursors to α-amino acids and esters (R = H, Me, Et, Ph) *(267)*.

$$RCH{=}C(COR')(NHC(O)Me) \xrightarrow{H_2} RCH_2{-}CH(COR')(NHC(O)Me) \quad (7.97)$$

7.5: References

1. Bottomley, F. *React. Coord. Ligands* **1989**, *2*, 115.
2. Masek, J. *Inorg. Chim. Acta Rev.* **1969**, *3*, 99.
3. Eisenberg, R.; Meyer, C. D. *Acc. Chem. Res.* **1975**, *8*, 26.
4. Masek, J. *Chem. Listy* **1980**, *74*, 751.
5. Johnson, B. F. G.; Haymore, B. L.; Dilworth, J. R. In *Comprehensive Coordination Chemistry*: Wilkinson, G.; Gillard, R. D.; McCleverty, J. A., Eds.; Pergamon: Oxford, 1987; Vol. 2, pp 99-118.
6. Pandey, K. K. *Coord. Chem. Rev.* **1983**, *51*, 69.
7. Eisenberg, R.; Hendrickson, D. E. *Adv. Catal.* **1979**, *28*, 79.
8. McCleverty, J. A. *Chem. Rev.* **1979**, *79*, 53.
9. Bottomley, F.; Darkwa, J.; White, P. S. *J. Chem. Soc. Dalton Trans.* **1985**, 1435.
10. Bottomley, F.; Darkwa, J.; White, P. S. *J. Chem. Soc., Chem. Commun.* **1982**, 1039.
11. Mason, J.; Mingos, D. M. P.; Sherman, D.; Wardle, R. W. M. *J. Chem. Soc., Chem. Commun.* **1984**, 1223.

12. Mingos, D. M. P.; Sherman, D. J.; Bott, S. G. *Transition Met. Chem. (London)* **1987**, *12(5)*, 471.
13. Mingos, D. M. P.; Sherman, D. J.; Williams, I. D. *Transition Met. Chem. (London)* **1987**, *12(6)*, 493.
14. Schoonover, M. W.; Baker, E. C.; Eisenberg, R. *J. Am. Chem. Soc.* **1979**, *101*, 1880.
15. Georgiou, S.; Wight, C. A. *J. Phys. Chem.* **1990**, *94*, 4935.
16. Evans, W.; Zink, J. I. *J. Am. Chem. Soc.* **1981**, *103*, 2635.
17. Georgiou, S.; Wight, C. A. *Chem. Phys. Lett.* **1986**, *132*, 511.
18. Herberhold, M.; Kremnitz, W.; Trampisch, H.; Hitam, R. B.; Rest, A. J.; Taylor, D. J. *J. Chem. Soc., Dalton Trans.* **1982**, 1261.
19. Geiger, W. E.; Rieger, P. H.; Tulyathan, B.; Rausch, M. D. *J. Am. Chem. Soc.* **1984**, *106*, 7000.
20. Satija, S. K.; Swanson, B. I.; Crichton, O.; Rest, A. J. *Inorg. Chem.* **1978**, *17*, 1737.
21. Crichton, O.; Rest, A. J. *J. Chem. Soc., Dalton Trans.* **1978**, 202.
22. Crichton, O.; Rest, A. J. *J. Chem. Soc., Dalton Trans.* **1978**, 208.
23. Basolo, F. *Polyhedron* **1990**, *9*, 1503.
24. Enemark, J. H.; Feltham, R. D.; Riker-Nappier, J.; Bizot, K. F. *Inorg. Chem.* **1975**, *14*, 624.
25. Kirchner, R. M.; Marks, T. J.; Kristoff, J. S.; Ibers, J. A. *J. Am. Chem. Soc.* **1973**, *95*, 6602.
26. Marks, T. J.; Kristoff, J. S. *J. Organomet. Chem.* **1972**, *42*, C91.
27. Adams, R. D.; Cotton, F. A. *J. Am. Chem. Soc.* **1973**, *95*, 6594.
28. Herrmann, W. A.; Bauer, C. *J. Organomet. Chem.* **1981**, *204*, C21.
29. Fjeldsted, D. O. K.; Stobart, S. R.; Zaworotko, M. J. *J. Am. Chem. Soc.* **1985**, *107*, 8258.
30. Adams, D. M. *Metal-Ligand and Related Vibrations*; Edward Arnold: London, U.K., 1967; pp 84-111 and 268-271.
31. Foffani, A.; Poletti, A.; Cataliotti, R. *Spectrochim. Acta* **1968**, *24A*, 1437.
32. Haymore, B. L.; Ibers, J. A. *Inorg. Chem.* **1975**, *14*, 3060.
33. Horrocks, W. D., Jr.; Mann, R. H. *Spectrochim. Acta* **1965**, *21*, 399.
34. Fairey, M. B.; Irving, R. J. *Spectrochim. Acta* **1964**, *20*, 1757.
35. Gutmann, V. *The Donor Acceptor Approach to Molecular Interaction*; Plenum: New York, 1978.
36. Mu, X. H.; Kadish, K. M. *Inorg. Chem.* **1988**, *27*, 4720.

37. Hunter, A. D.; Legzdins, P. *Organometallics* **1986**, *5*, 1001.
38. Kadish, K. M.; Cornillon, J. L.; Yao, C. L.; Malinski, T.; Gritzner, G. *J. Electroanal. Chem. Interfacial Electrochem.* **1987**, *235*, 189.
39. Titov, V. A.; Pashkin, V. A. *Zh. Obshch. Khim.* **1987**, *57*, 865. CA108(25):221848h.
40. Onaka, S. *Inorg. Chem.* **1980**, *19*, 2132.
41. Rausch, M. D.; Mintz, E. A.; Macomber, D. W. *J. Org. Chem.* **1980**, *45*, 689.
42. Ginzburg, A. G.; Setkina, V. N.; Kursanov, D. N. *Izv. Akad. Nauk SSSR, Ser. Khim.* **1985**, 447. CA103(13):105099n.
43. Lokshin, B. V.; Rusach, E. B.; Kolobova, N. E.; Makarov, Y. V.; Ustynyuk, N. A.; Zdanovich, V. I.; Zhakaeva, A. Z; Setkina, V. N. *J. Organomet. Chem.* **1976**, *108*, 353.
44. Crease, A. E.; Legzdins, P. *J. Chem. Soc., Dalton Trans.* **1973**, 1501.
45. Legzdins, P.; Rettig, S. J.; Sanchez, L. *Organometallics* **1988**, *7*, 2394.
46. Lee, K. E.; Arif, A. M.; Gladysz, J. A. *Chem. Ber.* **1991**, *124*, 309.
47. Lee, K. E.; Arif, A. M.; Gladysz, J. A. *Inorg. Chem.* **1990**, *29*, 2885.
48. Lokshin, B. V.; Kazaryan, S. G.; Ginzburg, A. G. *Izv. Akad. Nauk SSSR, Ser. Khim.* **1987**, 948. CA108(11):94715k.
49. Stevens, R. E.; Guettler, R. D.; Gladfelter, W. L. *Inorg. Chem.* **1990**, *29*, 451.
50. Legzdins, P.; Nurse, C. R.; Rettig, S. J. *J. Am. Chem. Soc.* **1983**, *105*, 3727.
51. Delgado, E.; Jeffery, J. C.; Simmons, N. D.; Stone, F. G. A. *J. Chem. Soc., Dalton Trans.* **1986**, 869.
52. Stevens, R. E.; Gladfelter, W. L. *J. Am. Chem. Soc.* **1982**, *104*, 6454.
53. Mronga, N.; Weller, F.; Dehnicke, K. *Z. Anorg. Allg. Chem.* **1983**, *502*, 35.
54. Sellmann, D.; Keller, J.; Moll, M.; Beck, H. P.; Milius, W. *Z. Naturforsch., B: Anorg. Chem., Org. Chem.* **1986**, *41B*, 1551.
55. Sellmann, D.; Keller, J.; Moll, M.; Campana, C. F.; Haase, M. *Inorg. Chim. Acta* **1988**, *141*, 243.
56. Bhaduri, S.; Johnson, B. F. G.; Savory, C. J.; Segal, J. A.; Walter, R. H. *J. Chem. Soc., Chem. Commun.* **1974**, 809.
57. McPhail, A. T.; Knox, G. R.; Robertson, C. G.; Sim, G. A. *J. Chem. Soc. (A)* **1971**, 205.

58. Hitam, R. B.; Rest, A. J.; Herberhold, M.; Kremnitz, W. *J. Chem. Soc., Chem. Commun.* **1984**, 471.
59. Gladfelter, W. L. *Adv. Organomet. Chem.* **1985**, *24*, 41.
60. Adams, R. D.; Horváth, I. T. *Prog. Inorg. Chem.* **1985**, *33*, 127.
61. Collins, M. A.; Johnson, B. F. G.; Lewis, J.; Mace, J. M.; Morris, J.; McPartlin, M.; Nelson, W. J. H.; Puga, J.; Raithby, P. R. *J. Chem. Soc., Chem. Commun.* **1983**, 689.
62. Attard, J. P.; Johnson, B. F. G.; Lewis, J.; Mace, J. M.; Raithby, P. R. *J. Chem. Soc., Chem. Commun.* **1985**, 1526.
63. Fjare, D. E.; Gladfelter, W. L. *J. Am. Chem. Soc.* **1984**, *106*, 4799.
64. Gibson, C. P.; Dahl, L. F. *Organometallics* **1988**, *7*, 543.
65. Feasey, N. D.; Knox, S. A. R. *J. Chem. Soc., Chem. Commun.* **1982**, 1062.
66. Sellmann, D.; Binker, G. *Z. Naturforsch., B: Chem. Sci.* **1987**, *42*, 341.
67. Legzdins, P.; Rettig, S. J.; Ross, K. J.; Veltheer, J. E. *J. Am. Chem. Soc.* **1991**, *113*, 4361.
68. McGilligan, B. S.; Arnold, J.; Wilkinson, G.; Hussain-Bates, B.; Hursthouse, M. B. *J. Chem. Soc., Dalton Trans.* **1990**, 2465.
69. McCleverty, J. A. *J. Mol. Catal.* **1981**, *13*, 309.
70. Bottomley, F. *Acc. Chem. Res.* **1978**, *11*, 158.
71. Bottomley, F.; Brooks, W. V. F.; Clarkson, S. G.; Tong, S.-B. *Chem. Commun.* **1973**, 919.
72. Tatsumi, T.; Sekizawa, K.; Tominaga, H. *Bull. Chem. Soc. Jpn.* **1980**, *53*, 2297.
73. Grundy, K. R.; Reed, C. A.; Roper, W. R. *J. Chem. Soc., Chem. Commun.* **1970**, 1501.
74. La Monica, G.; Freni, M.; Cenini, S. *J. Organomet. Chem.* **1974**, *71*, 57.
75. Wilson, R. D.; Ibers, J. A. *Inorg. Chem.* **1979**, *18*, 336.
76. Sellmann, D.; Seubert, B.; Moll, M.; Knoch, F. *Angew. Chem., Int. Ed. Engl.* **1988**, *27*, 1164.
77. Swinehart, J. H. *Coord. Chem. Rev.* **1967**, *2*, 385.
78. Wasielewska, E. *Inorg. Chim. Acta* **1986**, *113*, 115.
79. Manoharan, P. T.; Gray, H. B. *Inorg. Chem.* **1966**, *5*, 823.
80. Banyai, I.; Bodi, Z.; Dozsa, L.; Beck, M. *Magy. Kem. Foly.* **1988**, *94*, 394. CA111(2): 13077r.

81. Butler, A. R.; Glidewell, C.; Reglinski, J.; Waddon, A. *J. Chem. Res., Synop.* **1984**, *(9)*, 279.

82. Fiedler, J.; Masek, J. *Inorg. Chim. Acta* **1985**, *105*, 83.

83. Dozsa, L.; Kormos, V.; Beck, M. T. *Inorg. Chim. Acta* **1984**, *82*, 69.

84. Dozsa, L.; Kormos, V.; Beck, M. *Magy. Kem. Foly.* **1983**, *89*, 97. CA98(22):186366a.

85. Katz, N. E.; Blesa, M. A.; Olabe, J. A.; Aymonino, P. J. *J. Inorg. Nucl. Chem.* **1980**, *42*, 581.

86. Katho, A.; Bodi, Z.; Dozsa, L.; Beck, M. T. *Inorg. Chim. Acta* **1984**, *83*, 145.

87. Katho, A.; Bodi, Z.; Dozsa, L.; Beck, M. *Magy. Kem. Foly.* **1983**, *89*, 102. CA98(22):186367b.

88. Butler, A. R.; Glidewell, C.; Chaipanich, V.; McGinnis, J. *J. Chem. Soc., Perkin Trans. 2* **1986**, 7.

89. Butler, A. R.; Calsy-Harrison, A. M.; Glidewell, C.; Soerenson, P. E. *Polyhedron* **1988**, *7*, 1197.

90. Morando, P. J.; Borghi, E. B.; De Schteingart, L. M.; Blesa, M. A. *J. Chem. Soc., Dalton Trans.* **1981**, 435.

91. Gol'tsov, Y. G.; Zhilinskaya, V. V. *Russ. J. Inorg. Chem.* **1989**, *63*, 1596.

92. (a) Butler, A. R.; Calsy-Harrison, A. M.; Glidewell, C.; Johnson, I. L. *Inorg. Chim. Acta* **1988**, *146*, 187. (b) Chevalier, A. A.; Gentil, L. A.; Olabe, J. A. *J. Chem. Soc., Dalton Trans.* **1991**, 1959.

93. Walsh, J. L.; Bullock, R. M.; Meyer, T. J. *Inorg. Chem.* **1980**, *19*, 865.

94. Chakravarty, A. R.; Chakravorty, A. *J. Chem. Soc., Dalton Trans.* **1982**, 1765.

95. Bottomley, F.; White, P. S.; Mukaida, M.; Shimura, K.; Kakihana, H. *J. Chem. Soc., Dalton Trans.* **1988**, 2965.

96. Chakravarty, A. R.; Chakravorty, A. *J. Chem. Soc., Dalton Trans.* **1983**, 961.

97. Stershic, M. T.; Keefer, L. K.; Sullivan, B. P.; Meyer, T. J. *J. Am. Chem. Soc.* **1988**, *110*, 6884.

98. Bottomley, F.; Brooks, W. V. F.; Paez, D. E.; White, P. S.; Mukaida, M. *J. Chem. Soc., Dalton Trans.* **1983**, 2465.

99. Fenske, R. F.; Milletti, M. C. *Organometallics* **1986**, *5*, 1243.

100. Mueller, J.; de Oleveira, G. M.; Pickardt, J. *J. Organomet. Chem.* **1987**, *329*, 241.

101. Johnson, B. F. G.; Bhaduri, S.; Connelly, N. G. *J. Organomet. Chem.* **1972**, *40*, C36.

102. Rhodes, M. R.; Barley, M. H.; Meyer, T. J. *Inorg. Chem.* **1991**, *30*, 629.

103. Murphy, W. R., Jr.; Takeuchi, K.; Barley, M. H.; Meyer, T. J. *Inorg. Chem.* **1986**, *25*, 1041.

104. Müller, J.; Dorner, H.; Kohler, F. H. *Chem. Ber.* **1973**, *106*, 1122.

105. Gallop, M. A.; Rickard, C. E. F.; Roper, W. R. *J. Organomet. Chem.* **1984**, *269*, C21.

106. Boyar, E. B.; Dobson, A.; Robinson, S. D.; Haymore, B. L.; Huffman, J. C. *J. Chem. Soc., Dalton Trans.* **1985**, 621.

107. Haymore, B.; Huffman, J. C.; Dobson, A.; Robinson, S. D. *Inorg. Chim. Acta* **1982**, *65*, L231.

108. Becker, P. N.; Bergman, R. G. *J. Am. Chem. Soc.* **1983**, *105*, 2985.

109. Becker, P. N.; Bergman, R. G. *Organometallics* **1983**, *2*, 787.

110. Becker, P. N.; White, M. A.; Bergman, R. G. *J. Am. Chem. Soc.* **1980**, *102*, 5676.

111. Brunner, H.; Loskot, S. *J. Organomet. Chem.* **1973**, *61*, 401.

112. Jorgensen, K. A.; Hoffmann, R. *J. Am. Chem. Soc.* **1986**, *108*, 1867.

113. McCleverty, J. A.; Nennes, C. W.; Wolochowicz, I. *J. Chem. Soc., Dalton Trans.* **1986**, 743.

114. Klein, H.-F.; Karsch, H. H. *Chem. Ber.* **1976**, *109*, 1453.

115. Middleton, A. R.; Wilkinson, G. *J. Chem. Soc., Dalton Trans.* **1981**, 1898.

116. Middleton, A. R.; Wilkinson, G. *J. Chem. Soc., Dalton Trans.* **1980**, 1888.

117. Jones, C. J.; McCleverty, J. A.; Rothin, A. S. *J. Chem. Soc., Dalton Trans.* **1985**, 405.

118. Weiner, W. P.; Bergman, R. G. *J. Am. Chem. Soc.* **1983**, *105*, 3922.

119. Seidler, M. D.; Bergman, R. G. *Organometallics* **1983**, *2*, 1897.

120. Diel, B. N. *J. Organomet. Chem.* **1985**, *284*, 257.

121. Chang, J.; Seidler, M. D.; Bergman, R. G. *J. Am. Chem. Soc.* **1989**, *111*, 3258.

122. Ghedini, M.; Longeri, M.; Neve, F.; Lanfredi, A. M. M.; Tiripichio, A. *J. Chem. Soc., Dalton Trans.* **1989**, 1217.

123. Ghedini, M.; Lanfredi, A. M. M.; Neve, F.; Tiripicchio, A. *J. Chem. Soc., Chem. Commun.* **1987**, 847.
124. Foa, M.; Cassar, L. *J. Organomet. Chem.* **1971**, *30*, 123.
125. Clamp, S.; Connelly, N. G.; Howard, J. A. K.; Manners, I.; Payne, J. D.; Geiger, W. E. *J. Chem. Soc., Dalton Trans.* **1984**, 1659.
126. Cameron, M.; Gowenlock, B. G.; Vasapollo, G. *Chem. Soc. Rev.* **1990**, *19*, 355.
127. Danilina, L. I.; Iretskii, A. V.; Kukushkin, Y. N. *Russ. J. Inorg. Chem.* **1985**, *30*, 1073.
128. Iretskii, A. V.; Kukushkin, V. Y. *Zh. Obsch. Khim.* **1986**, *56*, 480. CA105(16):144914k.
129. Nagao, H.; Mukaida, M.; Shimizu, K.; Howell, F. S.; Kakihana, H. *Inorg. Chem.* **1986**, *25*, 4312.
130. Wah, H. L. K.; Postel, M.; Tomi, F. *Inorg. Chem.* **1989**, *28*, 233.
131. Laing, K. R.; Roper, W. R. *J. Chem. Soc., Chem. Commun.* **1968**, 1568.
132. Grundy, K. R.; Laing, K. R.; Roper, W. R. *J. Chem. Soc., Chem. Commun.* **1970**, 1500.
133. Chen, T. L.; Venkatasubramanian, P. N.; Basolo, F. *J. Phys. Chem.* **1984**, *88*, 477.
134. English, R. *Polyhedron* **1989**, *8*, 2565.
135. Bhaduri, S. A.; Bratt, I.; Johnson, B. F. G.; Khair, A.; Segal, J. A.; Walters, R.; Zuccaro, C. *J. Chem. Soc., Dalton Trans.* **1981**, 234.
136. Solar, J. P.; Mares, F.; Diamond, S. E. *Catal. Rev.-Sci. Eng.* **1985**, *27*, 1.
137. Tomi, F.; Wah, H. L. K.; Postel, M. *New J. Chem.* **1988**, *12*, 289.
138. (a) Leising, R. A.; Takeuchi, K. J. *J. Am. Chem. Soc.* **1988**, *110*, 4079. (b) Leising, R. A.; Kubow, S. A.; Takeuchi, K. J. *Inorg. Chem.* **1990**, *29*, 4569.
139. Gordon, C. M.; Feltham, R. D.; Turner, J. J. *J. Phys. Chem.* **1991**, *95*, 2889.
140. Tovrog, B. S.; Diamond, S. E.; Mares, F. *J. Am. Chem. Soc.* **1979**, *101*, 270.
141. Tovrog, B. S.; Diamond, S. E.; Mares, F. *US Patent 4322562 A*, 1982. CA97(11):91716e.
142. Andrews, M. A.; Chang, T. C. T.; Cheng, C. W. F. *Organometallics* **1985**, *4*, 268.
143. Ercolani, C.; Pennesi, G. *Inorg. Chim. Acta* **1985**, *101*, L41.

144. Tovrog, B. S.; Mares, F.; Diamond, S. E. *J. Am. Chem. Soc.* **1980**, *102*, 6616.
145. Mares, F.; Diamond, S. E *Fundam. Res. Homogeneous Catal.* **1984**, *4*, 55.
146. Tovrog, B. S.; Mares, F.; Diamond, S. E. *Chem. Ind. (Dekker)* **1981**, *5*, 139. CA98(25):215001k.
147. Andrews, M. A.; Cheng, C.-W. F. *J. Am. Chem. Soc.* **1982**, *104*, 4268.
148. Diamond, S. E.; Mares, F.; Szalkiewicz, A.; Muccigrosso, D. A.; Solar, J. P. *J. Am. Chem. Soc.* **1982**, *104*, 4266.
149. Bhaduri, S.; Sarma, K. R.; Narayan, B. A. *Transition Met. Chem. (Weinheim)* **1981**, *6*, 206.
150. Andrews, M. A.; Kelly, K. P. *J. Am. Chem. Soc.* **1981**, *103*, 2894.
151. Bäckvall, J.-E.; Heumann, A. *J. Am. Chem. Soc.* **1986**, *108*, 7107.
152. Filatov, M. J.; Gritsenko, O. V.; Zhidomirov, G. M. *J. Mol. Catal.* **1989**, *54*, 462.
153. Gusevskaya, E. V.; Beck, I. E.; Stepanov, A. G.; Likholobov, V. A.; Nekipelov, V. M.; Yermakov, Y. I.; Zamaraev, K. I. *J. Mol. Catal.* **1986**, *37*, 177.
154. Zavorokhina, Z. M.; Levchenko, L. V. *Izv. Akad. Nauk Kaz. SSR, Ser. Khim* **1983**, *(5)*, 69. CA99(26):224155m.
155. Beck, I. E.; Gusevskaya, E. V.; Stepanov, A. G.; Likholobov, V. A.; Nekipelov, V. M.; Yermakov, Y. I.; Zamaraev, K. I. *J. Mol. Catal.* **1989**, *50*, 167.
156. Legzdins, P.; Phillips, E. C.; Rettig, S. J.; Sanchez, L.; Trotter, J.; Yee, V. C. *Organometallics* **1988**, *7*, 1877.
157. Herberhold, M.; Kremnitz, W.; Razavi, A.; Schöllhorn, H.; Thewalt, U. *Angew. Chem., Int. Ed. Engl.* **1985**, *24*, 601.
158. Bottomley, F.; Sutin, L. *Adv. Organomet. Chem.* **1988**, *28*, 339.
159. Palocsay, F. A.; Rund, J. V. *Inorg. Chem.* **1969**, *8*, 696.
160. Innorta, G.; Torroni, S.; Foffani, A. *J. Organomet. Chem.* **1974**, *66*, 459.
161. Ileperuma, O. A.; Feltham, R. D. *Inorg. Chem.* **1977**, *16*, 1876.
162. Maejima, T.; Miki, E.; Tanaka, M.; Tezuka, H.; Mizumachi, K.; Ishimori, T. *Bull. Chem. Soc. Jpn.* **1990**, *63*, 1596.
163. Nikol'skii, A. B.; Ershov, A. Y.; Vasilevskii, I. V.; Suvorov, A. V.; Yurkova, O. E. *Koord. Khim.* **1980**, *6*, 987. CA93(12):121012j.
164. Pandey, K. K.; Agarwala, U. C. *Indian J. Chem., Sect. A* **1981**, *20A*, 74.

165. Pandey, K. K.; Agarwala, U. C. *Inorg. Chem.* **1981**, *20*, 1308.

166. Hursthouse, M. B.; Walker, N. P. C.; Warrens, C. P.; Woollins, J. D. *J. Chem. Soc., Dalton Trans.* **1985**, 1043.

167. Masuda, H.; Miyokawa, K.; Masuda, I. *Thermochim. Acta* **1985**, *84*, 337.

168. Stochel, G. *Zesz. Nauk. Uniw. Jagiellon., Pr. Chem.* **1985**, *29*, 97. CA103(14):113221m.

169. Nikol'skii, A. B.; Popov, A. M. *Dokl. Akad. Nauk SSSR* **1980**, *250*, 902. CA93(2):16835g.

170. Mohai, B.; Horvath, A.; Honti, P. E. *J. Therm. Anal.* **1986**, *31*, 157.

171. Horvath, A.; Mohai, B. *J. Inorg. Nucl. Chem.* **1980**, *42*, 195.

172. Uehara, A.; Kitayama, M.; Suzuki, M.; Tsuchiya, R. *Inorg. Chem.* **1985**, *24*, 2184.

173. Georgiou, S.; Wight, C. A. *J. Chem Phys.* **1989**, *90*, 1694.

174. Kubota, M.; Chan, M. K.; Boyd, D. C.; Mann, K. R. *Inorg. Chem.* **1987**, *26*, 3261.

175. Herberhold, M.; Trampisch, H. *Inorg. Chim. Acta* **1983**, *70*, 143.

176. Fischer, E. O.; Strametz, H. *J. Organomet. Chem.* **1967**, *10*, 323.

177. Seyferth, K.; Rosenthal, K; Kuehn, G.; Taube, R. *Z. Anorg. Allg. Chem.* **1984**, *513*, 57.

178. Sellmann, D.; Ludwig, W.; Huttner, G.; Zsolnai, L. *J. Organomet. Chem.* **1985**, *294*, 199.

179. Lloyd, M. K.; McCleverty, J. A. *J. Organomet. Chem.* **1973**, *61*, 261.

180. Soong, S. L.; Hain, J. H., Jr.; Millar, M.; Koch, S. A. *Organometallics* **1988**, *7*, 556.

181. Roustan, J. L.; Ansari, N. Ahmed, F. R. *Inorg. Chim. Acta* **1987**, *129*, L11.

183. Greenwood, N. N.; Earnshaw, A. *Chemistry of the Elements*; Pergamon: Oxford, U. K., 1984; p 512.

184. Bencze, L.; Kohan, J.; Mohai, B. *Acta Chim. Hung.* **1983**, *113*, 183.

185. Mohai, B.; Gyoryova, K.; Bencze, L. *J. Therm. Anal.* **1979**, *17*, 159.

186. Gyoryova, K.; Mohai, B.; Bencze, L. *Thermochim. Acta* **1979**, *33*, 169.

187. Baev, A. K.; Gubar, Y. L.; Gasanov, K. S. *Russ. J. Phys. Chem.* **1982**, *56*, 1490.

188. Hightower, J. W.; Van Leirsburg, D. A. In *The Catalytic Chemistry of Nitrogen Oxides*; Klimisch, R. L.; Larson, J. G., Eds.; Plenum: New York, 1975; p 63.

189. Olevskii, S. S.; Tolstikhina, A. L.; Repko, V. P.; Sergeev, M. S. *Poverkhnost* **1983**, *(3)*, 108. CA98(22):189196f.

190. Gasanov, K. S.; Strizhkov, B. V.; Kozyrkin, B. I.; Kurbanov, T. K.; Koshchienko, A. V. *Izv. Akad. Nauk SSSR, Neorg. Mater.* **1983**, *19*, 1682.

191. Moser, W. R. In *The Catalytic Chemistry of Nitrogen Oxides*; Klimisch, R. L.; Larson, J. G., Eds.; Plenum: New York, 1975; pp 33-43.

192. Ambach, E.; Beck, W. *Z. Naturforsch., B: Anorg. Chem., Org. Chem.* **1985**, *40B*, 288.

192. Gwost, D.; Caulton, K. G. *Inorg. Chem.* **1974**, *13*, 414.

193. Gargano, M.; Giannoccaro, P.; Rossi, M.; Sacco, A.; Vasapollo, G. *Gazz. Chim. Ital.* **1975**, *105*, 1279.

194. Nazarenko, Y. P.; Zhilinskaya, V. V.; Tananaiko, Z. Y. *Russ. J. Inorg. Chem.* **1980**, *25*, 1369.

195. Kaduk, J. A.; Tulip, T. H.; Budge, J. R.; Ibers, J. A. *J. Mol. Catal.* **1981**, *12*, 239.

196. Kubota, M.; Evans, K. J.; Koerntgen, C. A.; Marsters, J. C., Jr. *J. Mol. Catal.* **1980**, *7*, 481.

197. Sun, K. S.; Kong, K. C.; Cheng, C. H. *Inorg. Chem.* **1991**, *30*, 1998.

198. Cheng, C. H.; Sun, K. S. *Inorg. Chem.* **1990**, *29*, 2547.

199. Fang, W. P.; Cheng, C. H. *J. Chem. Soc., Chem. Commun.* **1986**, 503.

200. Mol, J. C.; Moulijn, J. A. *Adv. Catal.* **1975**, *24*, 131.

201. Zuech, E. A. *J. Chem. Soc., Chem. Commun.* **1968**, 1182.

202. Zuech, E. A.; Hughes, W. B.; Kubicek, D. H.; Kittleman, E. T. *J. Am. Chem. Soc.* **1970**, *92*, 528.

203. Hughes, W. B. *J. Am. Chem. Soc.* **1970**, *92*, 532.

204. Grubbs, R. H.; Hoppin, C. R. *J. Chem. Soc., Chem. Commun.* **1977**, 634.

205. Seyferth, K.; Taube, R. *J. Mol. Catal.* **1985**, *28*, 53.

206. Seyferth, K.; Taube, R. *J. Organomet. Chem.* **1984**, *268*, 155.

207. Masek, J.; Fiedler, J.; Klima, J.; Seyferth, K.; Taube, R. *Collect. Czech. Chem. Commun.* **1982**, *47*, 1721.

208. Seyferth, K.; Taube, R. *J. Organomet. Chem.* **1982**, *229*, 275.

209. Taube, R.; Seyferth, K. *Z. Chem.* **1974**, *14*, 410.

210. Leconte, M.; Taarit, Y. B.; Bilhou, J. L.; Basset, J. M. *J. Mol. Catal.* **1980**, *8*, 263.

211. Crease, A. E.; Egglestone, H.; Taylor, N. *J. Organomet. Chem.* **1982**, *238*, C5.

212. Keller, A. *J. Mol. Catal.* **1991**, *64*, 171.

213. Keller, A.; Szterenber, L. *J. Mol. Catal.* **1989**, *57*, 207.

214. Keller, A. *J. Mol. Catal.* **1989**, *53*, L9.

215. Keller, A. *Inorg. Chim. Acta* **1988**, *149*, 165.

216. Keller, A. *Transition Met. Chem. (London)* **1987**, *12*, 320.

217. Keller, A. *Inorg. Chim. Acta* **1987**, *133*, 207.

218. Keller, A. *J. Organomet. Chem.* **1990**, *393*, 389.

219. Keller, A. *J. Organomet. Chem.* **1991**, *407*, 237.

220. Keller, A. *J. Organomet. Chem.* **1990**, *385*, 285.

221. Grubbs, R. H.; Swetnick, S.; Su, S. C.-H. *J. Mol. Catal.* **1977/8**, *3*, 11.

222. Sen, A. *Acc. Chem. Res.* **1988**, *21*, 421.

223. Sen, A.; Thomas, R. R. *Organometallics* **1982**, *1*, 1251.

224. Legzdins, P.; Martin, D. T. *Organometallics* **1983**, *2*, 1785.

225. Ballivet-Tkatchenko, D.; Bremard, C. *J. Chem. Soc., Dalton Trans.* **1983**, 1143.

226. Ballivet-Tkatchenko, D.; Billard, C.; Revillon, A. *J. Polym. Sci., Polym. Chem. Ed.* **1981**, *19*, 1697.

227. Ballivet-Tkatchenko, D.; Nickel, B.; Rassat, A.; Vincent-Vaucquelin, J. *Inorg. Chem.* **1986**, *25*, 3479.

228. Ballivet-Tkatchenko, D.; Esselin, C.; Goulon, J. *J. Phys.* **1986**, *C8*, 343.

229. Legzdins, P.; Richter-Addo, G. B.; Einstein, F. W. B.; Jones, R. H. *Organometallics* **1990**, *9*, 431.

230. Honeychuck, R. V.; Bonnesen, P. V.; Farahi, J.; Hersh, W. H. *J. Org. Chem.* **1987**, *52*, 5293.

231. Faller, J. W.; Ma, Y. *J. Am. Chem. Soc.* **1991**, *113*, 1579.

232. Mortreaux, A.; Bavay, J. C.; Petit, F. *Nouv. J. Chim.* **1980**, *4(11)*, 671.

233. Huchette, D.; Nicole, J.; Petit, F. *Tetrahedron Lett.* **1979**, *(12)*, 1035.

234. Leroy, E.; Huchette, D.; Mortreux, A.; Petit, F. *Nouv. J. Chim.* **1980**, *4(3)*, 173.

235. El Murr, N.; Tirouflet, J. *Fundam. Res. Homogeneous Catal.* **1979**, *3*, 1007.

236. Tkatchenko, I. *J. Mol. Catal.* **1978**, *4(2)*, 163.

237. Piazza, G.; Innorta, G. *J. Organomet. Chem.* **1982**, *240*, 257.

238. Ballivet-Tkatchenko, D.; Riveccie, M.; Murr, N. E. *Inorg. Chim. Acta* **1978**, *30*, L289.

239. Ballivet-Tkatchenko, D.; Riveccie, M.; El Murr, N. *J. Am. Chem. Soc.* **1979**, *101*, 2763.

240. Gadd, G. E.; Poliakoff, M.; Turner, J. J. *Organometallics* **1987**, *6*, 391.

241. Sanchez-Delgado, R. A.; Wilkinson, G. *J. Chem. Soc., Dalton Trans.* **1977**, 804.

242. Tajima, Y.; Kuniuka, E. *J. Polym. Sci.* **1967**, *B5*, 221.

243. Su, C. C.; Ueng, C. H.; Lin, W. H.; Gi, M. J.; Lii, K. H.; Ting, J. S.; Jan, S. C.; Hodgson, K. O. *Proc. Natl. Sci. Counc., Repub. China* **1982**, *6B*, 45.

244. Schnurpfeil, D.; Lauterbach, G.; Seyferth, K.; Taube, R. *J. Prakt. Chem.* **1984**, *326*, 1025. CA103(17):141765k.

245. Gonzales, O.; Schaefer, R.; Schnurpfeil, D.; Seyferth, K.; Taube, R.; Mohai, B. *J. Prakt. Chem.* **1983**, *325*, 981. CA100(7):50823z.

246. Schurpfeil, D.; Dittmar, U.; Kramer, K.; Megdiche, R. *Z. Phys. Chem. (Leipzig)* **1988**, *269*, 794.

247. Ungar, R. K.; Baird, M. C. *J. Chem. Soc., Chem. Commun.* **1986**, 643.

248. Legzdins, P.; Wassink, B.; Einstein, F. W. B.; Willis, A. C. *J. Am. Chem. Soc.* **1986**, *108*, 317.

249. Legzdins, P.; Richter-Addo, G. B.; Wassink, B.; Einstein, F. W. B.; Jones, R. H.; Willis, A. C. *J. Am. Chem. Soc.* **1989**, *111*, 2097.

250. Goldhaber, A.; Vollhardt, K. P. C.; Walborsky, E. C.; Wolfgruber, M. *J. Am. Chem. Soc.* **1986**, *108*, 516.

251. Jones, C. J.; McCleverty, J. A.; Rothin, A. S. *J. Chem. Soc., Dalton Trans.* **1985**, 401.

252. Fischer, E. O. *Pure Appl. Chem.* **1970**, *24*, 407.

253. Fischer, E. O.; Beck, H.-J. *Angew. Chem., Int. Ed. Engl.* **1970**, *9*, 72.

254. Hegedus, L. S.; Perry, R. J. *J. Org. Chem.* **1985**, *50*, 4955.

255. Xu, Y.; Zhou, B. *J. Org. Chem.* **1987**, *52*, 974.

256. Ito, K.; Nakanishi, S.; Otsuji, Y. *Chem. Lett.* **1988**, *(3)*, 473.

257. Ito, K.; Nakanishi, S.; Otsuji, Y. *Chem. Lett.* **1987**, *(10)*, 2103.

258. Roustan, J. L.; Bisnaire, M.; Park, G.; Guillaume, P. *J. Organomet. Chem.* **1988**, *356*, 195.

259. Davies, S. G.; Smallridge, A. J.; Ibbotson, A. *J. Organomet. Chem.* **1990**, *386*, 195.

260. Gambarotta, S.; Alper, H. *J. Organomet. Chem.* **1981**, *212*, C23.

261. Falicky, S.; Alper, H. *J. Chem. Soc., Chem. Commun.* **1987**, 1039.

262. Dolcetti, G.; Hoffman, N. W.; Collman, J. P. *Inorg. Chim. Acta* **1972**, *6*, 531.

263. Collman, J. P.; Hoffman, N. W.; Morris, D. E. *J. Am. Chem. Soc.* **1969**, *91*, 5659.

264. Cenini, S.; Ugo, R.; Porta, F. *Gazz. Chim. Ital.* **1981**, *111 (7-8)*, 293.

265. Legzdins, P.; Martin, J. T.; Einstein, F. W. B.; Jones, R. H. *Organometallics* **1987**, *6*, 1826.

266. Roustan, J. L. A.; Forgues, A.; Merour, J. Y.; Venayak, N. D.; Morrow, B. A. *Can. J. Chem.* **1983**, *61*, 1339.

267. Zwick, B. D.; Arif, A. M.; Patton, A. T.; Gladysz, J. A. *Angew. Chem., Int. Ed. Engl.* **1987**, *26*, 910.

8

Nitric Oxide and the Environment

The Earth's atmosphere (Greek 'atmos' = vapor, 'sphaira' = ball) is a giant gas-phase photochemical reactor with the Sun as the major light source. The total mass of the atmosphere is *ca.* 5 x 10^8 kg, and while half of this mass is located less than 5.5 km from the Earth's surface, almost all (99%) is located below 30 km *(1)*. A schematic drawing of the main regions of the atmosphere is shown in Figure 8.1.

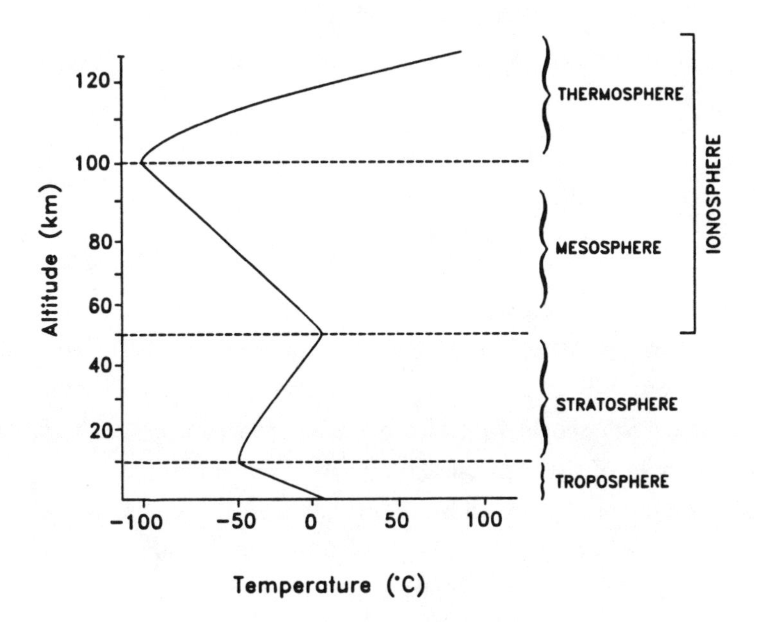

Figure 8.1. The main regions of the earth's surface.

The movement of atmospheric species may either be by horizontal mixing (which occurs mostly in the troposphere) or by vertical mixing. The type of chemical species in each region roughly depends on (i) the height and temperature gradient of that region, and (ii) the type of solar radiation (i.e., the magnitude of λ) that penetrates to that region. Shorter wavelengths penetrate less deeply, so that their chemical effects are not felt at lower altitudes.

In this chapter, the role that nitric oxide and other nitrogen oxides play in regulating the chemistry of the troposphere, stratosphere and ionosphere will be discussed briefly. In presenting this material, it is understood that the nitrogen oxides constitute only a fraction of the chemical compounds contained in the Earth's atmosphere, and the reader is advised to consult more comprehensive treatments on atmospheric chemistry for other aspects of this fascinating, but rather complex, area of research *(1-3)*.

8.1: The Troposphere

The troposphere is the atmospheric region that is in direct contact with the earth's surface (< 12 km height). The composition of the atmosphere in this region is tabulated in Table 8.1.

8.1.1: Non-Urban NO_x Inventories

The concentrations of the nitrogen oxides ($NO_x = NO + NO_2$), sulfur dioxide and other trace gases vary with geographical regime. These regimes and their areas are listed in Table 8.2. Studies of the chemical compositions of the atmospheres in direct contact with these regions (excluding urban areas, of course) should help to provide an indication of the background (natural) levels of these chemicals. It is known, for example, that forest regions are responsible for emissions of unsaturated, saturated and oxidized hydrocarbons as well as N_2O *(5,6)*. Marshlands (a favorable environment for anaerobic bacteria) are responsible for emissions of methane, ammonia, hydrogen and sulfides. Ocean surfaces, on the other hand, emit halogenated compounds and aerosols, as well as those compounds emitted by marshlands. NO_x emissions derive from grasslands (soils and plants) and from urban areas *(5)*.

Table 8.1 The Constitution of the Atmosphere Near The Earth's Surface

Species	Concentration (% vol)
N_2	78
O_2	21
Ar	0.9
CO_2	0.01 - 0.1
Ne, He, Kr, Xe	25×10^{-4}
H_2	0.5×10^{-4}
N_2O	0.3×10^{-4}
CH_4	1.6×10^{-4}
NO_2	0.2×10^{-6}
O_3	$0 - 10^{-4}$
SO_2	$0 - 10^{-4}$

Data taken from reference 4.

Other 'natural' sources of NO_x are forest fires and lightning. The contribution by biomass burning to the global N budget is estimated at 21%, and is similar to the 20% contributed by lightning *(7)*. For comparison, the contribution by the burning of fossil fuels is estimated at 37%. Indeed, the production of NO by cloud-to-ground lightning is the major non-biological source of fixed nitrogen *(8,9)*, and it results from the reaction between N_2 and O_2 in a lightning stroke, where temperatures may reach up to *ca.* 30,000 K. It has been estimated that on average, on the order of 8×10^{25} molecules of NO are produced per lightning stroke, contributing to a total global budget of 1.2×10^{13} g N/yr due to lightning *(7)*. However, it must also be noted that in the vicinity of a lightning thunderstorm, the concentration of NO is *ca.* 0.82 ppbv, and when this is compared to sub-urban concentrations of between 3 - 10 ppbv and urban concentrations of >

10 ppbv, it becomes apparent that only the NO_x budget in the non-polluted geographical areas are significantly affected by lightning.

Table 8.2 The Geographical Regimes of the Earth's Surface

Regime	Area (cm^2)	% of surface
Oceans	3.3×10^{18}	66.3
Forests	5.7×10^{17}	11.4
Grasslands	3.8×10^{17}	7.6
Steppes and Mountains	2.6×10^{17}	5.2
Deserts	2.4×10^{17}	4.8
Urban areas	1.9×10^{17}	3.8
Marshlands	4.0×10^{16}	0.8

Data taken from reference 5.

8.1.2: Urban NO_x Inventories

Nearly all of the NO_x production in urban areas is as a result of high temperature combustion devices, via the oxidation of N_2 or organic nitrogen compounds *(10)*. Indeed, the high temperatures present in combustion devices favor the formation of NO_x, and this is illustrated in Figure 8.2.

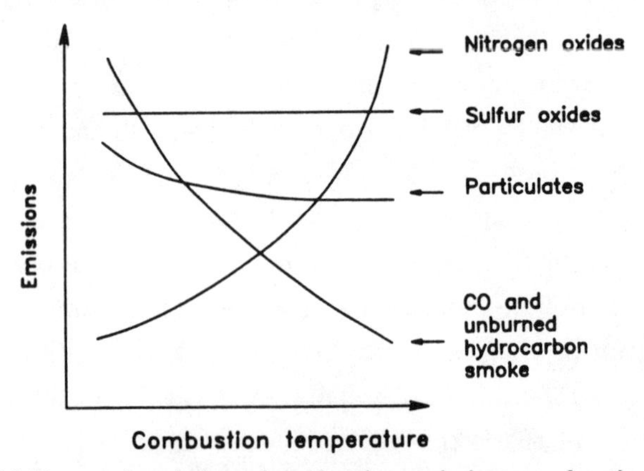

Figure 8.2 Expected variation of combustion emission as a function of peak combustion temperature (redrawn from reference 2).

A listing of emission levels and residence times of some common pollutants are displayed in Table 8.3.

Table 8.3 Common Combustion Pollutants: Estimated Annual Emissions and Residence Times

Pollutant	Estimated annual Emission (Tg/yr $= 10^6$ metric tons/yr)		Estimated Atmospheric Residence Time
	Pollution	Natural	
NO_x (as NO_2)	75	180	2 - 5 days
N_2O	3	340	20 - 100 yrs
SO_2	212	20	1 - 4 days
CO_2	22 x 10^3	10^6	2 - 4 yrs

Data taken from reference 2.

It is immediately obvious from the data contained in Table 8.3 that only sulfur dioxide has a pollution level higher than that of its background level. However, this does not mean that the concerns over NO_x emissions are unwarranted. This is especially so, since levels of NO_x are very high in the vicinity of their sources. For example, in the city of Los Angeles, automobile exhausts account for a daily production of 1,000 tons of NO_x per day; a figure that represents about two-thirds of the total NO_x emissions for that area *(11)*. Stationary sources such as boilers, gas turbines, ship berthing and refinery heaters account for the remaining third.

NO_x compounds are primary pollutants, and increased levels of these pollutants affect the levels of other chemical species such as ozone and peroxyacetylnitrate (secondary pollutants) *(1)*. The term *smog*, originally coined to describe a combination of smoke and fog, is now used to describe any smoky or hazy pollution of the atmosphere. Southern California is an area that is enclosed by high mountainous regions, and suffers from an

Figure 8.3. Los Angeles, California, on a smoggy day (above) and on a clear day after a heavy rainfall (below). [Photographs courtesy of South Coast Air Quality Management.]

overwhelmingly high NO_x production. These two features contribute to the smoggy conditions so frequently observed in this area, and the photographs displayed in Figure 8.3 illustrate this point well. The smog-induced low visibility is not only caused by the brown color of NO_2, but also by airborne particulates less than 2.5 microns in diameter (e.g., ammonium nitrates and sulfates). Thus, whereas the levels of primary pollutants such as combustion gases (e.g., NO and CO_2) are of some concern, it is the presence of secondary pollutants that are directly responsible for most of the smog, haze, eye irritation and material damage.

Evidence has now been accumulated that in the polluted marine environment, the presence of NO_2 results in the formation of airborne nitrosyl halides via the reaction shown below *(12-14)*.

$$2\ NO_2 + NaX \longrightarrow XNO + NaNO_3$$

$$(X = Cl, Br)$$

These XNO compounds may then either undergo photolysis during daytime conditions to yield X atoms which then reduce ozone levels (vide infra), or may hydrolyze to yield HNO_2 which is a major source of the OH radical.

Without question, it is the realization that the increased NO_x levels in urban areas may indeed have disastrous consequences on human health that has resulted in tremendous advances in transition-metal-mediated conversions of NO_x into less harmful substances. The catalytic reduction of NO is one such conversion, and has been dealt with elsewhere in this book (Sections 5.1 and 7.4). The advances made in NO_x control technology for passenger cars also reflect society's concern over high NO_x levels resulting from automobile exhausts. In the 1966 -1970 period, the attachment of a retrofit NO_x device to gasoline vehicles resulted in a 25% NO_x reduction in emission. The attachment of a three-way catalyst to gasoline vehicles in 1980 resulted in a 70 - 90% reduction in such emissions *(11)*. The success of these catalytic converters in automobiles depends on the choice of metal catalyst, and several reviews are available that treat this subject in depth *(15-17)*.

In the next section, the tropospheric chemistry of NO_x will be discussed. Particular attention will be focused on how NO_x levels affect ozone levels and acidic deposition.

8.1.3: Tropospheric NO_x Chemistry

The troposphere (Greek 'tropos' = turning) is characterized by strong vertical mixing which is generally on the order of a few days for individual molecules, but may reduce to a few minutes in a thunderstorm *(1)*. The ozone present in the stratosphere limits the solar radiation available to the troposphere to $\lambda > 280$ nm. In the troposphere, O_3, NO_2 and H_2CO are the most important species that are photochemically labile, and they all lead eventually to the production of the OH or HO_2 radicals. Indeed, the reactions of OH radicals appear to dominate the chemistry of the troposphere via reactions 8.1 - 8.8 (M = third body, e.g. N_2, O_2) *(3)*.

$$NO_2 + h\nu \longrightarrow NO + O \tag{8.1}$$

$$O + O_2 + M \longrightarrow O_3 + M \tag{8.2}$$

$$O_3 + h\nu \longrightarrow O_2 + O \tag{8.4}$$

$$H_2CO + h\nu \longrightarrow H + HCO \tag{8.5}$$

$$H + O_2 + M \longrightarrow HO_2 + M \tag{8.6}$$

$$HCO + O_2 \longrightarrow HO_2 + CO \tag{8.7}$$

$$HO_2 + NO \longrightarrow NO_2 + OH \tag{8.8}$$

Indeed, the photolysis of NO_2 is the only known source of O_3 in the troposphere (eqs 8.1 and 8.2) *(1)*, and in urban areas, eq 8.8. represents the major supplier of OH *(5)*. A photostationary state exists between NO_2, NO and O_3, since NO reacts with O_3 to regenerate NO_2, i.e.,

$$NO_2 + h\nu \longrightarrow NO + O \tag{8.1}$$

$$O + O_2 + M \longrightarrow O_3 + M \quad (8.2)$$

$$NO + O_3 \longrightarrow NO_2 + O_2 \quad (8.9)$$

---

no net reaction

and this photostationary state is expressed by $[O_3] = k[NO_2]/[NO]$. In other words, the presence of other species capable of oxidizing NO to NO_2 effectively increases net O_3 production *(18-20)*. It must be noted that the direct reaction of NO with O_2 in an unpolluted atmosphere is very slow and in reality, the efficient conversion of NO to NO_2 is accomplished by reactive oxidizing species such as O_3, OH and peroxy radicals (RO_2) which result from the breakdown of volatile organics *(21)*.

$$R + O_2 \longrightarrow RO_2 \quad (8.10)$$

$$RO_2 + NO \longrightarrow RO + NO_2 \quad (8.11)$$

What this means, of course, is that NO_x polluted areas will also experience high ozone levels which result in eye and lung irritations and increased health risks.

Due to the photostationary state that exists for NO, NO_2 and O_3, the reactions 8.1 and 8.9 do not constitute net removal processes for NO_x. The only true removal (depletion) processes are,

$$NO_2 + OH + M \longrightarrow HNO_3 + M \quad (8.12)$$

$$NO_2 + HO_2 + M \longrightarrow HO_2NO_2 + M \quad (8.13)$$

$$NO_2 + MeCO(O_2) \longrightarrow MeCO(O_2)NO_2 \quad (8.14)$$

and these are schematically represented in Figure 8.4 *(21)*.

The major pathway for NO_x depletion appears to be its reaction with OH in the presence of a third body to form HNO_3 (eq 8.12)*(22)*. Other reactions to produce RO_2NO_2 compounds and peroxyacetyl nitrate (PAN, also known as ethaneperoxoic nitric anhydride (EPNA)) are also important *(22)*.

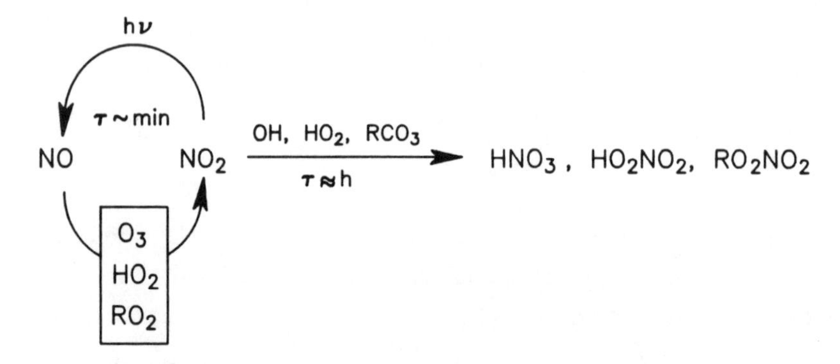

Figure 8.4. The NO/NO_2 conversion and NO_2 removal processes in the troposphere (Redrawn from reference 21).

The atmospheric chemistry of the nitrate radical and of PAN in the presence of NO-NO_2-air mixtures have been reviewed recently *(23-25)* and will not be discussed here. A schematic drawing outlining the essential aspects of tropospheric NO_x chemistry is shown in Figure 8.5.

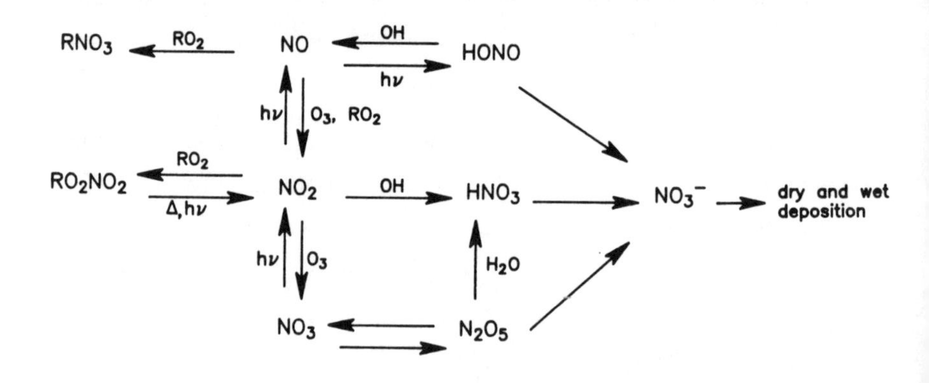

Figure 8.5. Tropospheric NO_x Chemistry (Redrawn from reference 26).

8.1.4: Effects of Increased NO_x Levels on Climate and Health.

It is well known that the earth's natural atmosphere acts as a greenhouse to keep the earth warm, and without it there would be no life. Concern over the destruction of the 'natural' atmosphere arises from three principal viewpoints (i) increased production of greenhouse gases, (ii) depletion of the protective ozone layer, and (iii) increased levels of secondary pollutants such as ozone and PAN which cause eye and lung irritation.

8.1.4.1: Greenhouse gases

Radiation (thermal) trapping by atmospheric absorbers such as water vapor, clouds, and carbon dioxide contribute to the 'normal' warming of the atmosphere. Increasing the concentration of combustion gases such as CO_2 should, to a first approximation, result in an 'unnatural' warming of the atmosphere, and many predictions have been made to this effect. For example, it has been estimated that the cumulative surface temperature rise caused by increases in concentrations of CO_2, CH_4, N_2O, O_3 and CFCs (chlorofluorocarbons) by the year 2030 will be as shown in Figure 8.6 *(27)*.

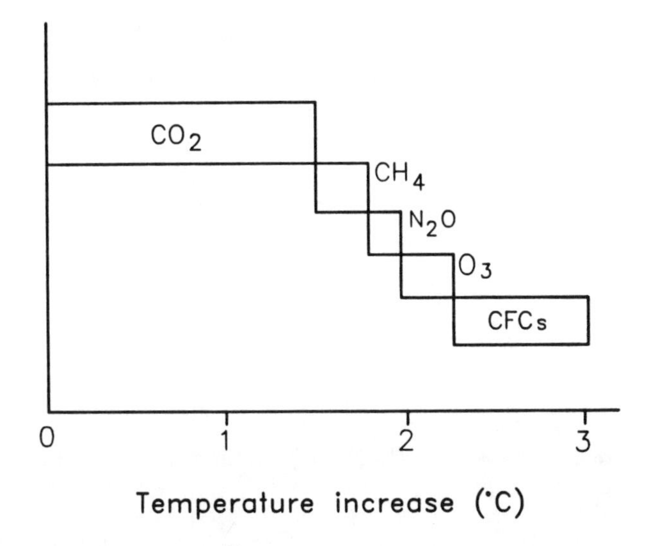

Figure 8.6. Estimated cumulative temperature rise caused by pollutants by the year 2030. (Adapted from reference 27).

Increased NO_x levels (i.e., NO + NO_2) do not appear to contribute directly to the greenhouse effect. However, it must also be remembered that only 0.04% of all nitrogen is in a combined form (i.e., other than N_2). Of this 0.04%, 99.6% is N_2O *(28)*. The annual emission of N_2O is estimated at *ca.* 30 million tonnes: anthropogenic sources account for a quarter of this amount *(27)*. Natural emissions derive from the soil and water as a result of incomplete microbial nitrification and denitrification processes *(1)*, whereas artificial sources derive from the burning of fossil fuels *(27)*. The global concentration of N_2O appears to be rising by about 0.2 - 0.3% per year, mainly as a result of combustion processes *(27)*. Its major sink is its reaction with oxygen in the stratosphere (eq 8.18) *(27)*.

8.1.4.2: Acid Rain

'Natural' rain is slightly acidic (pH of *ca.* 5.6) due to the reaction of CO_2 and H_2O vapor to form carbonic acid. As previously noted in Figures 8.4 and 8.5, the true removal processes for NO_x involve its reactions to form HNO_3. Acidic depositions due to the presence of NO_x may be broadly classified into two categories. *Dry depositions* (non-rain events) generally involve the deposition of species such as NO_2, HNO_3 and NH_4NO_3 onto vegetation and soils, whereas *wet depositions* involve the incorporation of NO_2^- or NO_3^- into rain droplets and are referred to as acid rain. The pH range of acid rain is shown in Figure 8.7.

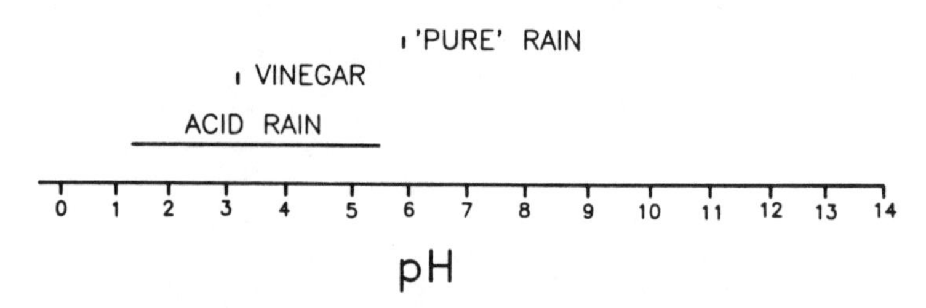

Figure 8.7. The typical pH range of acid rain (adapted from reference 2).

In areas with high NO_x levels, the acid rain is caused mainly by HNO_3, and in areas of high SO_2 levels the acid rain is caused mainly by H_2SO_4. For example, in southwestern U.S.A., acid rain is normally 80% HNO_3, whereas in some regions of eastern U.S.A., it is 65% H_2SO_4 and 30% HNO_3 *(2)*, although the latter levels show seasonal variations. For instance, it has been determined that in the Adirondack mountains of eastern U.S.A., the SO_4^{2-} aerosols are the predominant atmospheric acidic species in the summer, whereas HNO_3 predominates in other seasons *(29)*.

Freshwater aquatic systems and vegetation suffer most from acidic deposition. Indeed, at low pH levels (< *ca.* 5) many fish cannot reproduce and survive. Furthermore, the more acidic the aquatic body is, the greater the incorporation of metal ions in it. Acidification may also cause the leeching of soil, thus decreasing its fertility. Indeed, out of 5,000 lakes in southern Norway, 1,750 had lost all their fish as of 1983, and about 900 others were seriously affected by acid rain *(30)*. The concern over acid rain in different parts of the world has been expressed in a number of scientific papers *(31-35)*.

Interestingly, it has been found that stone material from the corroding surfaces of buildings release NO to the atmosphere *(36)*. This is because a large fraction of the endothilic microflora of building stones is made up of nitrifying bacteria which, in addition to oxidizing NH_4^+ to HNO_2 and HNO_3, produce NO during the NH_4^+ to NO_3^- biogenic oxidation *(36)*.

8.1.4.3: Health concerns

Densely populated areas are subject to smoggy conditions, which normally reflect a high concentration of aerosol particles, ozone and an variety of eye irritants such as aliphatic aldehydes (from ozone attack on alkenes) and peroxynitrates. While NO_2 (or SO_2) is known to cause respiratory problems and PAN to cause eye irritation, ozone is known to cause coughing, chest discomfort and increased asthma attacks especially in the young, sick and elderly *(2)*. It is known that asthmatics bronchoconstrict upon inhalation of NO_2 at near-ambient urban levels *(37)*. Laboratory tests on rats indicate that inhalation of NO_2 causes lesions at the airways and bronchopulmonary junctions, in effect damaging the lung *(38)*. Not unexpectedly, short-term high NO_2 exposure of 20 ppm for 1 h has been found to be morphologically similar to long-term low exposure of 0.5 ppm for one month *(39)*. Furthermore, NO_2 initiates lipid peroxidation in biological systems *(40)*, and its inhalation has been associated with the biosynthesis of carcinogenic nitrosamines in rabbits *(41)*. Also, NO_2 has been found to stimulate the nitration of polycyclic aromatic hydrocarbons to produce complexes that display mutagenicity, genotoxicity and carcinogenicity. [It must be remembered that nitroarenes are also emitted directly from diesel engines, kerosene heaters and gas burners *(42)*!]

The deleterious effects of nitrogen oxides on health have been compiled in an excellent text on the subject *(43)*. It must be emphasized again, however, that the increased levels of the secondary pollutants such as PAN, O_3, HNO_3 and oxygenated hydrocarbons result from the photochemical conversion of primary pollutants such as NO_x, SO_2 and hydrocarbons. Consequently, the levels of these eye and lung irritants are highest in the presence of intense sunlight (e.g., at noon).

8.1.5: Indoor Air Pollution

Although pollutants such as SO_2 and O_3 generally have higher outdoor concentrations, others such as NO and NO_2 (NO_x) can have considerably higher indoor concentrations *(44)*. In residences with gas cookers/heaters, the indoor concentrations of NO_2 can reach very high levels, often exceeding outdoor concentrations by at least an order of magnitude. For example, in a recent study in London (U. K.), it was noted that the NO_2 concentration varied as a function of the combustion device used (Table 8.4.) *(44)*.

Table 8.4 Increase in [NO_2] as a Function of Combustion Device.

Room	Device	Increase in [NO_2] (ppb)
Kitchen	gas cooker	19.83 ± 9.64
Living room	kerosene heater	20.8 ± 9.78
Bedroom	kerosene heater	45.09 ± 9.38

Such studies in Europe and in the United States have been prompted by indications that there might be a direct association of prevalence of respiratory illnesses in primary school children and the use of gas cookers/kerosene heaters in the home *(45)*. Tobacco smoking also contributes to NO_2 contamination. Indeed, NO_2 levels in restaurant (where smoking is allowed) and bars may reach 63 ppb and 21 ppb, respectively, compared to background levels of *ca.* 4 ppb *(44)*. However, contributions from combustion processes in the kitchens of restaurant may be responsible for the higher levels observed.

8.2: The Stratosphere

The stratosphere (Greek 'stratus' = layered) occupies the 15 - 55 km region above the earth (Figure 8.1) and is important to aircraft scientists. It is in the stratosphere that the major part of the earth's ozone shield is formed *(28)*. [Less than 10% of all ozone is located in the troposphere *(3)*.] It is this ozone that absorbs solar *uv* radiation between 240 and 290 nm, radiation that is lethal to unicellular organisms and the surface cells of higher plants and animals *(1)*. Although O_3 and atomic O dominate the chemistry of the stratosphere via the reactions

$$O_2 + h\nu \longrightarrow O + O \tag{8.15}$$

$$O + O_2 + M \longrightarrow O_3 + M \tag{8.16}$$

$$O_3 + h\nu \longrightarrow O_2 + O \tag{8.17}$$

NO_x compounds play a crucial role in the odd-oxygen balance in this region *(1,3,5,46)*.

The major source of NO_x in the stratosphere is by the vertical mixing of tropospheric N_2O,

$$N_2O + O \longrightarrow NO + NO \tag{8.18}$$

whereas the NO_x sink is by reaction with OH in the presence of a third body,

$$NO_2 + OH + M \longrightarrow HONO_2 + M \tag{8.19}$$

A summary of stratospheric NO_x chemistry is presented in Figure 8.8 *(47)*.

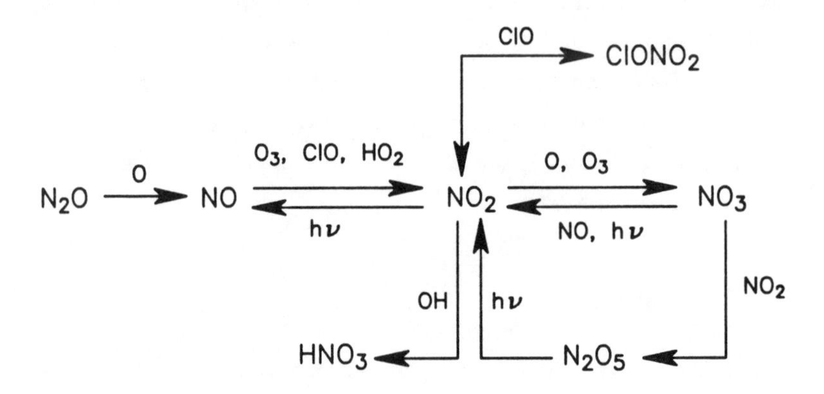

Figure 8.8. Stratospheric NO_x chemistry (adapted from reference 47).

Interestingly, it has been proposed that HNO_3 cloud formation in the cold Antarctic stratosphere is the major cause for the springtime 'ozone hole' via the formation of HNO_3-H_2O aerosols *(48)*. The proposed mechanism of such action involves the removal of odd nitrogen due to the conversion of N_2O_5 and $ClONO_2$ to HNO_3 vapor and then to the HNO_3-H_2O aerosols. Consequently, the depleted concentrations of odd nitrogen species lead to large OH concentrations which then serve to convert HCl to ClOX rapidly, which in turn is directly responsible for the ozone destruction.

Radical species such as H, OH, NO and Cl (denoted by X in eqs 8.20 and 8.21) play an important role in the catalytic destruction of ozone *(3)* via

$$X + O_3 \longrightarrow XO + O_2 \quad (8.20)$$

$$XO + O \longrightarrow X + O_2 \quad (8.21)$$

net reaction

$$O + O_3 \longrightarrow O_2 + O_2 \quad (8.22)$$

In the lower atmosphere, X = NO dominates and is the major factor contributing to the loss of odd oxygen. Although all the X = H, HO_x, NO_x and Cl_x species are present in the natural "unpolluted" atmosphere, the Cl_x species are supplemented by human activities in the troposphere (e.g., by the escape into the stratosphere and subsequent photolysis of the tropospherically inert chlorofluorocarbons) *(27)*. Also, the direct emissions into the stratosphere of NO_x, CO_2, H_2O and unburned hydrocarbons by supersonic jets (such as the Concorde flying at 17 km) do give rise to concerns over the chemical balance in the stratosphere.

8.3: The Ionosphere

Charged particles represent only a minute fraction of atmospheric chemical species. However, due to the fact that most of the sun's ionizing radiation (extreme *uv* and X-rays) is absorbed at *ca.* 60 km, it is not surprising that ionic species (including 'free' electrons) are most abundant at altitudes above 60 km. The region above this level is appropriately termed the ionosphere, and may be conveniently divided into sub-regions as shown in Figure 8.9. The upper F region ($>$ 200 km) experiences the sun's radiation of all λs, whereas the lower F region (150 - 200 km) experiences radiation characterized by λ = 20 - 91 nm. The E region (100 - 150 km) experiences radiation characterized by λ = 80 - 103 nm and X-rays (1 - 10 nm). The D region (64 - 112 km) experiences radiation of longer wavelengths (λ = 103 nm), X-rays (0.2 - 0.8 nm) and Lyman α at λ = 121.6 nm *(1)*. However, it must be noted that ionization does not stop at night (i.e., in the absence of sunlight), it just slows down since other light sources such as starlight are also available during this time.

The distribution of chemical species in the ionosphere may be roughly correlated with their masses. The heavier ions (e.g., the cluster ions) are thus found at lower altitudes, whereas the lighter ions (e.g., He^+) are located at higher altitudes. Although the emphasis in this chapter is on the atmospheric chemistry of nitrogen oxides, it must be borne in mind that atmospheric chemistry is indeed very complex, and factors such as temperature, geographical area and presence of other chemical species determine the concentrations and reactions of the nitrogen oxides *(1,49,50)*. Many of the atmospheric reactions of the nitrogen oxides have been studied under simulated conditions in laboratories with a view to understanding some aspects of this chemistry *(51-59)*.

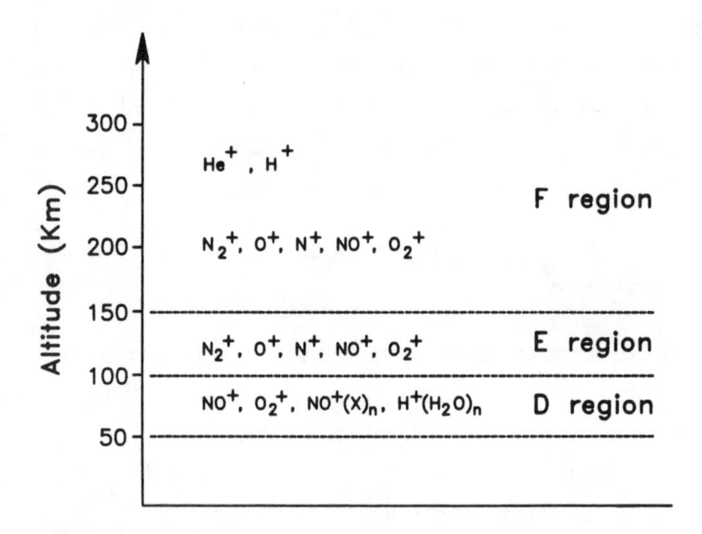

Figure 8.9 Regions of the ionosphere and their constituent species (adapted from reference 1).

8.3.1: The F Region

In the F region, O^+ is the dominant primary ion, formed via

$$O + h\nu \longrightarrow O^+ + e \tag{8.23}$$

Other species are then formed via the reaction of O^+ with other molecules, as in the following charge-transfer reactions

$$O^+ + O_2 \longrightarrow O + O_2^+ \tag{8.24}$$

$$O^+ + N_2 \longrightarrow O + N_2^+ \tag{8.25}$$

or in the NO^+-forming reactions

$$O^+ + N_2 \longrightarrow N + NO^+ \tag{8.26}$$

$$N_2^+ + O \longrightarrow NO^+ + N \tag{8.27}$$

reaction 8.27 being the major daytime source of NO^+ *(60,61)*. Generally speaking, sunrise conditions favor the increase in NO^+ and O_2^+ levels via eqs 8.23, 8.28 and 8.29:

$$O^+ + N \longrightarrow NO^+ \tag{8.28}$$

$$O^+ + O \longrightarrow O_2^+ \tag{8.29}$$

The only sink of NO^+ in the F region is by its temperature-dependent dissociative recombination reaction

$$NO^+ + e \longrightarrow N + O \tag{1.7}$$

(the term 'recombination' refers to the reverse of ionization) *(58,61,63)*.

There are two important consequences arising from these photochemical, charge-transfer and recombination reactions. The first is that they occasionally produce excited-state atoms and molecules (usually designated by *) that emit radiation which is observable as auroral emissions and airglows. Examples of such auroral emitters are NO, NO^+, O, N and CO_2 *(55,64-67)*, e.g.,

$$N^+ + O_2 \longrightarrow (NO^+)^* + O \tag{8.30}$$

$$NO^+ + e \longrightarrow N^* + O \tag{8.31}$$

$$N_2^+ + O \longrightarrow NO^+ + N^* \tag{8.32}$$

Furthermore, molecular species such as NO are excited by sunlight, and the subsequent radiative decay of NO* has been found to be the strongest dayglow feature in the 200 - 300 nm spectral range. However, it is observable only by high-flying rockets due to atmospheric absorption. At night (in the absence of sunlight) chemiluminescence of N and NO_2 contribute to nightglow *(1)*, e.g.,

$$NO + O \longrightarrow NO_2^* \longrightarrow NO_2 + h\nu \tag{8.33}$$

$$NO^+ + e \longrightarrow N^* + O \longrightarrow N + O + h\nu \tag{8.34}$$

The second consequence is that ionization and dissociative recombination reactions serve as sources and sinks for electrons, respectively. Thus, although electron densities would be expected to be higher in daytime than nighttime due to increased ionizations, greater electron-densities could also mean the increased occurrence of some dissociative recombination reactions.

Consequently, ionic ratios such as $[NO^+]/[O_2^+]$ may be expected to fluctuate accordingly *(68)*.

8.3.2: The E Region

The NO^+ and O_2^+ molecular ions are the most important chemical species in the E region (100 - 150 km), although metallic ions such as Na^+, Al^+ and Fe^+ may also be present *(69)*. In the E region, the $[NO^+]/[O_2^+]$ ratio may sometimes be as large as 10 *(69)*. The chemistry of NO^+ is generally dominated by the reactions,

$$NO^+ + e \longrightarrow N + O \qquad (1.7)$$

$$N + O_2 \longrightarrow NO + O \qquad (8.35)$$

$$O_2^+ + N_2 \longrightarrow NO + NO^+ \qquad (8.36)$$

$$NO + O_2 + \longrightarrow NO^+ + O_2 \qquad (8.37)$$

Other molecular ions that play a role in NO_x chemistry of the E region include N_2 at high altitude (eq 8.27) or at low altitude (eq 8.28)

$$N_2^+ + O \longrightarrow NO^+ + N \qquad (8.27)$$

$$N_2^+ + O_2 \longrightarrow N_2 + O_2^+ \qquad (8.38)$$

8.3.3: The D Region

In the D region, the main ionizable component is NO (λ = 121.6 nm, Lyman-α), although excited O_2 ionization (102.7 - 111.8 nm) contributes to some extent *(70)*. Consequently, any increase in NO concentration in this region would, simplistically speaking, be expected to result in a concomitant increase in electron density.

The D region has the lowest temperatures in the entire atmosphere, and may be divided into two sub-regions, with a "boundary" at *ca.* 82 km. In these two sub-regions, complex positive ions dominate the chemistry, in addition to some negative ions such as O_2^-, NO_3^-, O^- and free electrons *(1)*. At all D region heights, $[NO^+] \geq [O_2^+]$ *(71)*.

In the upper D region (> 82 km), the dominant ions during daytime are NO^+ (*ca.* 80 %) and O_2^+ (*ca.* 20%). However, there is a strong variability due to other factors such as temperature and pressure *(1,72)*. In the lower D region (< 80 km), cluster ions of the form $NO^+(L)_n$ and $O_2^+(L)_n$ ($L = N_2$, CO_2) and proton hydrates $H^+(H_2O)_n$ (n = 2 - 8) occur in addition to NO^+ and O_2^+ *(72)*. The formation of the cluster ions may be perceived to occur via *(71,73)*:

$$NO^+ + L + M \longrightarrow NO^+(L) + M \tag{8.39}$$

$$NO^+(L) + H_2O \longrightarrow NO^+(H_2O) + L \tag{8.40}$$

$$NO^+(H_2O) + H_2O \longrightarrow NO^+(H_2O)_2 \tag{8.41}$$

$$NO^+(H_2O)_2 + H_2O \longrightarrow H^+(H_2O)_2 + HNO_2 \tag{8.42}$$

The $[NO^+(H_2O)_n]/[NO^+]$ ratio is temperature dependent. However, below the 82 km boundary, the $H^+(H_2O)_n$ cluster ions are the dominant species *(74)*. Interestingly, the concentration of free electrons changes drastically within a few kilometers of the boundary, with high electron concentration at higher levels. Two factors appear to be responsible for the removal of electrons at lower altitudes: (a) formation of negative ions by reaction of free electrons with neutral molecules, and (b) reactions of electrons with proton hydrates (dissociative recombination),

$$H^+(H_2O)_n + e \longrightarrow H + n\,H_2O \tag{8.43}$$

a reaction that is apparently entropy-driven *(75)*.

During some winter days, a five- to ten-fold increase in the concentration of NO above 'normal' values is observed in the 75 - 97 km range (e.g., [NO] = 6×10^6 cm^{-3} at 150 km vs 2×10^7 cm^{-3} at 80 km *(75)*) which is often coupled with a lowering of the inner D boundary from 82 km to *ca.* 74 - 77 km *(72,75,76)*. Such a phenomenon has been broadly termed the *winter anomaly*. The mechanisms of the processes that give rise to this phenomenon are very complex and will not be discussed further here.

8.3.4: The Ionosphere and Radio Propagation

It is now generally accepted that the charged species responsible for radiowave reflection are electrons and not atomic, molecular or cluster ions. However, these latter species (including neutral atoms or molecules) play a central role in the distribution and concentration of these electrons *(1)*. In the ionosphere (> 70 km), the electron concentrations are high enough to exert an effect on radiowave reflection, absorption or propagation *(72)*. Indeed, low-frequency radiowaves (of up to 4.4 MHz) are reflected by the E region (at *ca.* 100 km). Higher frequency radiowaves normally penetrate the E region and are reflected by the F region at higher altitudes. Interestingly, the D region is more involved with absorption (rather than reflection) of radio waves *(1,72)*. It thus becomes apparent that any geographical or chemical process that leads to increased ionization in the D region will lead to increased electron density in this region and will result in an 'interference' of radio reflection by the upper ionospheric regions. For instance, (a) seasonal variations in NO concentration in the D region (such as the winter anomaly described previously) will affect electron concentrations and thus affect radio propagation, and (b) periods of sunlight (e.g., daytime) should increase ionization of NO to NO^+ and free electrons. Consequently, daytime absorption of radiowaves prevents normal reflection of these waves at higher altitudes.

8.3.5: Reliability of Measurements of Ionospheric NO_x

A common method of measurement of constituent species in the ionosphere is by the use of ion mass spectrometers aboard rockets and spacecraft. However, the measurements obtained may not always be reflective of the 'natural' ionosphere, since the rockets and spacecraft are themselves local sources of atmospheric species. For example, in a recent flight by the shuttle Spacelab-2, the frequencies of increased bursts in the measured fluxes of NO^+ and H_2O were found to coincide with the frequencies of thruster firings *(77)*. Incidentally, the thrusters employed monomethylhydrazine and dinitrogen tetroxide.

8.4: NO_x on Other Planets?

Throughout this book, an attempt has been made to link the different aspects of the chemistry of NO from laboratory experiments to nature's way of synthesizing NO and then transforming it into its derivatives. As a final comment in this chapter, it must be stressed that the chemistry of nitrogen oxides may not be limited to the chemistry observed on this planet. It has been suggested that NO_x chemistry may also be present on other planets such as Venus, where lightning storms in its clouds produce NO via the reaction of CO_2 (95%) and N_2 (5%) *(78)*. It is clear that with the development of modern space technology, more information will become available as to the role of NO and its compounds in the universe as a whole.

8.5: References

1. Wayne, R. P. *Chemistry of Atmospheres*; Oxford: Oxford, 1985.
2. Stern, A. C.; Boubel, R. W.; Turnerm, D. B.; Fox, D. L. In *Fundamentals of Air Pollution*, 2nd ed.; Academic: Orlando, Florida, 1984.
3. Graedel, T. E.; Hawkins, D. T.; Claxton, L. D. In *Atmospheric Chemical Compounds*; Academic: Orlando, Florida, 1986.
4. *The World Environment 1972-82: A Report by UNEP*; Holdgate, M. W.; Kassas, M.; White, G. F., Eds; Tycooly International: Dublin, Ireland, 1982.
5. Graedel, T. E. In *The Handbook of Environmental Chemistry*; Hutzinger, O., Ed.; Springer-Verlag: Berlin, 1980; Vol 2A, pp 107-143.
6. Schmidt, J.; Seiler, W.; Conrad, R. *J. Atmos. Chem.* **1988**, *6*, 95.
7. Sisterson, D. L.; Liaw, Y. P. *J. Atmos. Chem.* **1990**, *10*, 83.
8. Gladyshev, G. P. *Dokl. Akad. Nauk SSSR* **1983**, *271*, 341.
9. Mancinelli, R. L.; McKay, C. P. *Origins Life Evol. Biosphere* **1988**, *18*, 311.

10. *Atmospheric Pollution: Proceedings of the 14th. International Colloquium, Paris, France, 5-8 May, 1980*; Benarie, M. M., Ed.; Elsevier: Amsterdam, 1980; Vol. 8, pp 173-186.
11. *Staff Report on the Hearing on Controls of Oxides of Nitrogen*; South Coast Air Quality Management District: CA, 1986.
12. Finlayson-Pitts, B. J.; Johnson, S. N. *Atmos. Environ.* **1988**, *22*, 1107.
13. Sturges, W. T. *Atmos. Environ.* **1989**, *23*, 1167.
14. Finlayson-Pitts, B. J. *Nature (London)* **1983**, *306 (5944)*, 676.
15. Moser, W. R. In *The Catalytic Chemistry of Nitrogen Oxides*; Klimisch, R. L.; Larson, J. G., Eds.; Plenum: New York, 1975; pp 33-43.
16. Dwyer, F. G. *Catal. Rev.* **1972**, *6*, 261.
17. Wei, J. *Adv. Catal.* **1975**, *24*, 57.
18. *Tropospheric Ozone*; Isaksen, I. S. A., Ed.; D. Reidel: Holland, 1988.
19. Liu, S. C.; Trainer, M. *J. Atmos. Chem.* **1988**, *6*, 221.
20. Kondo, Y.; Muramatsu, H.; Matthews, W. A.; Toriyama, N.; Hirota, M. *J. Atmos. Chem.* **1988**, *6*, 235.
21. Wagner, H. G.; Zellner, R. *Angew. Chem., Int. Ed. Engl.* **1979**, *18*, 663.
22. Roberts, J. M. *Atmos. Environ.* **1990**, *24A*, 243.
23. Wayne, R. P.; Barnes, I.; Biggs, P.; Burrows, J. P.; Canosa-Mas, C. E.; Hjorth, J.; Bras, G. L.; Moorgat, G.K.; Perner, D.; Poulet, G.; Restelli, G.; Sidebottom, H. *Atmos. Environ.* **1991**, *25A*, 1.
24. Tuazon, E. C.; Carter, W. P. L.; Atkinson, R. *J. Phys. Chem.* **1991**, *95*, 2434.
25. Cantrell, C. A.; Davidson, J. A.; Shetter, R. E.; Anderson, B. A.; Calvert, J. G. *J. Phys. Chem.* **1987**, *91*, 6017.
26. Cox, R. A. In *Tropospheric Ozone*; Isaksen, I. S. A., Ed.; D. Reidel: Holland, 1988; pp 263-292.
27. *The State of the World Environment*; U. N. Environment Programme, 1989.
28. Söderlund, R.; Rosswall, T. In *The Handbook of Environmental Chemistry*; Hutzinger, O., Ed.; Springer-Verlag: Berlin, 1982; Vol 1B, pp 61-81.
29. Kelly, T. J.; McLaren, S. E.; Kadlecek, J. A. *Atmos. Environ.* **1989**, *23*, 1315.
30. *The State of the Environment*; U. N. Environment Programme, 1983.

31. Dikaiakos, J. G.; Tsitouris, C. G.; Siskos, P. A.; Melissos, D. A.; Nastos, P. *Atmospheric Environment* **1990**, *24B*, 171.
32. Clarke, A. G.; Lambert, D. R.; Willison, M. J. *Atmospheric Environment* **1990**, *24B*, 159.
33. Lippmann, M. *Environ. Health Perspect.* **1985**, *63*, 63.
34. Goyer, R. A. *Environ. Health Perspect.* **1985**, *63*, 3.
35. Franklin, C. A.; Burnett, R. T.; Paolini, R. J. P.; Raizenne, M. E. *Environ. Health Perspect.* **1985**, *63*, 155.
36. Baumgärtner, M.; Remde, A.; Bock, E.; Conrad, R. *Atmospheric Environment* **1990**, *24B*, 87.
37. Utell, M. J. *Environ. Health Perspect.* **1985**, *63*, 39.
38. Kubota, K.; Murakami, M.; Takenaka, S.; Kawai, K.; Kyono, H. *Environ. Health Perspect.* **1987**, *73*, 157.
39. Hayashi, Y.; Kohno, T.; Ohwada, H. *Environ. Health Perspect.* **1987**, *73*, 135.
40. Sagai, M.; Ichinose, T. *Environ. Health Perspect.* **1987**, *73*, 179.
41. Kosaka, H.; Uozumi, M.; Nakajima, T. *Environ. Health Perspect.* **1987**, *73*, 153.
42. Tokiwa, H.; Nakagawa, R.; Horikawa, K.; Ohkubo, A. *Environ. Health Perspect.* **1987**, *73*, 191.
43. *Nitrogen Oxides and Their Effects on Health*; Lee, S. D., Ed.; Ann Arbor Science: Michigan, 1980.
44. Godish, T. *Indoor Air Pollution Control*; Lewis: Michigan, 1989.
45. Melia, R. J. W.; Chinn, S.; Rona, R. J. *Atmospheric Environment* **1990**, *24B*, 177.
46. *Atmospheric Ozone: Proceedings of the Quadrennial Ozone Symposium Held in Halkidiki, Greece, 3-7 September, 1984*; Zeretos, C. S.; Ghazi, A., Eds.; D. Reidel: Holland, 1985; Chapter 3.
47. Pyle, J. A.; Zavody, A. M. *J. Atmos. Chem.* **1988**, *6 (3)*, 201.
48. Crutzen, P. J.; Arnold, F. *Nature* **1986**, *324*, 651.
49. Schunk, R. W. *Pure Appl. Geophys.* **1988**, *127*, 255.
50. Atkinson, R.; Baulch, D. L.; Cox, R. A.; Hampson, R. F., Jr.; Kerr, J. A.; Troe, J. *Int. J. Chem. Kinet.* **1989**, *21*, 115.
51. Kuo, C. H.; Wyttenbach, T.; Beggs, C. G.; Kemper, P. R.; Bowers, M. T. *J. Chem. Phys.* **1990**, *92*, 4849.
52. Chambaud, G.; Rosmus, P. *Chem. Phys. Lett.* **1990**, *165*, 429.

53. Kuo, C. H.; Beggs, C. G.; Kemper, P. R.; Bowers, M. T.; Leahy, D. J.; Zare, R. N. *Chem. Phys. Lett.* **1989**, *163(4-5)*, 291.
54. O'Keefe, A.; Parent, D.; Mauclaire, G.; Bowers, M. T. *J. Chem Phys.* **1984**, *80*, 4901.
55. Smith, M. A.; Bierbaum, V. M.; Leone, S. R. *Chem. Phys. Lett.* **1983**, *94*, 398.
56. Torr, M. R.; Torr, D. G. *Planet. Space Sci.* **1979**, *27*, 1233.
57. Ringer, G.; Gentry, W. R. *J. Chem Phys.* **1979**, *71*, 1902.
58. Mul, P. M.; McGowan, J. W. *J. Phys. B* **1979**, *12*, 1591.
59. Fhadil, H. A.; Numan, A. T.; Shuttleworth, T.; Hasted, J. B. *Int. J. Mass Spectrom. Ion Processes* **1985**, *65*, 307.
60. Breig, E. L.; Hanson, W. B.; Hoffman, J. H. *J. Geophys. Res.* **1984**, *89(A4)*, 2359.
61. Torr, D. G.; Torr, M. R.; Brinton, H. C.; Brace, L. H.; Spencer, N. W.; Hedin, A. E.; Hanson, W. B.; Hoffman, J. H.; Nier, A. O. *J. Geophys. Res.* **1979**, *84 (A7)*, 3360.
62. Dachev, T.; Serafimov, K.; Velinov, P.; Spasov, C. *Dokl. Bolg. Akad. Nauk* **1989**, *42*, 87. CA112(16):143031c.
63. Torr, M. R.; Torr, D. G. *J. Geophys. Res.* **1979**, *84*, 4316.
64. Picard, R. H.; Winick, J. R.; Sharma, R. D.; Zachor, A. S.; Espy, P. J.; Harris, C. R. *Adv. Space Res.* **1987**, *7 (10)*, 23.
65. Winick, J. R.; Picard, R. H.; Sharma, R. D.; Joseph, R. A.; Wintersteiner, P. P. *Adv. Space Res.* **1987**, *7 (10)*, 17.
66. Rusch, D. W.; Gerard, J. C. *J. Geophys. Res.* **1980**, *85 (A3)*, 1285.
67. Serafimov, K. *Adv. Space Res.* **1984**, *4*, 139.
68. Sridharan, R.; Raghavarao, R. *J. Atmos. Terr. Phys.* **1985**, *47*, 1081.
69. Kopp, E.; Andre, L.; Smith, L. G. *J. Atmos. Terr. Phys.* **1985**, *47*, 301.
70. Krankowsky, D.; Laemmerzahl, P.; Goetzelmann, A.; Friedrich, M.; Torkar, K. M. *J. Atmos. Terr. Phys.* **1987**, *49*, 809.
71. Taubenheim, J.; Prasad, B. S. N. *Adv. Space Res.* **1985**, *5*, 107.
72. Smirnova, N. V.; Ogloblina, O. F.; Vlaskov, V. A. *Pure Appl. Geophys.* **1988**, *127*, 353.
73. Smith, D.; Adams, N. G.; Grief, D. *J. Atmos. Terr. Phys.* **1977**, *39(4)*, 513.
74. Chakrabarty, P.; Chakrabarty, D. K. *J. Geophys. Res.* **1979**, *84 (A7)*, 3403.

75. Koshelev, V. V. *J. Atmos. Terr. Phys.* **1987**, *49,* 81.

76. Prasad, B. S. N.; Mohanty, S. *Indian J. Radio Space Phys.* **1978**, *7(5)*, 224.

77. Grebowsky, J. M.; Taylor, H. A.; Pharo, M. W., III; Reese, N. *Planet. Space Sci.* **1987**, *35,* 1463.

78. Watson, A. J.; Donahue, T. M.; Stedman, D. H.; Knollenberg, R. G.; Ragent, B.; Blamont, J. *Geophys. Res. Lett.* **1979**, *6(9)*, 743.

Index